PE Environmental
Practice

R. Wane Schneiter, PhD, PE

PPI®

PPI2PASS.COM
A **KAPLAN** COMPANY

Report Errors for This Book

PPI is grateful to every reader who notifies us of a possible error. Your feedback allows us to improve the quality and accuracy of our products. Report errata at **ppi2pass.com**.

NFPA 70®, *National Electrical Code®*, and NEC® are registered trademarks of the National Fire Protection Association, Inc., Quincy, MA 02169. *National Electrical Safety Code®* and NESC® are registered trademarks of the Institute of Electrical and Electronics Engineers, Inc., New York, NY 10016.

PE ENVIRONMENTAL PRACTICE

Current release of this edition: 5

Release History

date	edition number	revision number	update
Apr 2020	1	3	Minor corrections.
Jul 2020	1	4	Minor corrections.
Mar 2021	1	5	Minor corrections.

PPI
ppi2pass.com

ISBN: 978-1-59126-576-4

Topics

Water

Air

Solid & Hazard. Waste

Site Assessment & Remediation

Environmental Health & Safety

Assoc. Eng. Principles

Table of Contents

Topic VI: Associated Engineering Principles

Preface and Acknowledgments

The purpose of *PE Environmental Practice* is to prepare you for the Principles and Practice of Engineering (PE) exam for environmental engineering. The PE exam is developed, administered, and scored by the National Council of Examiners for Engineering and Surveying (NCEES). State and other local licensing boards use the PE exam as a way of assessing uniformly the preparation and competency of engineers practicing within their jurisdictions.

The NCEES PE environmental exam is a computer-based test (CBT). During the test, you may refer to the electronic *NCEES PE Environmental Reference Handbook* (*NCEES Handbook*, for short), use an on-screen calculator, or write your calculations on a reusable notepad. To help you prepare, the problems in *PE Environmental Practice* have been written with the goal of strengthening your mastery of exam knowledge areas and familiarizing you with the location of relevant information in the *NCEES Handbook*.

Because NCEES keeps exam problems confidential, actual NCEES exam problems cannot be included in any study guide, sample PE exam, or other publication, including publications and other materials available from NCEES. However, NCEES does make public the knowledge areas and topics that will be covered on the PE environmental exam. The problems in *PE Environmental Practice* are organized into these same knowledge areas (topics) and topics (chapters).

PE Environmental Practice is just one of a collection of PPI study tools designed to get you ready for your exam day. The problems in this book are organized into chapters that correspond to the chapters of *PE Environmental Review*, a complete review guide for the PE environmental exam. In the online PPI Learning Hub, you will find electronic versions of these books, diagnostic exams, full-length practice exams, and a quiz generator with hundreds more exam-like practice problems. Use the PPI Learning Hub to integrate these tools with a custom study plan that takes into account your exam date, how much time you have for study, and the subject areas you're already strong in and those you need more review in. To learn how, visit **ppi2pass.com**.

Many thanks go to the PPI editorial operations and product management staff. In particular, I wish to thank Megan Synnestvedt, senior product manager; Steve Shea, product manager; Tyler Hayes and Scott Marley, senior copy editors; Robert Genevro and Scott Rutherford, copy editors; Bradley Burch, production editor; Tom Bergstrom, illustrator and cover designer; Richard Iriye, typesetter; Ellen Nordman, publishing systems specialist; Sam Webster, product data operations manager; Cathy Schrott, editorial operations manager; and Grace Wong, director of editorial operations. In addition, sincere thanks to Kelly Winquist, proofreader, and Anil Ancharya, calculation checker.

Despite our best efforts, as you work problems, you may discover errors or an alternative, more efficient way to solve a problem. Please bring such discoveries to PPI's attention through **ppi2pass.com/errata**. Valid submitted errors will be incorporated into future printings of this book.

R. Wane Schneiter, PhD, PE

How to Use This Book

PE Environmental Practice is written for one purpose, and one purpose only: to get you ready for the NCEES PE environmental exam. Use it along with the other PPI PE environmental study tools to assess, review, and practice until you pass your exam.

ASSESS

To pinpoint the subject areas where you need more study, use the diagnostic exams on the PPI Learning Hub (**ppi2pass.com**). How you perform on these diagnostic exams will tell you which topics you need to spend more time on and which you can review more lightly. Table 1 lists the subject areas on the current exam specifications and the number of questions on the exam for each area.

REVIEW

PE Environmental Review, the companion book to *PE Environmental Practice*, is a complete review for the PE environmental exam. Its topics and chapters correspond neatly to the PE Environmental exam specifications. In turn, the chapters in *PE Environmental Practice* pair exactly with the chapters in *PE Environmental Review*, making it easy to find practice problems on a subject area you have just studied. If you don't fully understand the solution to a practice problem in *PE Environmental Practice*, just turn to the corresponding chapter in *PE Environmental Review* to review the subject area.

PRACTICE

Learn to Use the *NCEES PE Environmental Handbook*

Download a PDF of the *NCEES PE Environmental Reference Handbook* (*NCEES Handbook*) from the NCEES website. As you solve the problems in this book, use the *NCEES Handbook* as your reference. Although you could print out the *NCEES Handbook* and use it that way, it will be better for your preparations if you use it in PDF form on your laptop or computer. This is how you will be referring to it and searching in it during the actual exam.

A searchable electronic copy of the *NCEES Handbook* is the only reference you will be able to use during the exam, so it is critical that you get to know what it includes and how to find what you need efficiently. Even if you know how to find the equations and data you need more quickly in other references, take the time to search for them in *NCEES Handbook*. Get to know the terms and section titles used in the *NCEES Handbook* and use these as your search terms.

A step-by-step solution is provided for each problem in *PE Environmental Practice*. In these solutions, wherever an equation, a figure, or a table is used from the *NCEES Handbook*, the section heading from the *NCEES Handbook* is given in blue. See example below.

For average baffling, the baffling factor BF is $= 0.5$. [EPA Baffling Factors]

$T_{\text{theoretical}}$ = hydraulic residence time corrected for hydraulic efficiency, min

Secondary Drinking Water Standards

$$BF = \frac{T10}{T_{\text{theoretical}}}$$

$$T_{\text{theoretical}} = \frac{T10}{BF}$$

$$= \frac{48 \text{ min}}{0.5}$$

$$= 96 \text{ min}$$

Getting to know the content in the *NCEES Handbook* will save you valuable time on the exam.

Access the PPI Learning Hub

Although *PE Environmental Review* and *PE Environmental Practice* can be used on their own, they are designed to work with the PPI Learning Hub. At the PPI Learning Hub, you can access

- a personal study plan, keyed to your exam date, to help keep you on track

- diagnostic exams to help you identify the subject areas where you are strong and where you need more review

- a quiz generator containing hundreds of additional exam-like problems that cover all knowledge areas on the PE environmental exam

Table 1 *PE Environmental Exam Specifications*

I. Water: 21 to 35 questions

A. principles (hydraulics and fluid mechanics; chemistry; biology and microbiology; fate and transport; sampling and measurement methods; hydrology and hydrogeology; codes, standards, regulations, and guidelines): 3 to 5 questions

B. wastewater (sources of pollution; minimization and prevention; treatment technologies and management; collection systems; residuals (sludge) management; water reuse): 7 to 11 questions

C. stormwater (sources of pollution; treatment technologies and management; collection systems): 2 to 4 questions

D. potable water (source water quality; treatment technologies and management; distribution systems; residuals management—solid, liquid, and gas): 7 to 11 questions

E. water resources (sources of pollution; watershed management and planning; source supply and protection): 2 to 4 questions

II. Air: 14 to 22 questions

A. principles (sampling and measurement methods; codes, standards, regulations, and guidelines; chemistry; fate and transport; atmospheric science and meteorology): 7 to 11 questions

B. pollution control (sources of pollution; characterization, calculations, and inventory of emissions; treatment and control technologies; pollution minimization and prevention): 7 to 11 questions

III. Solid and Hazardous Waste: 11 to 18 questions

A. principles (chemistry; fate and transport; codes, standards, regulations, and guidelines; risk assessment; sampling and measurement methods; minimization, reduction, and recycling; mass and energy balance; hydrology, hydrogeology, and geology): 5 to 8 questions

B. municipal and industrial solid waste (storage, collection, and transportation systems; treatment and disposal technologies and management): 4 to 6 questions

C. hazardous, medical, and radioactive waste (storage, collection, and transportation systems; treatment and disposal technologies and management): 2 to 4 questions

IV. Site Assessment and Remediation: 12 to 19 questions

A. principles (codes, standards, regulations, and guidelines; chemistry and biology; hydrology and hydrogeology; sampling and measurement methods): 5 to 8 questions

B. applications (site assessment and characterization; risk assessment; fate and transport; remediation alternative identification; remediation technologies and management): 7 to 11 questions

V. Environmental Health and Safety: 7 to 11 questions

A. principles (health and safety; security, emergency plans, and incident response procedures; codes, standards, regulations, and guidelines): 3 to 5 questions

B. applications (industrial hygiene; exposure assessments; indoor air quality): 4 to 6 questions

VI. Associated Engineering Principles: 5 to 9 questions

A. principles (statistics; sustainability): 2 to 4 questions

B. applications (engineering economics, project management, mass and energy balance; data management): 3 to 5 questions

- NCEES-like, computer-based practice exams to familiarize you with the exam day experience and let you hone your time management and test-taking skills

- electronic versions of *PE Environmental Review* and *PE Environmental Practice*

For more about the PPI Learning Hub, visit PPI's website at **ppi2pass.com**.

Be Thorough

Really do the work.

Time and again, customers ask us for the easiest way to pass the exam. The short answer is pass it the first time you take it. Put the time in. Take advantage of the problems provided and practice, practice, practice! Take the practice exams and time yourself so you will feel comfortable during the exam. When you are prepared you will know it. Yes, the reports in the PPI Learning Hub will agree with your conclusion but, most importantly, if you have followed the PPI study plan and done the work, it is more likely than not that you will pass the exam.

Some people think they can read a problem statement, think about it for 10 seconds, read the solution, and then say, "Yes, that's what I was thinking of, and that's what I would have done." Sadly, these people find out too late that the human brain makes many more mistakes under time pressure and that there are many ways to get messed up in solving a problem even if you understand the concepts. It may be in the use of your calculator, like using log instead of ln or forgetting to set the angle to radians instead of degrees. It may be rusty math, like forgetting exactly how to factor a polynomial. Maybe you can't find the conversion factor you need, or don't remember what joules per kilogram is in SI base units.

For real exam preparation, you'll have to spend some time with a stubby pencil. You have to make these mistakes during your exam prep so that you do not make them during the actual exam. So do the problems—all of them. Do not look at the solutions until you have sweated a little.

About the Problems in This Book

Each practice problem in this book stands alone. The problems can be solved in any order, and there is never a need to retrieve information from other problems. Flipping a page back and forth in order to solve a problem may seem like a trivial matter to those reading this book in print or in sequence, but for those reading it on a web browser or an e-reader it is a serious inconvenience. This book has been designed to make it unnecessary.

As a result, where two or more problems are based on similar scenarios, the data that is given in each problem, including any tables or illustrations, will be similar. Flipping quickly through the pages of the print version may give the initial impression that some problems have been accidentally repeated, but a closer look will show that each problem is asking for something different and requires a different path to its solution.

Topic I Water

Chapter

Fluid Properties and Statics

Content in blue refers to the *NCEES Handbook.*

PRACTICE PROBLEMS

1. A gauge measures pressure in a system at 58 psi. The absolute pressure when the actual atmospheric pressure in the system equals standard atmospheric pressure is most nearly

(A) 36 psi

(B) 44 psi

(C) 58 psi

(D) 73 psi

2. The equilibrium concentration of oxygen in a pressurized tank partially filled with water is 1.2×10^{-4} expressed as a mole fraction. Most nearly, what is the equilibrium concentration of oxygen gas in the water expressed in mg/L? [Periodic Table of Elements]

(A) 59 mg/L

(B) 69 mg/L

(C) 120 mg/L

(D) 210 mg/L

3. A manometer filled with water is attached to a bulb filled with air. The other end of the manometer is open to the atmosphere. The height of the water in the manometer is 12 cm. The pressure at the bulb is most nearly

(A) 1.2 kPa

(B) 10 kPa

(C) 100 kPa

(D) 120 kPa

4. Water is impounded to a depth of 27 m behind a concrete dam. The face of the dam exposed to the water is vertical. Most nearly, what is the force of the water on the dam?

(A) 1300 kN

(B) 1600 kN

(C) 2400 kN

(D) 3600 kN

SOLUTIONS

1. Standard atmospheric pressure is 14.7 psia. [Normal Temperature and Pressure Air Density]

The absolute pressure is

Pressure Field in a Static Liquid

$$\begin{array}{l}\dfrac{\text{absolute}}{\text{pressure}} = \dfrac{\text{atmospheric}}{\text{pressure}} + \dfrac{\text{gauge pressure}}{\text{reading}}\end{array}$$

$$= 14.7 \ \dfrac{\text{lbf}}{\text{in}^2} + 58 \ \dfrac{\text{lbf}}{\text{in}^2}$$

$$= 72.7 \ \text{psi} \quad (73 \ \text{psi})$$

The answer is (D).

2. Calculate the molecular weight of water. [Periodic Table of Elements]

$$M_{\text{water}} = (2)\left(1 \ \dfrac{\text{g}}{\text{mol}}\right) + 16 \ \dfrac{\text{g}}{\text{mol}} = 18 \ \text{g/mol}$$

Assume 1 L of water and a water density of 1000 g/L. Calculate the moles of water, N_{water}.

$$N_{\text{water}} = \dfrac{1000 \ \dfrac{\text{g}}{\text{L}}}{18 \ \dfrac{\text{g}}{\text{mol}}} = 56 \ \text{mol/L}$$

From the equation for the mole fraction, x, of oxygen, the number of moles of oxygen is

Ideal Gas Mixtures

$$x_i = N_i/N = \dfrac{N_{\text{oxygen}}}{N_{\text{oxygen}} + N_{\text{water}}}$$

$$1.2 \times 10^{-4} = \dfrac{N_{\text{oxygen}}}{N_{\text{oxygen}} + 56 \ \dfrac{\text{mol}}{\text{L}}}$$

$$N_{\text{oxygen}} = 0.0067 \ \text{mol/L}$$

Calculate the concentration of oxygen in mg/L.

$$\left(0.0067 \ \dfrac{\text{mol}}{\text{L}}\right)\left(32 \ \dfrac{\text{g}}{\text{mol}}\right)\left(10^3 \ \dfrac{\text{mg}}{\text{g}}\right)$$

$$= 214 \ \text{mg/L} \quad (210 \ \text{mg/L})$$

The answer is (D).

3. At ambient conditions, the density of air is 1.204 kg/m^3. [Normal Temperature and Pressure Air Density]

At ambient conditions, the density of water is 1000 kg/m^3. [Properties of Water (SI Metric Units)]

Assume atmospheric pressure at the open end of the manometer tube is equal to the standard atmospheric pressure of 101.3 kPa. [Normal Temperature and Pressure Air Density]

Calculate the pressure difference.

P = pressure
ρ = fluid density
g = acceleration of gravity
h = height

Manometers

$$P_o = P_2 + (\rho_2 - \rho_1)gh$$

$$= 101.3 \ \text{kPa} + \left(1000 \ \dfrac{\text{kg}}{\text{m}^3} - 1.204 \ \dfrac{\text{kg}}{\text{m}^3}\right)\left(9.81 \ \dfrac{\text{m}}{\text{s}^2}\right)$$

$$\times (12 \ \text{cm})\left(10^{-2} \ \dfrac{\text{m}}{\text{cm}}\right)\dfrac{1 \ \text{kPa}}{10^3 \ \dfrac{\text{kg}}{\text{m}\cdot\text{s}^2}}$$

$$= 102.5 \ \text{kPa} \quad (100 \ \text{kPa})$$

The answer is (C).

4. Calculate the hydrostatic pressure on a vertical plane surface.

p = pressure, kN
ρ = density of water = 1000 kg/m^3
g = acceleration of gravity = 9.81 m/s^2
h = water depth = 27 m

$$p = \tfrac{1}{2}\rho g(h_1 + h_2)$$

$$= \left(\dfrac{1}{2}\right)\left(1000 \ \dfrac{\text{kg}}{\text{m}^3}\right)\left(9.81 \ \dfrac{\text{m}}{\text{s}^2}\right)(0 \ \text{m} + 27 \ \text{m})\dfrac{1 \ \text{kN}}{1000 \ \dfrac{\text{kg}\cdot\text{m}}{\text{s}^2}}$$

$$= 132.5 \ \dfrac{\text{kN}}{\text{m}^2}$$

Calculate the force acting on 1 m of width.

$$F = pA = \left(132.5 \ \dfrac{\text{kN}}{\text{m}^2}\right)(1 \ \text{m})(27 \ \text{m}) = 3578 \ \text{kN} \quad (3600 \ \text{kN})$$

The answer is (D).

2 Fluid Flow, Dynamics, Hydraulics

Content in blue refers to the NCEES Handbook.

PRACTICE PROBLEMS

1. A 4 in galvanized iron pipe conveys 380 gpm of water over a distance of 1200 ft. The water is at 70°F. Head loss due to flow is most nearly

- (A) 11 ft
- (B) 66 ft
- (C) 120 ft
- (D) 160 ft

2. A stirred tank reactor with a volume of 6.5 m³ was designed for a hydraulic residence time of 12 min. Dye tracer studies reveal that the actual residence time for the reactor is 9.3 min. The effective volume of the reactor is most nearly

- (A) 1.5 m³
- (B) 5.0 m³
- (C) 5.8 m³
- (D) 8.4 m³

3. A tank is 3 m long and 3 m wide, with a water depth of 3.5 m above the discharge outlet. The discharge outlet has a protruding pipe entrance with a 75 mm diameter. Most nearly, how much time is required to decrease the water depth to 0.5 m above the discharge outlet?

- (A) 0.76 min
- (B) 2.0 min
- (C) 22 min
- (D) 34 min

4. A continuous flow-through process tank is filled from an open-end pipe above the water surface. The water discharges through a 3 in pipe entrance protruding near the bottom of the tank. The water surface is maintained at a constant elevation. The flow rate into the tank is 83 gpm. Most nearly, what is the head loss through the discharge pipe?

- (A) 0.01 ft
- (B) 0.2 ft
- (C) 0.5 ft
- (D) 2 ft

5. A sharp-edged flow-restricting orifice with a 2 cm opening is placed inside a 5 cm pipe. The orifice coefficient is most nearly

- (A) 0.58
- (B) 0.61
- (C) 0.63
- (D) 0.65

6. A suspension of spherical particles in water is evaluated for clarification. The characteristics of the suspension are given.

$$\text{mean particle density} = 2.1 \text{ g/cm}^3$$
$$\text{mean particle diameter} = 0.042 \text{ mm}$$
$$\text{water temperature} = 20°C$$

The terminal settling velocity of a mean particle in the suspension is most nearly

- (A) 0.3 mm/s
- (B) 0.5 mm/s
- (C) 1 mm/s
- (D) 3 mm/s

SOLUTIONS

1. Find the cross-sectional area of the pipe.

A = pipe cross sectional area, ft^2
D = pipe diameter = 4 in

$$A = \frac{\pi D^2}{4} = \frac{\pi (4 \text{ in})^2}{(4)\left(12 \frac{\text{in}}{\text{ft}}\right)^2} = 0.087 \text{ ft}^2$$

Use the continuity equation, and solve for the flow velocity.

v = flow velocity
Q = flow rate = 380 gpm

Continuity Equation

$$Q = Av$$

$$v = \frac{Q}{A} = \frac{\left(380 \frac{\text{gal}}{\text{min}}\right)\left(0.134 \frac{\text{ft}^3}{\text{gal}}\right)}{(0.087 \text{ ft}^2)\left(60 \frac{\text{sec}}{\text{min}}\right)}$$

$$= 9.75 \text{ ft/sec}$$

The kinematic viscosity of water at 70°F is 1.059×10^{-5} ft^2/sec. [Properties of Water (English Units)]

Find the Reynolds number.

Reynolds Number (Newtonian Fluid)

$$Re = vD/v$$

$$= \frac{\left(9.75 \frac{\text{ft}}{\text{sec}}\right)(4 \text{ in})}{\left(1.059 \times 10^{-5} \frac{\text{ft}^2}{\text{sec}}\right)\left(12 \frac{\text{in}}{\text{ft}}\right)}$$

$$= 3.06 \times 10^5$$

Find the friction factor. [Moody (Stanton) Diagram]

ϵ = roughness = 0.0005 (average value)
ϵ/D = relative roughness = $\dfrac{0.0005 \text{ ft}}{(4 \text{ in})\left(\frac{1 \text{ ft}}{12 \text{ in}}\right)}$

$$= 0.0015$$
$$f = 0.022$$

Use the Darcy-Weisbach equation to find the head loss due to flow.

$$h_f = f \frac{L}{D} \frac{v^2}{2g}$$

$$= (0.022)\left[\frac{(1200 \text{ ft})\left(12 \frac{\text{in}}{\text{ft}}\right)}{4 \text{ in}}\right]\left[\frac{\left(9.75 \frac{\text{ft}}{\text{sec}}\right)^2}{(2)\left(32.2 \frac{\text{ft}}{\text{sec}^2}\right)}\right]$$

$$= 116.6 \text{ ft} \quad (120 \text{ ft})$$

The answer is (C).

2. The effective volume is

V_e = effective volume, m^3
V = dimensional volume, m^3
t = actual residence time, min
θ = theoretical residence time, min

$$V_e = V\frac{t}{\theta} = (6.5 \text{ m}^3)\left(\frac{9.3 \text{ min}}{12 \text{ min}}\right) = 5.0 \text{ m}^3$$

The answer is (B).

3. Find the tank cross-sectional area.

A_t = tank cross-sectional area, m^2

$$A_t = (3 \text{ m})(3 \text{ m}) = 9 \text{ m}^2$$

D_o = discharge outlet diameter = 75 mm
A_0 = discharge outlet area, m^2

$$A_0 = \frac{\pi D_o^2}{4} = \frac{\pi (75 \text{ mm})^2\left(10^{-3} \frac{\text{m}}{\text{mm}}\right)^2}{4} = 0.0044 \text{ m}^2$$

[Minor Losses in Pipe Fittings, Contractions, and Expansions]

C_d = coefficient of discharge = 0.8
t = time to empty tank, min
z = water depth at beginning and end = 3.5 m, 0.5 m
g = acceleration of gravity = 9.81 m/s^2

$$t = \frac{2A_t(\sqrt{z_1} - \sqrt{z_2})}{C_d A_0 \sqrt{2g}}$$

$$= \frac{(2)(9 \text{ m}^2)(\sqrt{3.5 \text{ m}} - \sqrt{0.5 \text{ m}})}{(0.8)(0.0044 \text{ m}^2)\sqrt{(2)\left(9.81 \dfrac{\text{m}}{\text{s}^2}\right)}\left(60 \dfrac{\text{s}}{\text{min}}\right)}$$

$$= 22 \text{ min}$$

The answer is (C).

4. Use the continuity equation and solve for the velocity.

v = velocity, ft/sec

A = pipe cross sectional area, ft^2

Q = flow rate = 83 gpm

Continuity Equation

$$Q = A\text{v}$$

$$= \left(\frac{\pi d^2}{4}\right)\text{v}$$

$$\text{v} = \frac{4Q}{\pi d^2}$$

$$= \frac{(4)\left(\dfrac{83 \dfrac{\text{gal}}{\text{min}}}{\left(7.48 \dfrac{\text{gal}}{\text{ft}^3}\right)\left(60 \dfrac{\text{sec}}{\text{min}}\right)}\right)}{\pi\left(\dfrac{3 \text{ in}}{12 \dfrac{\text{in}}{\text{ft}}}\right)^2}$$

$$= 3.8 \text{ ft/sec}$$

Find the head loss through the discharge pipe.

$h_{f,\text{fitting}}$ = head loss through the exit pipe, ft

C = exit coefficient for a protruding pipe = 0.8

Minor Losses in Pipe Fittings, Contractions, and Expansions

$$h_{f,\text{fitting}} = C\frac{\text{v}^2}{2g}$$

$$= (0.8)\left(\frac{\left(3.8 \dfrac{\text{ft}}{\text{sec}}\right)^2}{(2)\left(32.2 \dfrac{\text{ft}}{\text{sec}^2}\right)}\right)$$

$$= 0.18 \text{ ft} \quad (0.2 \text{ ft})$$

The answer is (B).

5. Find the coefficients of velocity and contraction for a sharp-edged orifice. [Orifices]

C_v = coefficient of velocity = 0.98
C_c = coefficient of contraction = 0.62

Calculate the orifice coefficient.

C = orifice coefficient

A_0 = orifice opening cross-sectional area

$$= \pi(2 \text{ cm})^{2/4}$$

A_1 = pipe cross-sectional area

$$= \pi(5 \text{ cm})^{2/4}$$

Orifices

$$C = \frac{C_\text{v} C_c}{\sqrt{1 - C_c^2(A_0/A_1)^2}}$$

$$= \frac{(0.98)(0.62)}{\sqrt{1 - (0.62)^2\left(\dfrac{(2 \text{ cm})^2}{(5 \text{ cm})^2}\right)^2}}$$

$$= 0.61$$

The answer is (B).

6. Find the properties of water at 20°C. [Properties of Water (SI Metric Units)]

ρ_f = water density = 998.2 kg/m^3

μ = water absolute viscosity = 0.001002 kg/m·s

Use the equation for the Reynolds number. [Reynolds Number (Newtonian Fluid)]

v_t = terminal settling velocity, mm/s

ρ_p = particle density = 2.1 g/cm^3

d = particle diameter = 0.042 mm

C_D = drag coefficient

Re = Reynolds number

Water

General Spherical

$$\mathrm{Re} = \frac{v_t \rho d}{\mu}$$

For laminar flow, $C_D = 24/\mathrm{Re}$. Find the terminal settling velocity for spherical particles.

General Spherical

$$v_t = \sqrt{\frac{4g(\rho_p - \rho_f)d}{3C_D\rho_f}}$$

$$v_t^2 = \frac{4g(\rho_p - \rho_f)d}{3\left(\dfrac{24}{\mathrm{Re}}\right)\rho_f}$$

$$= \frac{4g(\rho_p - \rho_f)d^2 v_t \rho_f}{72\mu\rho_f}$$

$$v_t = \frac{4g(\rho_p - \rho_f)d^2}{72\mu}$$

$$= \frac{(4)\left(9.81\ \dfrac{\mathrm{m}}{\mathrm{s}^2}\right)\left(2100\ \dfrac{\mathrm{kg}}{\mathrm{m}^3} - 998.2\ \dfrac{\mathrm{kg}}{\mathrm{m}^3}\right)}{\begin{array}{c}\times(4.2\times10^{-5}\ \mathrm{m})^2\left(1000\ \dfrac{\mathrm{mm}}{\mathrm{m}}\right)\\\hline(72)\left(0.001002\ \dfrac{\mathrm{kg}}{\mathrm{m}\cdot\mathrm{s}}\right)\end{array}}$$

$$= 1.07\ \mathrm{mm/s} \quad (1\ \mathrm{mm/s})$$

Check the Reynolds number to confirm laminar flow.

General Spherical

$$\mathrm{Re} = \frac{v_t \rho d}{\mu}$$

$$= \frac{\left(1.07\ \dfrac{\mathrm{mm}}{\mathrm{s}}\right)\left(998.2\ \dfrac{\mathrm{kg}}{\mathrm{m}^3}\right)(4.2\times10^{-5}\ \mathrm{m})}{\left(0.001002\ \dfrac{\mathrm{kg}}{\mathrm{m}\cdot\mathrm{s}}\right)\left(1000\ \dfrac{\mathrm{mm}}{\mathrm{m}}\right)}$$

$$= 0.04 \quad [\mathrm{Re} \le 1.0;\ \text{flow is laminar}]$$

The answer is (C).

3 Chemistry

Content in blue refers to the NCEES Handbook.

PRACTICE PROBLEMS

1. Chlorine gas is used for the disinfection of a municipal drinking water supply. The chlorine gas reacts with the water to form hypochlorous acid (HOCl) at a concentration of 4 mg/L. The water temperature is 15°C. The pK_a for hypochlorous acid at 15°C is 7.63. The hydrogen ion concentration in the water from the dissociation of the hypochlorous acid is most nearly

(A) 2.3×10^{-8} M

(B) 1.3×10^{-6} M

(C) 7.5×10^{-5} M

(D) 7.6×10^{-5} M

2. Chlorine gas reacts with water to form hypochlorous acid (HOCl). Which of the following statements is true regarding the dissociation of HOCl in water at equilibrium?

(A) $[HOCl] = [OCl^-] + [H^+]$

(B) $[HOCl] = [OCl^-]$

(C) $[HOCl] = [H^+]$

(D) $[OCl^-] = [H^+]$

3. Chlorine gas is used for the disinfection of a municipal drinking water supply. The chlorine gas reacts with the water to form hypochlorous acid (HOCl) at a concentration of 4 mg/L. The water temperature is 15°C. If the hypochlorite concentration is 1.3×10^{-6} M, the percent ionization of the hypochlorous acid in the water is most nearly

(A) 1.3%

(B) 13%

(C) 74%

(D) 100%

4. Calcium ions (Ca^{+2}) and carbonate ions $\left(CO_3^{-2}\right)$ are present in a water sample at concentrations of 25 mg/L

and 15 mg/L, respectively. The water temperature is 16°C. The solubility product constant, K_{SP}, for $CaCO_3$ is 5.0×10^{-9} at $T_1 = 25°C$ (298K). Solubility product constants at different absolute temperatures, T_1 and T_2, are related by

$$\ln \frac{K_{SP,T2}}{K_{SP,T1}} = \frac{-\Delta H°(T_1 - T_2)}{R T_1 T_2}$$

R is the ideal gas constant. The standard enthalpies, $\Delta H°$, for the reactants and product are

$$\Delta H° \ Ca^{2+} = -543.0 \ kJ/mol$$
$$\Delta H° \ CO_3^{2-} = -676.3 \ kJ/mol$$
$$\Delta H° \ CaCO_3 = -1207.0 \ kJ/mol$$

The solubility product for $CaCO_3$ at 16°C is most nearly

(A) 5.8×10^{-9}

(B) 1.6×10^{-7}

(C) 2.5×10^{-4}

(D) 6.3×10^{-4}

5. What effect do temperature and salinity have on the dissolved oxygen (DO) concentration in water?

(A) Increasing temperature raises DO; increasing salinity raises DO.

(B) Increasing temperature lowers DO; increasing salinity raises DO.

(C) Increasing temperature raises DO; increasing salinity lowers DO.

(D) Increasing temperature lowers DO; increasing salinity lowers DO.

6. Bench scale aeration tests are performed on a sample of water that is being evaluated for air stripping to remove VOCs. The target VOC is present in the water sample before aeration at 990 μg/L. The tests produce the following data.

elapsed time of aeration (s)	target VOC effluent concentration (μg/L)
30	497
60	251
90	124
120	63
180	14
240	3

What is the reaction order?

(A)   zero

(B)   first

(C)   second

(D)   pseudo-first

7. Bench scale aeration tests are performed on a sample of water that is being evaluated for air stripping to remove VOCs. The target VOC is present in the water sample before aeration at 990 μg/L. The tests produce the following data.

elapsed time of aeration (s)	target VOC effluent concentration (μg/L)
30	497
60	251
90	124
120	63
180	14
240	3

Most nearly, what is the value of the mass transfer coefficient?

(A)   0.023 s^{-1}

(B)   0.092 s^{-1}

(C)   0.21 s^{-1}

(D)   0.82 s^{-1}

8. Bench scale aeration tests are performed on a sample of water that is being evaluated for air stripping to remove VOCs. The target VOC is present in the water sample before aeration at 990 μg/L. If the value of the mass transfer coefficient is 0.023 s^{-1}, the time required

to reduce the target VOC concentration to 1.0 μg/L is most nearly

(A)   260 s

(B)   300 s

(C)   390 s

(D)   470 s

9. Wastewater samples are prepared and incubated at 20°C for 5 d for BOD analysis. Standard 300 mL BOD bottles are used. Sample dilutions and initial and final dissolved oxygen concentrations are summarized in the table shown.

bottle	sample volume (mL)	initial dissolved oxygen (mg/L)	final dissolved oxygen (mg/L)
1	5	9.3	7.7
2	10	9.2	6.6
3	15	9.1	5.2
4	20	9.1	4.1
5	30	8.9	0.8

The BOD$_5$ at 20°C is most nearly

(A)   71 mg/L

(B)   77 mg/L

(C)   80 mg/L

(D)   96 mg/L

10. Wastewater samples are prepared and incubated for BOD analysis. The reaction rate coefficient at 20°C is 0.40 d^{-1} (base e). If the BOD$_5$ is 77 mg/L, the ultimate BOD is most nearly

(A)   82 mg/L

(B)   89 mg/L

(C)   93 mg/L

(D)   110 mg/L

11. Wastewater samples are prepared and incubated for BOD analysis. The temperature correction coefficient is 1.047, and the reaction rate coefficient at 20°C is 0.40 d^{-1} (base e). If the ultimate BOD is 89 mg/L, the BOD$_7$ at 15°C is most nearly

(A)   73 mg/L

(B)   80 mg/L

(C)   85 mg/L

(D)   98 mg/L

12. Calcium ion (Ca^{+2}) and carbonate ion (CO_3^{-2}) are present in a water sample at concentrations of 25 mg/L and 15 mg/L, respectively. The water temperature is 16°C. The reaction quotient for calcium carbonate dissociation or precipitation is most nearly

(A) 5.8×10^{-9}

(B) 1.6×10^{-7}

(C) 2.5×10^{-4}

(D) 6.3×10^{-4}

13. The solubility product constant, K_{SP}, is 5.8×10^{-9} for the reaction

$$CaCO_3 \rightarrow Ca^{2+} + CO_3^{2-}$$

The reaction quotient, Q, is 2.1×10^{-8}. Is the calcium carbonate, $CaCO_3$, precipitating, dissociating, or at equilibrium?

(A) precipitating

(B) dissociating

(C) at equilibrium

(D) unable to determine from the information provided

SOLUTIONS

1. Find the molecular weight of HOCl. [Periodic Table of Elements]

$$1\,\frac{g}{mol} + 16\,\frac{g}{mol} + 35.5\,\frac{g}{mol} = 52.5\ g/mol$$

The molarity of HOCl in the water is

$$\left(4\,\frac{mg}{L}\right)\left(\frac{1\ mol}{52.5\ g}\right)\left(\frac{1\ g}{1000\ mg}\right) = 7.6 \times 10^{-5}\ mol/L$$

In aqueous solution, hypochlorous acid partially dissociates into hypochlorite OCl^-.

$$HOCl \rightleftharpoons OCl^- + H^+$$
$$\frac{[H^+][OCl^-]}{[HOCl]} = 10^{-7.63}$$
$$[H^+] = [OCl^-] = x$$
$$[HOCl] = 7.6 \times 10^{-5}\,\frac{mol}{L} - x$$

$$\frac{x^2}{7.6 \times 10^{-5}\,\dfrac{mol}{L} - x} = 10^{-7.63} = 2.3 \times 10^{-8}$$

Solve for x using the quadratic formula.

$$x = 1.3 \times 10^{-6}\,\frac{mol}{L} = 1.3 \times 10^{-6}\ M$$

The answer is (B).

2. The equilibrium equation for the dissociation of HOCl is

$$HOCl \leftrightarrow OCl^- + H^+$$

At equilibrium, $[OCl^-] = [H^+]$.

The answer is (D).

3. Find the molecular weight of HOCl. [Periodic Table of Elements]

$$1\,\frac{g}{mol} + 16\,\frac{g}{mol} + 35.5\,\frac{g}{mol} = 52.5\ g/mol$$

The molarity of HOCl in the water is

$$\left(4\,\frac{mg}{L}\right)\left(\frac{1\ mol}{52.5\ g}\right)\left(\frac{1\ g}{1000\ mg}\right) = 7.6 \times 10^{-5}\ mol/L$$

Water

Subtract the molarity of OCl^- from the molarity of the $HOCl$ in the water to find the molarity of $HOCl$ that is ionized.

$$[HOCl] = 7.6 \times 10^{-5} \frac{mol}{L} - OCl^-$$

$$= 7.6 \times 10^{-5} \frac{mol}{L} - 1.3 \times 10^{-6} \frac{mol}{L}$$

$$= 7.5 \times 10^{-5} \ mol/L$$

The percent ionization of $HOCl$ is

$$\left(1 - \frac{7.5 \times 10^{-5} \frac{mol}{L}}{7.6 \times 10^{-5} \frac{mol}{L}}\right)(100\%) = 1.3\%$$

The answer is (A).

4. In aqueous solution, calcium carbonate dissociates into calcium and carbonate as shown.

$$CaCO_3 \rightarrow Ca^{+2} - CO_3^{-2}$$

The standard enthalpy, $\Delta H°$, for the reaction is equal to $\Delta H°$ products $- \sum \Delta H°$ reactants. [Heats of Reaction]

$$-543.0 \frac{kJ}{mol} + \left(-676.3 \frac{kJ}{mol}\right)$$
$$-\left(-1207.0 \frac{kJ}{mol}\right) = -12.3 \ kJ/mol$$

Use the equation for solubility product constants given in the problem, and solve for $K_{SP,T2}$.

$K_{SP,T1} =$ solubility product at temperature 1
$$= 5.0 \times 10^{-9}$$

$K_{SP,T2} =$ solubility product at temperature 2

$\Delta H° =$ standard enthalpy for the reaction
$$= -12.3 \ kJ/mol$$

$E_a =$ activation energy $= \Delta H°$
$T_1 =$ temperature 1 $= 298K$
$T_2 =$ temperature 2 $= 289K$
$R =$ universal gas constant $= 8.314 \ J/mol\cdot K$

$$\ln \frac{K_{SP,T1}}{K_{SP,T2}} = \frac{\Delta H°(T_1 - T_2)}{RT_1 T_2}$$

$$\ln \frac{K_{SP,T2}}{K_{SP,T1}} = -\frac{\Delta H°(T_1 - T_2)}{RT_1 T_2}$$

$$\ln \frac{K_{SP,T2}}{5.0 \times 10^{-9}} = \frac{-\left(-12.3 \frac{kJ}{mol}\right) \times (298K - 289K) \times \left(1000 \frac{J}{kJ}\right)}{\left(8.314 \frac{J}{mol\cdot K}\right)(298K)(289K)}$$

$$= 0.15$$

$$K_{SP,T2} = (5.0 \times 10^{-9})(e^{0.15})$$

$$= 5.8 \times 10^{-9}$$

The answer is (A).

5. Increasing temperature lowers DO, and increasing salinity lowers DO. [Dissolved-Oxygen Concentration in Water]

For example,

- From 10°C to 20°C, the DO concentration in fresh water is lowered from 11.28 mg/L to 9.08 mg/L ($\Delta DO = 2.2$ mg/L).

- Increasing salinity from 5 parts per thousand (ppth) to 20 ppth for water at a constant temperature of 20°C lowers the DO from 8.81 mg/L to 8.07 mg/L ($\Delta DO = 0.74$ mg/L).

- The DO of fresh water at 10°C is 11.28 mg/L, but decreases to 8.07 mg/L at 20°C and 20 ppth salinity ($\Delta DO = 3.21$ mg/L).

The answer is (D).

6. Check each reaction order equation against the data in the table to see which equation plots a straight line.

elapsed time of aeration (s)	target VOC effluent concentration (μg/L)
0	990
30	497
60	251
90	124
120	63
180	14
240	3

Check zero order.

Zero-Order Irreversible Reaction Kinetics
$$C_A = C_{A0} - kt$$
$$C_A - C_{A0} = C = -kt$$

Because k is not constant, the reaction is not zero order.

Check first order.

First-Order Irreversible Reaction Kinetics
$$\ln(C_A/C_{A0}) = -kt$$
$$\frac{\ln(C_A/C_{A0})}{-t} = k$$

Because k is approximately constant, the reaction is probably first order.

Check second order.

Second-Order Irreversible Reaction Kinetics
$$1/C_A - 1/C_{A0} = kt$$
$$\frac{\dfrac{1}{C_A} - \dfrac{1}{C_{A0}}}{t} = k$$

Because k is not constant, the reaction is not second order.

The reaction is first order.

The answer is (B).

7. Check first order.

First-Order Irreversible Reaction Kinetics
$$\ln(C_A/C_{A0}) = -kt$$
$$\frac{\ln \dfrac{C_A}{C_{A0}}}{-t} = k$$

The equation for the mass transfer coefficient is

$$K_L a = \text{mass transfer coefficient}, \text{s}^{-1}$$

$$C_o = \text{initial concentration} = 990 \ \mu\text{g/L}$$

$$C = \text{concentration at each time increment}, \\ \mu\text{g/L}$$

$$t = \text{time}, \text{s}$$

$$K_L a = \frac{-\ln\left(\dfrac{C}{C_o}\right)}{t}$$

This equation applies because the reaction is first order.

t (s)	C (μg/L)	$\dfrac{-\ln(C/C_o)}{t}$
30	497	0.0230
60	251	0.0229
90	124	0.0231
120	63	0.0230
180	14	0.0237
240	3	0.0242

$$\text{average} = 0.0233 \ \text{s}^{-1}$$

$$K_L a = 0.023 \ \text{s}^{-1}$$

The answer is (A).

8. Check first order.

First-Order Irreversible Reaction Kinetics
$$\ln(C_A/C_{A0}) = -kt$$
$$\frac{\ln \dfrac{C_A}{C_{A0}}}{-t} = k$$
$$t = \frac{\ln \dfrac{C_A}{C_{A0}}}{k}$$
$$= \frac{-\ln\left(\dfrac{1.0 \ \dfrac{\mu\text{g}}{\text{L}}}{990 \ \dfrac{\mu\text{g}}{\text{L}}}\right)}{0.023 \ \text{s}^{-1}}$$
$$= 300 \ \text{s}$$

The answer is (B).

9. The final dissolved oxygen in bottle 1 is greater than 7.0 mg/L, and that in bottle 5 is less than 2.0 mg/L. Therefore, both of these bottles are excluded from the calculations.

Find the BOD for bottles 2, 3, and 4.

$$D_1 = \text{dissolved oxygen concentration of diluted sample} \\ \text{immediately after preparation, mg/L}$$

$$D_2 = \text{dissolved oxygen concentration of diluted sample} \\ \text{after 5-day incubation period, mg/L}$$

$$P = \text{fraction of wastewater sample volume to} \\ \text{total combined volume}$$

BOD Test Solution and Seeding Procedures
$$\text{BOD}, \text{mg/L} = \frac{D_1 - D_2}{P}$$

$$\text{BOD bottle 2} = \frac{9.2 \frac{mg}{L} - 6.6 \frac{mg}{L}}{\frac{10 \text{ mL}}{300 \text{ mL}}} = 78 \text{ mg/L}$$

$$\text{BOD bottle 3} = \frac{9.1 \frac{mg}{L} - 5.2 \frac{mg}{L}}{\frac{15 \text{ mL}}{300 \text{ mL}}} = 78 \text{ mg/L}$$

$$\text{BOD bottle 4} = \frac{9.1 \frac{mg}{L} - 4.1 \frac{mg}{L}}{\frac{20 \text{ mL}}{300 \text{ mL}}} = 75 \text{ mg/L}$$

The BOD at 20°C is

$$\text{BOD}_5 \; 20°C = \frac{78 \frac{mg}{L} + 78 \frac{mg}{L} + 75 \frac{mg}{L}}{3}$$

$$= 77 \text{ mg/L}$$

The answer is (B).

10. Use the equation for finding the amount of BOD exerted, and solve for the ultimate BOD.

y_t = the amount of BOD exerted at time t = 77 mg/L

k = BOD decay rate constant (base e)

   = 0.40 d^{-1} at 20°C

L = ultimate BOD, mg/L

t = time = 5 d

BOD Exertion

$$y_t = L(1 - e^{-kt})$$

$$L = \frac{y_t}{1 - e^{-kt}}$$

$$= \frac{77 \frac{mg}{L}}{1 - e^{(-0.40 \text{ d}^{-1})(5 \text{ d})}}$$

$$= 89 \text{ mg/L}$$

The answer is (B).

11. Calculate the rate coefficient at 15°C.

k_{15} = rate coefficient at 15°C, d^{-1}

k_{20} = rate coefficient at 20°C = 0.40 d^{-1}

θ = temperature correction coefficient, unitless = 1.047

Kinetic Temperature Corrections

$$k_T = k_{20}(\theta)^{T-20}$$

$$k_{15} = k_{20}(\theta)^{15-20}$$

$$= (0.40 \text{ d}^{-1})(1.047)^{15-20}$$

$$= 0.32 \text{ d}^{-1}$$

Use the equation for finding the amount of BOD exerted, solve for the BOD$_7$ at 15°C.

y_t = the amount of BOD exerted at time t, mg/L

k = BOD decay rate constant (base e)

   = 0.32 d^{-1} at 15°C

L = ultimate BOD = 89 mg/L

t = time = 7 d

BOD Exertion

$$y_t = L(1 - e^{-kt})$$

$$\text{BOD}_7 \text{ at } 15°C = \left(89 \frac{mg}{L}\right)\left(1 - e^{(-0.32 \text{ d}^{-1})(7 \text{ d})}\right)$$

$$= 79.5 \text{ mg/L} \quad (80 \text{ mg/L})$$

The answer is (B).

12. Find the reaction quotient for calcium carbonate dissociation or precipitation.

Q = reaction quotient

$[Ca^{+2}]$ = Ca^{+2} concentration, mol/L

$[CO_3^{-2}]$ = CO_3^{-2} concentration, mol/L

$$Q = [Ca^{+2}][CO_3^{-2}]$$

$$[Ca^{+2}] = \frac{\left(25 \frac{mg}{L}\right)\left(\frac{1 \text{ g}}{1000 \text{ mg}}\right)}{40 \frac{g}{mol}} = 6.25 \times 10^{-4}$$

$$[CO_3^{-2}] = \frac{\left(15 \frac{mg}{L}\right)\left(\frac{1 \text{ g}}{1000 \text{ mg}}\right)}{12 \frac{g}{mol} + (3)\left(16 \frac{g}{mol}\right)} = 2.5 \times 10^{-4}$$

$$Q = (6.25 \times 10^{-4})(2.5 \times 10^{-4})$$

$$= 1.56 \times 10^{-7} \quad (1.6 \times 10^{-7})$$

The answer is (B).

13. K_{sp} is written for the reaction $CaCO_3 \rightarrow Ca^{+2} + CO_3^{-2}$. Since Q is greater than K_{sp}, the reaction is proceeding from right to left and $CaCO_3$ is precipitating.

The answer is (A).

Biology and Microbiology

Content in blue refers to the *NCEES Handbook*.

PRACTICE PROBLEMS

1. What characteristic distinguishes an autotrophic organism from a heterotrophic organism?

(A) Autotrophs use both carbon dioxide and organic carbon for their carbon needs, whereas heterotrophs use only carbon dioxide.

(B) Autotrophs use only carbon dioxide for their carbon needs, whereas heterotrophs use both carbon dioxide and organic carbon.

(C) Autotrophs use only organic carbon for their carbon needs, whereas heterotrophs use only carbon dioxide.

(D) Autotrophs use only carbon dioxide for their carbon needs, whereas heterotrophs use only organic carbon.

2. Which of the following microorganisms are commonly used as indicator organisms to assess the presence of pathogens?

(A) enteroviruses

(B) flagellated protozoa

(C) *giardia* and *cryptosporidium*

(D) total and fecal coliform

3. A biological wastewater treatment system maintains an active biomass as mixed liquor suspended solids at a concentration of 1600 mg/L in a 30 m^3 reaction tank. The system is characterized by an endogenous decay rate coefficient of 0.045 d^{-1}. The rate of endogenous decay in the reaction tank is most nearly

(A) 2.2 mg/m^3·d

(B) 53 mg/m^3·d

(C) 72 mg/L·d

(D) 350 mg/L·d

4. The following equations illustrate the difference in the amount of biomass produced by aerobic and anaerobic processes.

$$18 \text{ waste} + NH_4^+ + HCO_3^- \rightarrow C_5H_7O_2N + 6.5CH_4$$
$$+ 7.5CO_2 + 4H_2O$$

$$13 \text{ waste} + 3NH_4^+ + 10O_2 \rightarrow 3C_5H_7O_2N + 10HCO_3^-$$
$$+ CO_2 + 10H_2O$$

The ratio of biomass produced by the aerobic process to that produced by the anaerobic process is most nearly

(A) 1:3

(B) 3:1

(C) 4:1

(D) 8:1

SOLUTIONS

1. Autotrophs use only carbon dioxide for their carbon needs, whereas heterotrophs use only organic carbon.

The answer is (D).

2. Total and fecal coliform are commonly used as indicator organisms. Coliform bacteria are effective at providing evidence of recent fecal contamination of water from warm-blooded animals, a primary source of pathogens presenting a risk to human health.

The answer is (D).

3. Calculate the rate of endogenous decay.

r_d = rate of endogenous decay, mg/L·d
k_d = endogenous decay rate coefficient = 0.045 d^{-1}
X = suspended solids concentration = 1600 mg/L

$$r_d = -k_d X = -(0.045 \text{ d}^{-1})\left(1600 \; \frac{\text{mg}}{\text{L}}\right)$$
$$= -72 \; \frac{\text{mg}}{\text{L·d}}$$

The negative sign for the decay rate indicates a loss to the system. Because the result is a decay rate, the negative sign is implicit and can be ignored.

The answer is (C).

4. This problem can be solved by directly comparing the biomass produced in each equation. The upper equation is for the anaerobic process, and the lower equation is for the aerobic process. This is seen by the occurrence of free oxygen as a reactant in the lower equation. Biomass is represented by $C_5H_7O_2N$. The aerobic process produces 3 moles of biomass for 13 moles of waste and the anaerobic process produces 1 mole of biomass for 18 moles of waste.

The biomass yield for the aerobic process is

$$y = \frac{n_{\text{biomass}}}{n_{\text{waste}}} = \frac{3 \text{ mol biomass}}{13 \text{ mol waste}}$$
$$= 0.23 \text{ mol/mol}$$

The biomass yield for the anaerobic process is

$$y = \frac{n_{\text{biomass}}}{n_{\text{waste}}} = \frac{1 \text{ mol biomass}}{18 \text{ mol waste}}$$
$$= 0.056 \text{ mol/mol}$$

The ratio of aerobic to anaerobic biomass is

$$\frac{0.23 \; \dfrac{\text{mol}}{\text{mol}}}{0.056 \; \dfrac{\text{mol}}{\text{mol}}} = 4{:}1$$

The answer is (C).

5 Fate and Transport

Content in blue refers to the *NCEES Handbook*.

PRACTICE PROBLEMS

1. What are the characteristics of a fund pollutant?

(A) large assimilative capacity; does not accumulate in the environment; little or slow degradation

(B) large assimilative capacity; does not accumulate in the environment; readily degrade

(C) small assimilative capacity; accumulate in the environment; little or slow degradation

(D) small assimilative capacity; accumulate in the environment; readily degrade

2. Which stream self-purification zone coincides with the critical dissolved oxygen deficit of the dissolved oxygen sag curve?

(A) decomposition

(B) impacted

(C) recovery

(D) septic

3. A small natural lake occupies an area of 73 ha to an average depth of 30 m. The stream feeding the lake has an average flow of 0.21 m³/s with a total phosphorus concentration of 0.13 mg/L. The total phosphorus deposition rate in the lake is 10 m/year. The annual total phosphorus loading to the lake is most nearly

(A) 0.050 kg/ha·m·yr

(B) 0.39 kg/ha·m·yr

(C) 12 kg/ha·m·yr

(D) 23 kg/ha·m·yr

4. A small natural lake occupies an area of 73 ha to an average depth of 30 m. The stream feeding the lake has an average flow of 0.21 m³/s with a total phosphorus concentration of 0.13 mg/L. The total phosphorus deposition rate in the lake is 10 m/year and the total

phosphorous concentration in the lake is 0.062 mg/L. The annual total phosphorus mass deposited in the lake sediments is most nearly

(A) 410 kg/yr

(B) 450 kg/yr

(C) 820 kg/yr

(D) 860 kg/yr

5. A stream enters and exits a small natural lake. The average inflow and outflow rate is 0.21 m³/s. The total phosphorous concentration in the lake is 0.062 mg/L. The annual total phosphorus mass lost from the lake in stream outflow is most nearly

(A) 410 kg/yr

(B) 450 kg/yr

(C) 820 kg/yr

(D) 860 kg/yr

6. Analyses for BOD and dissolved oxygen are performed at several locations along a segment of river downstream from a municipal wastewater treatment plant outfall. The flow velocity in the river channel is 0.45 m/s and the average temperature of the river water is 15°C. River constants are $k_1 = 0.5$ d⁻¹ and $k_2 = 0.9$ d⁻¹ at 15°C. Typical data from representative monitoring stations along the river are summarized in the following table.

monitoring station	BOD$_u$ (mg/L)	dissolved oxygen (mg/L)
A (upstream)	12	8.7
B (discharge)	18	7.9
C (downstream)	27	6.2
D (downstream)	31	5.8
E (downstream)	16	7.1

The oxygen deficit at station D is most nearly

(A) 2.9 mg/L

(B) 4.3 mg/L

(C) 5.8 mg/L

(D) 7.9 mg/L

monitoring station	BOD$_u$ (mg/L)	dissolved oxygen (mg/L)
A (upstream)	12	8.7
B (discharge)	18	7.9
C (downstream)	27	6.2
D (downstream)	31	5.8
E (downstream)	16	7.1

7. Analyses for BOD and dissolved oxygen are performed at several locations along a segment of river downstream from a municipal wastewater treatment plant outfall. The flow velocity in the river channel is 0.45 m/s and the average temperature of the river water is 15°C. River constants are $k_1 = 0.5$ d^{-1} and $k_2 = 0.9$ d^{-1} at 15°C. Typical data from representative monitoring stations along the river are summarized in the following table.

monitoring station	BOD$_u$ (mg/L)	dissolved oxygen (mg/L)
A (upstream)	12	8.7
B (discharge)	18	7.9
C (downstream)	27	6.2
D (downstream)	31	5.8
E (downstream)	16	7.1

If the time to the critical oxygen sag point is 1.1 d, the oxygen deficit at the critical oxygen sag point is most nearly

(A) 3.2 mg/L

(B) 5.4 mg/L

(C) 7.6 mg/L

(D) 8.2 mg/L

8. Analyses for BOD and dissolved oxygen are performed at several locations along a segment of river downstream from a municipal wastewater treatment plant outfall. The flow velocity in the river channel is 0.45 m/s and the average temperature of the river water is 15°C. River constants are $k_1 = 0.5$ d^{-1} and $k_2 = 0.9$ d^{-1} at 15°C. Typical data from representative monitoring stations along the river are summarized in the following table. Assume an initial dissolved oxygen concentration at 15-degrees celcius of 10.07 mg/L.

The time to the critical oxygen sag point is most nearly

(A) 0.39 d

(B) 0.72 d

(C) 0.86 d

(D) 1.2 d

9. Analyses for BOD and dissolved oxygen are performed at several locations along a segment of river downstream from a municipal wastewater treatment plant outfall. The flow velocity in the river channel is 0.45 m/s and the average temperature of the river water is 15°C. River constants are $k_1 = 0.5$ d^{-1} and $k_2 = 0.9$ d^{-1} at 15°C. Typical data from representative monitoring stations along the river are summarized in the following table.

monitoring station	BOD$_u$ (mg/L)	dissolved oxygen (mg/L)
A (upstream)	12	8.7
B (discharge)	18	7.9
C (downstream)	27	6.2
D (downstream)	31	5.8
E (downstream)	16	7.1

If the time to the critical oxygen sag point is 1.1 d, the distance below the discharge at which the critical oxygen sag point occurs is most nearly

(A) 15 km

(B) 28 km

(C) 33 km

(D) 43 km

SOLUTIONS

1. Fund pollutants are substances or materials for which the assimilative capacity is relatively large. These pollutants are compounds that degrade to produce little accumulation in the environment over time. Examples of fund pollutants are human and animal waste.

Stock pollutants, on the other hand, are substances or materials for which the assimilative capacity is very small—essentially any discharge results in an unacceptable negative impact. These are compounds that are toxic and that accumulate with little or very slow degradation. Examples of compounds that may be considered stock pollutants are organochlorine pesticides and heavy metals.

The answer is (B).

2. The five stream self-purification zones are clean, decomposition, septic, recovery, and clean. The septic zone coincides with the critical dissolved oxygen deficit of the dissolved oxygen sag curve. Because of low dissolved oxygen conditions, the septic zone is characterized by the absence of fish and other higher order aquatic organisms. Insects such as worms and midges are present, as are mosquito and other insect larvae.

The answer is (D).

3. The annual total phosphorus loading is

$$\frac{\left(0.21\ \dfrac{\text{m}^3}{\text{s}}\right)\left(0.13\ \dfrac{\text{mg}}{\text{L}}\right)\left(1000\ \dfrac{\text{L}}{\text{m}^3}\right)}{}$$

$$\frac{\times\left(10^{-6}\ \dfrac{\text{kg}}{\text{mg}}\right)\left(86\,400\ \dfrac{\text{s}}{\text{d}}\right)}{(73\ \text{ha})(30\ \text{m})\left(\dfrac{1\ \text{yr}}{365\ \text{d}}\right)}$$

$$= 0.39\ \text{kg/ha·m·yr}$$

The answer is (B).

4. The annual total phosphorus mass deposited to sediments is

$$\frac{\left(0.21\ \dfrac{\text{m}^3}{\text{s}}\right)\left(0.13\ \dfrac{\text{mg}}{\text{L}}\right) - \left(0.21\ \dfrac{\text{m}^3}{\text{s}}\right)\left(0.062\ \dfrac{\text{mg}}{\text{L}}\right)}{\left(\dfrac{1\ \text{m}^3}{1000\ \text{L}}\right)\left(10^6\ \dfrac{\text{mg}}{\text{kg}}\right)\left(\dfrac{1\ \text{d}}{86\,400\ \text{s}}\right)\left(\dfrac{1\ \text{yr}}{365\ \text{d}}\right)}$$

$$= 450\ \text{kg/yr}$$

The answer is (B).

5. The annual total phosphorus mass lost from the lake in stream outflow is

$$\frac{\left(0.21\ \dfrac{\text{m}^3}{\text{s}}\right)\left(10^3\ \dfrac{\text{L}}{\text{m}^3}\right)\left(0.062\ \dfrac{\text{mg}}{\text{L}}\right)\left(10^{-6}\ \dfrac{\text{kg}}{\text{mg}}\right)}{\left(\dfrac{1\ \text{d}}{86\,400\ \text{s}}\right)\left(\dfrac{1\ \text{yr}}{365\ \text{d}}\right)}$$

$$= 410\ \text{kg/yr}$$

The answer is (A).

6. Assume the water temperature along the river course is constant and the water salinity is low.

From a table of values for dissolved-oxygen concentration, the saturated dissolved oxygen at 15°C and low salinity is 10.07 mg/L. [Dissolved-Oxygen Concentration in Water]

The oxygen deficit at station D is

$$10.07\ \frac{\text{mg}}{\text{L}} - 5.8\ \frac{\text{mg}}{\text{L}} = 4.27\ \text{mg/L}\quad(4.3\ \text{mg/L})$$

The answer is (B).

7. The initial dissolved oxygen deficit at the discharge point is

$$D_0 = 10.07\ \frac{\text{mg}}{\text{L}} - 7.9\ \frac{\text{mg}}{\text{L}} = 2.17\ \text{mg/L}$$

Use the Streeter Phelps equation to find the dissolved oxygen deficit.

$$D = \text{dissolved oxygen deficit}$$
$$\qquad\text{at sag points, mg/L}$$
$$L_0 = \text{BOD}_u\ \text{at discharge point} = 18\ \text{mg/L}$$
$$k_1 = \text{deoxygenation constant, base } e = 0.5\ \text{d}^{-1}$$
$$k_2 = \text{reaeration rate constant, base } e = 0.9\ \text{d}^{-1}$$
$$t = \text{time} = 1.1\ \text{d}$$

Stream Modeling

$$D = \frac{k_1 L_0}{k_2 - k_1}\left(e^{-k_1 t} - e^{-k_2 t}\right) + D_0 e^{-k_2 t}$$

$$= \left(\frac{(0.5\ \text{d}^{-1})\left(18\ \dfrac{\text{mg}}{\text{L}}\right)}{0.9\ \text{d}^{-1} - 0.5\ \text{d}^{-1}}\right)\left(e^{(0.5\ \text{d}^{-1})(1.1\ \text{d})} - e^{(0.9\ \text{d}^{-1})(1.1\ \text{d})}\right)$$

$$+ 2.17\ \frac{\text{mg}}{\text{L}}e^{(-0.9\ \text{d}^{-1})(1.1\ \text{d})}$$

$$= 5.43\ \text{mg/L}\quad(5.4\ \text{mg/L})$$

The answer is (B).

Water

8. The dissolved oxygen deficit at the discharge point is
[Dissolved-Oxygen Concentration in Water]

$$D_0 = 10.07 \ \frac{mg}{L} - 7.9 \ \frac{mg}{L} = 2.17 \ mg/L$$

Find the time to the critical oxygen standpoint.

t_c = time of critical oxygen sag points, d
D_0 = dissolved oxygen deficit at
          discharge point
L_0 = BOD_u at discharge point = 18 mg/L
k_1 = deoxygenation constant, base e (days^{-1})
k_2 = reaeration rate constant, base e (days^{-1})

Stream Modeling

$$t_c = \frac{1}{k_2 - k_1} \ln \left[\frac{k_2}{k_1} \left(1 - D_0 \frac{(k_2 - k_1)}{k_1 L_0} \right) \right]$$

$$= \left(\frac{1}{0.9 \ d^{-1} - 0.5 \ d^{-1}} \right) \ln \left[\left(\frac{0.9 \ d^{-1}}{0.5 \ d^{-1}} \right) \right.$$

$$\left. \times \left(1 - \frac{(2.17 \ mg/L)(0.9 \ d^{-1} - 0.5 \ d^{-1})}{(0.5 \ d^{-1}) \left(18 \ \frac{mg}{L} \right)} \right) \right]$$

$$= 1.22 \ d \quad (1.2 \ d)$$

The answer is (D).

9. The distance below the discharge of the critical oxygen sag point is

$$\left(0.45 \ \frac{m}{s} \right) (1.1 \ d) \left(86\,400 \ \frac{s}{d} \right) \left(\frac{1 \ km}{1000 \ m} \right)$$

$$= 42.8 \ km \quad (43 \ km)$$

The answer is (D).

6 Sampling and Measurement Methods

Content in blue refers to the *NCEES Handbook.*

PRACTICE PROBLEMS

1. The protocol for the collection and analysis of water samples are defined in a joint publication of the American Public Health Association (APHA), the American Water Works Association (AWWA), and the Water Environment Federation (WEF). What is the name of the publication?

(A) *Analytical Protocol for Wastewater and Water*

(B) *Sample and Analyses Methods for Water and Wastewater*

(C) *Standard Methods for the Examination of Water and Wastewater*

(D) *Water and Wastewater Sample Collection, Handling, Analysis, and Recording*

SOLUTIONS

1. The collection and analysis of water samples for many common parameters follows the protocol defined by *Standard Methods for the Examination of Water and Wastewater*, a joint publication of the American Public Health Association (APHA), the American Water Works Association (AWWA), and the Water Environment Federation (WEF).

The answer is (C).

Hydrology and Hydrogeology

Content in blue refers to the *NCEES Handbook*.

PRACTICE PROBLEMS

1. Rainfall records for four precipitation stations are summarized in the following table.

station	normal annual precipitation (cm)	annual precipitation for year indicated (cm)		
		year 1	year 2	year 3
A	27	26	24	31
B	31	29	27	30
C	26	33		29
D	29	27	26	28

Stations A, B, C, and D are located in close proximity to each other. What is the value for the missing record at Station C for year 2?

(A) 23 cm

(B) 26 cm

(C) 29 cm

(D) 31 cm

2. A watershed occupies a 30 ha site. 18 ha of the site have been cleared and are used for pasture land; 1 ha is occupied by farm buildings, a house, and paved surfaces; the remaining 11 ha are woodland. The average land slope is 2.1%. Because the site is upland from a residential development, the rainfall runoff from the site is collected in a catchment that discharges directly to a culvert. The overland flow distance to the catchment is 212 m. The weighted average runoff coefficient for the watershed is most nearly

(A) 0.07

(B) 0.25

(C) 0.34

(D) 0.62

3. 20 years of 24-hour annual peak discharge values are summarized as shown in the table.

year	annual peak discharge (m³/s)
1995	112
1996	94
1997	54
1998	49
1999	42
2000	51
2001	128
2002	103
2003	88
2004	96
2005	65
2006	79
2007	83
2008	71
2009	57
2010	89
2011	62
2012	53
2013	64
2014	92

The peak discharge expected for the average 2 yr, 24 h storm is most nearly

(A) 42 m³/s

(B) 46 m³/s

(C) 50 m³/s

(D) 75 m³/s

4. An urban parkland with a uniform slope of 0.0081 ft/ft experiences sheet flow runoff over a distance of 112 ft. The runoff coefficient of the urban parkland is 0.20. The rainfall intensity is 2.1 in/hr. The inlet time is most nearly

(A) 3 min

(B) 8 min

(C) 40 min

(D) 70 min

Water

5. A storm produces a 3.7 in of runoff with a time to peak of 8 hr. The peak discharge from the storm for a 42,000 acre watershed is most nearly

(A) 4.0×10^3 ft^3/sec

(B) 1.5×10^4 ft^3/sec

(C) 1.2×10^6 ft^3/sec

(D) 1.3×10^{10} ft^3/sec

6. The drainage area of a river is 11 839 km^2 with an average annual runoff of 144.4 m^3/s. The average annual precipitation is 1.08 m. Assume storage remains constant and groundwater inflow and outflow are equal. The evaporation and transpiration losses are most nearly

(A) 1.3×10^4 m^3

(B) 1.2×10^6 m^3

(C) 8.3×10^9 m^3

(D) 1.3×10^{10} m^3

7. 20 years of 24-hour annual peak discharge values are summarized as shown in the table.

year	annual peak discharge (m^3/s)
1995	112
1996	94
1997	54
1998	49
1999	42
2000	51
2001	128
2002	103
2003	88
2004	96
2005	65
2006	79
2007	83
2008	71
2009	57
2010	89
2011	62
2012	53
2013	64
2014	92

The peak discharge expected for the average 10 yr, 24 h storm is most nearly

(A)   65 m^3/s

(B)   75 m^3/s

(C)   81 m^3/s

(D)   108 m^3/s

8. A well constructed in a confined aquifer and screened through the entire aquifer thickness of 18 m was pumped at 0.75 m^3/min for 48 h. Time-drawdown observations at a well located 61 m and 100 m away were recorded and the data plotted. Using a Theis-type curve and the data plot, values for $h_o - h = 1.32$ m and $t = 47$ min were obtained. The transmissivity of the aquifer is most nearly

(A)   3.3 m^2/d

(B)   6.5 m^2/d

(C)   33 m^2/d

(D)   64 m^2/d

9. 10 min unit hydrographs developed for a 30 min storm are shown in the figure.

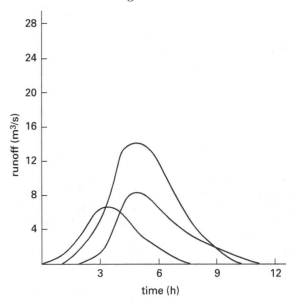

The peak discharge for the storm is most nearly

(A)   7 m^3/s

(B)   9 m^3/s

(C)   15 m^3/s

(D)   26 m^3/s

10. 10 min unit hydrographs developed for a 30 min storm are shown in the figure.

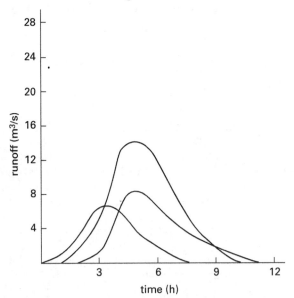

How long after the beginning of the storm does peak discharge occur?

(A) 3 h

(B) 5 h

(C) 7 h

(D) 9 h

11. 10 min unit hydrographs developed for a 30 min storm are shown in the figure.

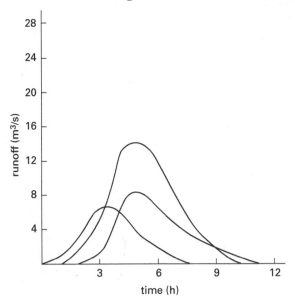

The total flow from the storm runoff is most nearly

(A) 7000 m³

(B) 18 000 m³

(C) 170 000 m³

(D) 420 000 m³

12. A watershed occupies a 30 ha site. 18 ha of the site have been cleared and are used for pasture land; 1 ha is occupied by farm buildings, a house, and paved surfaces; the remaining 11 ha are woodland. The average land slope is 2.1%. Because the site is upland from a residential development, the rainfall runoff from the site is collected in a catchment that discharges directly to a culvert. The overland flow distance to the catchment is 212 m. The 20 yr storm is characterized by the intensity duration curve presented in the figure.

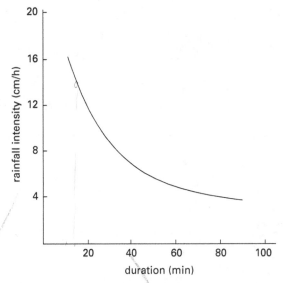

For a 15 min storm duration and average runoff coefficient of 0.18, the time of overland flow for the watershed is most nearly

(A) 0.6 min

(B) 1.8 min

(C) 5.0 min

(D) 18 min

13. Stormwater runoff from a fully developed urban area is collected for treatment prior to discharge into an estuary. The stormwater enters a basin used for treatment to remove suspended solids and for retention. The maximum runoff flow into the basin is 10 cfs for a 30 min period and the maximum allowed discharge from the basin is 1 cfs. The settling time in the basin is most nearly

(A) 0.50 hr

(B) 2.2 hr

(C) 4.5 hr

(D) 5.0 hr

SOLUTIONS

1. Calculate the differences in normal annual precipitation among the four stations.

$$\frac{26 \text{ cm} - 27 \text{ cm}}{26 \text{ cm}} \times 100\% = 3.8\%$$

$$\frac{26 \text{ cm} - 31 \text{ cm}}{26 \text{ cm}} \times 100\% = 19\%$$

$$\frac{26 \text{ cm} - 29 \text{ cm}}{26 \text{ cm}} \times 100\% = 12\%$$

Because the normal annual precipitation among stations varies by more than 10%, use the normal ratio method.

$$P_x = \frac{1}{3}\left[\left(\frac{N_x}{N_A}\right)P_A + \left(\frac{N_x}{N_B}\right)P_B + \left(\frac{N_x}{N_D}\right)P_D \right]$$

$$= \frac{1}{3}\left[\begin{array}{c} \left(\frac{26 \text{ cm}}{27 \text{ cm}}\right)(24 \text{ cm}) + \left(\frac{26 \text{ cm}}{31 \text{ cm}}\right)(27 \text{ cm}) \\ + \left(\frac{26 \text{ cm}}{29 \text{ cm}}\right)(26 \text{ cm}) \end{array} \right]$$

$$= 23 \text{ cm}$$

The answer is (A).

2. The typical runoff coefficient for each land use type is shown in the table.

land use	area (ha)	runoff coefficient range	typical runoff coefficient
pasture	18	0.05-0.45	0.13
developed	1	0.5-0.95	0.75
woodland	11	0.05-0.25	0.20

The weighted average runoff coefficient for the watershed is

$$\frac{(0.25)(18 \text{ ha}) + (0.73)(1 \text{ ha}) + (0.15)(11 \text{ ha})}{30 \text{ ha}} = 0.2\bar{3}$$

The answer is (B).

3. Rank the 20 years of 24-hour peak discharge values and calculate the cumulative frequency for each value.

rank	annual peak discharge (m^3/s)	cumulative frequency
1	42	0.025
2	49	0.075
3	51	0.125
4	53	0.175
5	54	0.225
6	57	0.275
7	62	0.325
8	64	0.375
9	65	0.425
10	71	0.475
11	79	0.525
12	83	0.575
13	88	0.625
14	89	0.675
15	92	0.725
16	94	0.775
17	96	0.825
18	103	0.875
19	112	0.925
20	128	0.975

$$\text{cumulative frequency} = \frac{\text{rank} - 0.5}{20}$$

$$\text{exceedence probability} = \frac{1}{\text{return period}}$$

The exceedence probability for the 2 yr storm is 1/2 or 0.5 and the corresponding cumulative frequency is 1/2 or 0.5.

From the table, a 0.5 cumulative frequency corresponds approximately to a peak discharge of 75 m^3/s.

The peak discharge expected for the average 2 yr, 24 h storm is 75 m^3/s.

The answer is (D).

4. Find the inlet time.

t_i = time of overland flow, min
L = distance of overland flow, ft
S = slope, ft/ft
i = rainfall intensity, in/hr
C = runoff coefficient

Time of Concentration

$$t_i = C(L/S\ i^2)^{1/3}$$

$$= (0.20)\left[\frac{112\ \text{ft}}{\left(0.0081\ \dfrac{\text{ft}}{\text{ft}}\right)\left(2.1\dfrac{\text{in}}{\text{hr}}\right)^2}\right]^{1/3}$$

$$= 2.9\ \text{min} \quad (3\ \text{min})$$

The answer is (A).

5. Find the peak discharge.

q_p = peak discharge, ft^3/sec
T_p = time to peak, hr
A = watershed area, mi^2

Unit Hydrographs

$$q_p = \frac{484AQ}{T_p}$$

$$= \frac{(484)(42{,}000\ \text{ac})(3.7\ \text{in})}{(8\ \text{hr})\left(640\ \dfrac{\text{ac}}{\text{mi}^2}\right)}$$

$$= 14{,}690\ \text{ft}^3/\text{sec} \quad (1.5 \times 10^4\ \text{ft}^3/\text{sec})$$

The answer is (B).

6. The water balance equation is a mass balance of water entering and leaving a hydrologic area. In this case, the equation can be written as

$$\text{water in} = \text{water out} + \text{evaporation/transpiration}$$

$$\text{evaporation/transpiration} = \text{water in} - \text{water out}$$

$$= \left(1.08\ \frac{\text{m}}{\text{yr}}\right)(11\,893\ \text{km}^2)\left(1000\ \frac{\text{m}}{\text{km}}\right)^2$$

$$- \left(144.4\ \frac{\text{m}^3}{\text{s}}\right)\left(3.15 \times 10^7\ \frac{\text{s}}{\text{yr}}\right)$$

$$= 8.3 \times 10^9\ \text{m}^3$$

The answer is (C).

7. Rank the 20 years of 24-hour peak discharge values, and calculate the cumulative frequency for each value.

rank	annual peak discharge (m^3/s)	cumulative frequency
1	42	0.025
2	49	0.075
3	51	0.125
4	53	0.175
5	54	0.225
6	57	0.275
7	62	0.325
8	64	0.375
9	65	0.425
10	71	0.475
11	79	0.525
12	83	0.575
13	88	0.625
14	89	0.675
15	92	0.725
16	94	0.775
17	96	0.825
18	103	0.875
19	112	0.925
20	128	0.975

$$\text{cumulative frequency} = \frac{\text{rank} - 0.5}{20}$$

$$\text{exceedence probability} = \frac{1}{\text{return period}}$$

The exceedence probability for the 10 yr storm is 1/10 or 0.1 and the corresponding cumulative frequency is 9/10 or 0.9.

From the table, a 0.9 cumulative frequency corresponds approximately to a peak discharge of 108 m^3/s.

The peak discharge expected for the average 10 yr, 24 h storm peak discharge is 108 m^3/s.

The answer is (D).

8. For a confined aquifer, use the Theim equation.

$$Q = \text{flow rate} = 0.75 \ m^3/min$$

$$T = \text{transmissivity}, m^2/d$$

$$h_1, h_2 = \text{heights of piezometer surface above}$$
$$\text{aquifer bottom}, h_2 - h_1 = 1.32 \ m$$

$$r_1, r_2 = \text{radii from pumping well} = 61 \ m, 100 \ m$$

$$Q = \frac{2\pi T(h_2 - h_1)}{\ln\left(\dfrac{r_2}{r_1}\right)}$$

$$
\begin{aligned}
T &= \frac{Q\ln\left(\dfrac{r_2}{r_1}\right)}{2\pi(h_2 - h_1)} \\[2mm]
&= \frac{\left(0.75 \ \dfrac{m^3}{min}\right)\left(1440 \ \dfrac{min}{d}\right)\ln\left(\dfrac{100 \ m}{61 \ m}\right)}{2\pi(1.32 \ m)} \\[2mm]
&= 64 \ m^2/d
\end{aligned}
$$

The answer is (D).

9. The figure shown is the synthesized hydrograph from the unit hydrographs given in the problem statement. It was constructed by summing the unit hydrograph runoff at 1 h increments.

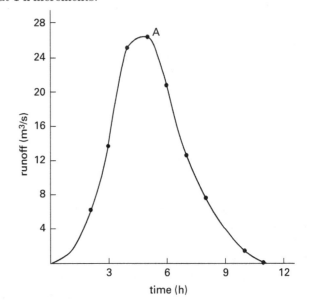

From the figure, the peak discharge occurs at point A and is 26 m^3/s.

The answer is (D).

10. The figure shown is the synthesized hydrograph from the unit hydrographs given in the problem statement. It was constructed by summing the unit hydrograph runoff at 1 h increments.

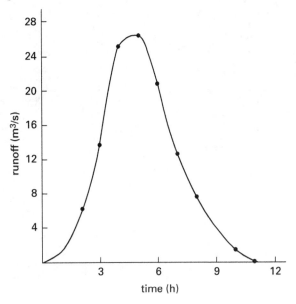

From the figure, the peak discharge occurs 5 h after the beginning of the storm.

The answer is (B).

11. The figure shown is the synthesized hydrograph from the unit hydrographs given in the problem statement. It was constructed by summing the unit hydrograph runoff at 1 h increments.

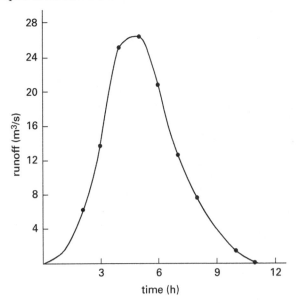

From the figure, the total runoff is equal to the area under the hydrograph curve. To find this area, integrate the hydrograph curve using 1 h time intervals.

$$
\begin{aligned}
&\text{runoff volume} \\
&= (1 \text{ h})\left(3600 \ \frac{\text{s}}{\text{h}}\right) \\
&\times \left(
\begin{aligned}
&0.5 \ \frac{\text{m}^3}{\text{s}} + 3.5 \ \frac{\text{m}^3}{\text{s}} + 9.5 \ \frac{\text{m}^3}{\text{s}} \\
&+ 19.5 \ \frac{\text{m}^3}{\text{s}} + 25.5 \ \frac{\text{m}^3}{\text{s}} + 23.5 \ \frac{\text{m}^3}{\text{s}} \\
&+ 16.0 \ \frac{\text{m}^3}{\text{s}} + 9.5 \ \frac{\text{m}^3}{\text{s}} + 5.5 \ \frac{\text{m}^3}{\text{s}} \\
&+ 3.0 \ \frac{\text{m}^3}{\text{s}} + 0.5 \ \frac{\text{m}^3}{\text{s}}
\end{aligned}
\right) \\
&= 419\,400 \text{ m}^3 \quad (420\,000 \text{ m}^3)
\end{aligned}
$$

The answer is (D).

12. From the figure, for a storm duration of 15 min, rainfall intensity, i, is 14 cm/hr.

$$
i = \frac{14 \ \frac{\text{cm}}{\text{hr}}}{2.54 \ \frac{\text{cm}}{\text{in}}} = 5.51 \text{ in/hr}
$$

Find the time of overland flow.

t_i = time of overland flow, min
C = runoff coefficient = 0.18
L = overland flow distance

$$= (212 \text{ m})\left(3.28 \ \frac{\text{ft}}{\text{m}}\right) = 696 \text{ ft}$$

S = slope = 2.1%

Time of Concentration

$$t_i = C\left(L/S \ i^2\right)^{1/3}$$

$$
\begin{aligned}
&= (0.18)\left(\frac{696 \text{ ft}}{0.021 \ \frac{\text{ft}}{\text{ft}}\left(5.51 \ \frac{\text{in}}{\text{hr}}\right)^2}\right)^{1/3} \\
&= 1.85 \text{ min} \quad (1.8 \text{ min})
\end{aligned}
$$

The answer is (B).

13. The total inflow volume to the basin is

Stokes' Law

$$\theta = V/Q$$
$$V = \theta/Q$$

$$\left(10\ \frac{\text{ft}^3}{\text{sec}}\right)(30\ \text{min})\left(60\ \frac{\text{sec}}{\text{min}}\right) = 18{,}000\ \text{ft}^3$$

The total outflow volume during the 30 min inflow period is

$$\left(1\ \frac{\text{ft}^3}{\text{sec}}\right)(30\ \text{min})\left(60\ \frac{\text{sec}}{\text{min}}\right) = 1800\ \text{ft}^3$$

The total retained volume is

$$18{,}000\ \text{ft}^3 - 1800\ \text{ft}^3 = 16{,}200\ \text{ft}^3$$

The settling time for the total retained volume is equal to the hydraulic residence time.

Stokes' Law

$$\theta = V/Q$$

$$= \frac{16{,}200\ \text{ft}^3}{\left(1\ \dfrac{\text{ft}^3}{\text{sec}}\right)\left(3600\ \dfrac{\text{sec}}{\text{hr}}\right)} = 4.5\ \text{hr}$$

The answer is (C).

8 Codes, Standards, Regulations, Guidelines

Content in blue refers to the *NCEES Handbook*.

PRACTICE PROBLEMS

1. What distinguishes National Primary Drinking Water Regulations from National Secondary Drinking Water Regulations?

(A) Each establishes enforceable drinking water quality standards. Primary contaminants do not transform to other regulated degradation products. Regulated secondary contaminants occur as parent compounds or elements transform into substances that threaten public health.

(B) Primary drinking water regulations define enforceable drinking water quality standards for protecting public health. Secondary drinking water standards are non-mandatory water quality standards for improving drinking water aesthetics such as taste, color, and odor.

(C) Primary and secondary standards apply to the same substances. Primary standards apply enforceable MCLs (maximum contaminant levels) and secondary standards apply enforceable MCLGs (maximum contaminant level goals), both to protect public health.

(D) Primary drinking water regulations address treatment and water quality standards for organic and inorganic chemical contaminants. Secondary drinking water regulations address treatment and water quality standards for biological contaminants.

2. The EPA regulations establish limits on concentrations of contaminants allowed in drinking water. Among other designations, these limits are expressed as MCL and TT. Which of the following statements are true regarding MCL and TT?

I. MCL and TT have distinct definitions, but are each established to control the highest level of a contaminant allowed in drinking water.

II. MCL refers to an enforceable standard. TT refers to a required treatment process for drinking water.

III. MCL defines the highest enforceable standard for controlling contaminants in drinking water. TT defines a non-enforceable treatment goal.

IV. MCL defines the contaminant level below which there is no known or expected risk. TT defines the treatment required to meet the MCL.

(A) I only

(B) I and II

(C) II and III

(D) I and IV

3. The EPA surface water treatment rules for systems using surface water or ground water under the direct influence of surface water require which of the following?

I. disinfection

II. filtration

III. watershed provisions to control cryptosporidium

IV. 99.9% Giardia and 99.99% virus removal/inactivation

(A) I only

(B) I and II only

(C) either I and II together or only III

(D) either I and II together or I, III, and IV together

4. Under which office of USEPA does the Office of Environmental Justice operate?

(A) Office of Enforcement and Compliance

(B) Office of International Activities

(C) Office of Policy

(D) Office of Solid Waste and Emergency Response

5. Where are USEPA's environmental justice goals implemented?

(A) They are implemented in the environmental impact statement (EIS) and environmental assessment (EA) process.

(B) They are implemented in the CERCLA remedial investigation (RI) and feasibility study (FS) process.

(C) They are implemented in the RCRA permitting process.

(D) They are implemented in international technologies cooperation activities.

SOLUTIONS

1. The EPA describes the National Primary Drinking Water Regulations as legally enforceable standards that limit the level of specific contaminants that can adversely affect public health and are known or anticipated to occur in water. These regulations take several forms.

- Maximum Contaminant Level (MCL): The maximum permissible level of a contaminant in water delivered to any user in a public water system. MCLs are enforceable standards.

- Maximum Contaminant Level Goal (MCLG): The level of a contaminant in drinking water below which there is no known or expected risk to health. MCLGs are non-enforceable public health goals.

- Treatment Technique (TT): A required process intended to reduce the level of a contaminant in drinking water.

- Maximum Residual Disinfectant Level (MRDL): The highest level of a disinfectant allowed in drinking water.

- Maximum Residual Disinfectant Level Goal (MRDLG): The level of a drinking water disinfectant below which there is no known or expected risk to health.

National Secondary Drinking Water Regulations are described by the EPA as non-mandatory water quality standards. The EPA does not enforce SMCLs (secondary maximum contaminant levels). They are established only as guidelines to assist public water systems in managing their drinking water for aesthetic considerations, such as taste, color, and odor. These contaminants are not considered to present a risk to human health at the SMCLs. States may choose to adopt them as enforceable standards. [Regulated Drinking Water Contaminants]

The answer is (B).

2. The maximum contaminant level (MCL) represents the highest level of a contaminant that is allowed in drinking water. MCLs consider best available treatment technology and cost, and they are enforceable. Maximum contaminant level goals (MCLGs) define the contaminant level below which there is no known or expected risk. MCLGs are not enforceable public health goals. [Regulated Drinking Water Contaminants]

TT refers to treatment technique, a required process intended to reduce the contaminant level in drinking water. Both MCL and TT are established to control the highest level of a contaminant allowed in drinking water.

The answer is (B).

3. The EPA surface water treatment rules require systems using surface water, or ground water under the direct influence of surface water, to either disinfect and filter the water or disinfect and meet the criteria for avoiding filtration by controlling cryptosporidium in the watershed and removing or inactivating 99.9% of Giardia and viruses. [Regulated Drinking Water Contaminants]

The answer is (D).

4. The Office of Environmental Justice operates under the USEPA Office of Enforcement and Compliance.

The answer is (A).

5. The USEPA's policy is to implement goals that promote environmental justice in the environmental impact statement (EIS) and environmental assessment (EA) process.

The answer is (A).

Wastewater Pollution, Minimization, Prevention

Content in blue refers to the *NCEES Handbook*.

PRACTICE PROBLEMS

1. A city of 87,000 people has an average daily per capita water demand of 100 gal/person-day. The projected total peak water demand for the entire population is most nearly

(A) 4 MGD

(B) 9 MGD

(C) 16 MGD

(D) 18 MGD

2. A municipality of generates 5.6 MGD of wastewater with an average untreated biochemical oxygen demand (BOD) of 263 mg/L. The population equivalent of the wastewater source is most nearly

(A) 12

(B) 28

(C) 62

(D) 72

3. Wastewater treatment guidelines for a planned community require that wastewater treatment capacity be provided based on four population equivalents (PE) per home. The community will eventually include 1750 homes. Assume that the typical person generates 0.2 lbm BOD/day-person. The approximate biochemical oxygen demand (BOD) loading expected at the wastewater plant from the community is most nearly

(A) 88 lbm BOD/day

(B) 1400 lbm BOD/day

(C) 2200 lbm BOD/day

(D) 35,000 lbm BOD/day

SOLUTIONS

1. Rearrange the equation for the ratio of peak to average water demand to find the projected total peak water demand.

Q_{peak} = peak total water demand, gal/day

Q_{ave} = average total water demand, gal/day

P = population served, thousands of people

$$\frac{Q_{\text{peak}}}{Q_{\text{ave}}} = \frac{18 + \sqrt{P}}{4 + \sqrt{P}}$$

$$Q_{\text{peak}} = Q_{\text{ave}}\left(\frac{18 + \sqrt{P}}{4 + \sqrt{P}}\right)$$

$$= \left(100 \ \frac{\text{gal}}{\text{person-day}}\right)(87{,}000 \ \text{people})$$

$$\times \left(\frac{18 + \sqrt{87}}{4 + \sqrt{87}}\right)\left(10^{-6} \ \frac{\text{MG}}{\text{gal}}\right)$$

$$= 17.8 \ \text{MGD} \quad (18 \ \text{MGD})$$

The answer is (D).

2. The population equivalent is

$P_{e,1000\text{s}}$ = population equivalent, thousands of people

Q = flow rate, gal/day

$$P_{e,1000\text{s}} = \frac{(\text{BOD})Q\left(8.345 \ \frac{\text{lbm-L}}{\text{MG·mg}}\right)}{\left(10^6 \ \frac{\text{gal}}{\text{MG}}\right)(1000 \ \text{people})\left(0.20 \ \frac{\text{lbm}}{\text{person-day}}\right)}$$

$$= \frac{\left(263 \ \frac{\text{mg}}{\text{L}}\right)\left(5.6 \times 10^6 \ \frac{\text{gal}}{\text{day}}\right)\left(8.345 \ \frac{\text{lbm-L}}{\text{MG-mg}}\right)}{\left(10^6 \ \frac{\text{gal}}{\text{MG}}\right)(1000 \ \text{people})\left(0.20 \ \frac{\text{lbm}}{\text{person-day}}\right)}$$

$$= 61.5 \quad (62)$$

The answer is (C).

Water

3. Use the equation for calculating the population equivalent and solve for the BOD loading rate.

$$\text{PE} = \frac{\text{BOD loading rate}}{\text{per capita BOD generation rate}}$$

$$\begin{aligned}
\text{BOD loading} \atop \text{rate} &= \left(\begin{array}{c}\text{per capita BOD}\\\text{generation rate}\end{array}\right)(\text{PE})\\
&= \left(0.2\ \frac{\text{lbm BOD}}{\text{day-person}}\right)\left(4\ \frac{\text{PE}}{\text{home}}\right)(1750\ \text{homes})\\
&= 1400\ \text{lbm BOD/day}
\end{aligned}$$

The answer is (B).

10 Wastewater Treatment and Management

Content in blue refers to the *NCEES Handbook*.

PRACTICE PROBLEMS

1. Complete mix activated sludge is selected to treat wastewater with the following characteristics.

influent flow	15,000 m³/day
influent BOD	168 mg/L
effluent BOD	20 mg/L
influent ammonia	36 mg/L
effluent ammonia	1 mg/L

The design constants for BOD oxidation are as shown.

maximum growth rate, μ_m	1.17 d^{-1}
half-velocity constant, K_s	0.60 mg/L
endogenous decay rate coefficient, k_d	0.081 d^{-1}

The corrected design constants for nitrification are as shown.

corrected maximum growth rate, m_m	0.54 d^{-1}
half-velocity constant, K_s	0.57 mg/L
endogenous decay rate coefficient, k_d	0.066 d^{-1}

Design is controlled by BOD oxidation or by nitrification, depending on which one has the greatest mean cell residence time. The mean cell residence times for BOD oxidation and nitrification are most nearly

(A) 0.93 d for BOD oxidation, 2.1 d for nitrification

(B) 1.9 d for BOD oxidation, 3.6 d for nitrification

(C) 2.1 d for BOD oxidation, 0.19 d for nitrification

(D) 2.6 d for BOD oxidation, 0.93 d for nitrification

2. A microbial system is known to follow the Monod kinetic model. For this system, the specific growth rate is 6 d^{-1} and the substrate concentration is 16 mg/L. The maximum specific growth rate is most nearly

(A) 3 d^{-1}

(B) 4 d^{-1}

(C) 9 d^{-1}

(D) 12 d^{-1}

3. A treatability study is conducted with wastewater samples using a series of five bench scale bioreactors. The bioreactors are operated without recycle, and no solids are wasted. Parameters monitored for each of the reactors are summarized as follows.

test reactor	influent total BOD$_5$ (mg/L)	effluent soluble BOD$_5$ (mg/L)	hydraulic residence time (d)	MLVSS concentration (mg/L)
1	275	8	3.5	137
2	275	15	1.9	128
3	275	21	1.6	135
4	275	32	1.3	130
5	275	44	1.0	124

The yield coefficient is most nearly

(A) 0.1 g/g

(B) 0.5 g/g

(C) 1.2 g/g

(D) 1.9 g/g

4. A treatability study is conducted with wastewater samples using a series of five bench scale bioreactors. The bioreactors are operated without recycle, and no solids are wasted. Parameters monitored for each of the reactors are summarized as shown.

test reactor	influent total BOD_5 (mg/L)	effluent soluble BOD_5 (mg/L)	hydraulic residence time (d)	MLVSS concentration (mg/L)
1	275	8	3.5	137
2	275	15	1.9	128
3	275	21	1.6	135
4	275	32	1.3	130
5	275	44	1.0	124

The endogenous decay rate coefficient is most nearly

(A) 0.007 d^{-1}

(B) 0.05 d^{-1}

(C) 0.8 d^{-1}

(D) 1.2 d^{-1}

5. A municipal wastewater requires nitrification for ammonia removal to 5 mg/L NH_3–N prior to discharge into a nearby stream. The complete mix activated sludge process has been selected for nitrification. The wastewater characteristics and design parameters are shown.

$NH_3 = 112$ mg/L as N

half-saturation constant $= 21.3$ mg/L

endogenous decay rate coefficient $= 0.045 \text{ d}^{-1}$

safety factor $= 2.5$

corrected maximum growth rate $= 0.61 \text{ d}^{-1}$

The design mean cell residence time for nitrification is most nearly

(A) 8 h

(B) 24 h

(C) 60 h

(D) 130 h

6. A municipal wastewater requires nitrification for ammonia removal to 5 mg/L NH_3–N prior to discharge into a nearby stream. The complete mix activated sludge process has been selected for nitrification. The

wastewater characteristics and design parameters are as shown.

$NH_3 = 112$ mg/L as N

mixed liquor volatile suspended solids $= 3200$ mg/L

growth rate constant $= 2.4 \text{ d}^{-1}$

half-saturation constant $= 21.3$ mg/L

endogenous decay rate coefficient $= 0.045 \text{ d}^{-1}$

yield coefficient $= 0.2$ g/g

safety factor $= 2.5$

design mean cell residence time $= \theta_c^d = 5.42$ d

biomass nitrifying fraction $= 0.17$

The hydraulic residence time for nitrification is most nearly

(A) 0.75 h

(B) 1.5 h

(C) 4.0 h

(D) 7.0 h

7. A municipal wastewater requires nitrification for ammonia removal to 5 mg/L NH_3–N prior to discharge into a nearby stream. The complete mix activated sludge process has been selected for nitrification. The wastewater characteristics and design parameters are as shown.

flow rate $= 10\,000$ m³/d

$NH_3 = 112$ mg/L as N

mixed liquor volatile suspended solids $= 3200$ mg/L

safety factor $= 2.5$

hydraulic residence time $= 4.0$ h

The bioreactor volume for nitrification is most nearly

(A) 320 m^3

(B) 620 m^3

(C) 1700 m^3

(D) 2900 m^3

8. An industrial wastewater treatment plant receives pulp and paper wastewater at $15\,000$ m³/d with a BOD_5 of 240 mg/L. The plant produces about 1400 kg/d of dry sludge as volatile solids and is able to satisfy effluent criteria for BOD_5 of 30 mg/L. The current aeration equipment is approaching the end of its design life and is scheduled for replacement with low-speed surface aerators. The ratio of BOD_u to BOD_5 is 1.38 and the ratio of O_2 to solids produced is 1.42 kg O_2/kg solids. The daily mass of dissolved oxygen required is most nearly

(A) 1200 kg O_2/d

(B) 2400 kg O_2/d

(C) 3200 kg O_2/d

(D) 4400 kg O_2/d

9. An industrial wastewater treatment plant receives pulp and paper wastewater at 15 000 m³/d with a BOD_5 of 240 mg/L. The plant produces about 1400 kg/d of dry sludge as volatile solids and is able to satisfy effluent criteria for BOD_5 of 30 mg/L. The current aeration equipment is approaching the end of its design life and is scheduled for replacement with low-speed surface aerators. The dissolved oxygen transfer rate under field conditions is 0.68 kg/kW·h. The power required at the aerator per 1000 kg of daily O_2 required is most nearly

(A) 30 kW

(B) 61 kW

(C) 83 kW

(D) 115 kW

10. The characteristics of an activated sludge process are as shown.

> flow rate = 5000 m³/d
> influent BOD_5 = 204 mg/L
> effluent BOD_5 = 20 mg/L
> soluble BOD: total BOD = 0.37
> observed yield coefficient = 0.36 g/g
> bioreactor hydraulic residence time = 6 h
> design mean cell residence time = 10 d

The rate of biomass produced in the bioreactor is most nearly

(A) 330 kg/d

(B) 350 kg/d

(C) 490 kg/d

(D) 520 kg/d

11. The characteristics of an activated sludge process are as shown.

> flow rate = 5000 m³/d
> mixed liquor suspended solids = 3100 mg/L
> wasted suspended solids = 15 000 mg/L
> recirculated solids flow rate = 1300 m³/d

The daily mass of solids recirculated is most nearly

(A) 1900 kg/d

(B) 13 000 kg/d

(C) 20 000 kg/d

(D) 26 000 kg/d

12. A complete mix-activated sludge process is being proposed to treat municipal wastewater at a design flow rate of 8000 m³/d. The influent soluble BOD must be reduced to an effluent total BOD_5 of 10 mg/L. The influent wastewater characteristics and design requirements are as shown.

soluble BOD_5	235 mg/L
effluent soluble BOD_5	$0.3 \times$ effluent total BOD_5
mixed liquor volatile suspended solids	3000 mg/L
temperature	20°C
growth rate constant	2.7 d^{-1}
half-saturation constant	63 mg/L
endogenous decay rate coefficient	0.05 d^{-1}
yield coefficient	0.23 g/g
design mean cell residence time	12 d

The required hydraulic residence time is most nearly

(A) 3.2 h

(B) 8.4 h

(C) 14 h

(D) 56 h

13. A complete mix-activated sludge process is being proposed to treat municipal wastewater at a design flow rate of 8000 m³/d. The influent soluble BOD must be reduced to an effluent total BOD_5 of 10 mg/L. The hydraulic residence time is 3.2 h. The required bioreactor volume is most nearly

(A) 1100 m³

(B) 2800 m³

(C) 4700 m³

(D) 26 000 m³

14. A complete mix-activated sludge process is being proposed to treat municipal wastewater at a design flow rate of 8000 m³/d. The influent soluble BOD of 235 mg/L must be reduced to an effluent total BOD$_5$ of 10 mg/L. The bioreactor volume is 1100 m³. The organic loading rate (volumetric) is most nearly

(A) 0.60 kg/m³·d

(B) 1.7 kg/m³·d

(C) 6.0 kg/m³·d

(D) 17 kg/m³·d

15. A complete mix-activated sludge process is being proposed to treat municipal wastewater at a design flow rate of 8000 m³/d. The influent soluble BOD of 235 mg/L must be reduced to an effluent total BOD$_5$ of 10 mg/L. The hydraulic residence time is 3.2 h and the mixed liquor volatile suspended solids concentration is 3000 mg/L. The organic loading rate (F:M) is most nearly

(A) 0.6 d^{-1}

(B) 1.7 d^{-1}

(C) 6.0 d^{-1}

(D) 17 d^{-1}

16. An anaerobic wastewater lagoon is selected to treat waste at a flow rate of 2300 m³/d with an influent BOD of 670 mg/L. Assume an acceptable winter loading rate of 0.122 kg BOD/m²·d and an acceptable summer loading rate of 0.130 kg BOD/m²·d. The pond depth is set at 4.0 m. The required surface area for the pond is most nearly

(A) 11 800 m²

(B) 12 600 m²

(C) 17 700 m²

(D) 18 900 m²

17. A food processing plant produces a high-strength wastewater that is to be pretreated using a super high-rate plastic media trickling filter prior to discharge to a municipal sewer. The plant produces a continuous flow of 1600 m³/d of wastewater at a temperature of 16°C. The wastewater BOD of 434 mg/L must be reduced to 150 mg/L in the first stage. The filter depth is 3.7 m and the recirculation factor is 2. The cross-sectional surface area required for the filter is most nearly

(A) 700 ft²

(B) 2000 ft²

(C) 4000 ft²

(D) 5200 ft²

18. A food processing plant produces a high-strength wastewater that is to be pretreated using a super high-rate plastic media trickling filter prior to discharge to a municipal sewer. The plant produces a continuous flow of 1600 m³/d of wastewater at a temperature of 16°C. The recirculation ratio is 2, and the filter cross-sectional surface area is 64 m². The hydraulic loading rate is most nearly

(A) 0.027 m³/m²·min

(B) 0.052 m³/m²·min

(C) 0.12 m³/m²·min

(D) 0.23 m³/m²·min

19. A food processing plant produces a high-strength wastewater that is to be pretreated using a super high-rate plastic media trickling filter prior to discharge to a municipal sewer. The plant produces a continuous flow of 1600 m³/d of wastewater at a temperature of 16°C. The wastewater has a BOD of 434 mg/L. The recirculation factor is 2, the filter cross-sectional surface area is 64 m², and the filter depth is 3.7 m. The organic loading rate is most nearly

(A) 1.2 kg/m³·d

(B) 1.6 kg/m³·d

(C) 2.9 kg/m³·d

(D) 10 kg/m³·d

20. A food processing plant produces a high-strength wastewater that is to be pretreated using a super high-rate plastic media trickling filter prior to discharge to a municipal sewer. The plant produces a continuous flow of 1600 m³/d of wastewater at a temperature of 16°C. The filter uses two distribution arms with a hydraulic loading rate of 0.052 m³/m² min and a dosing rate of 20 cm per pass of the recirculation arm. If a two-arm distributor is used, the distribution arm rotational speed is most nearly

(A) 0.08 rpm

(B) 0.13 rpm

(C) 1.7 rpm

(D) 2.6 rpm

21. A water supply is treated by direct filtration followed by chlorination to control Giardia Cysts. The free chlorine concentration is 2.2 mg/L, and the chlorinated water flows through a chlorine contact chamber with average baffling. The water temperature is 10°C and the pH is 6.5. The required hydraulic residence time for the chlorine contact chamber is most nearly

(A) 24 min

(B) 48 min

(C) 96 min

(D) 129 min

22. A wastewater contains phosphorus, PO_4^{3-}, at 15 mg/L. To precipitate the phosphorus, the stoichiometric dose required of ferric chloride, $FeCl_3$, is most nearly

(A) 10 mg/L

(B) 30 mg/L

(C) 60 mg/L

(D) 90 mg/L

23. Monitoring at a wastewater treatment plant reveals the daily flow variations summarized as shown in the table.

period	average flow (m^3/h)
0000–0200	275
0200–0400	389
0400–0600	621
0600–0800	1340
0800–1000	1383
1000–1200	1312
1200–1400	1098
1400–1600	1027
1600–1800	1084
1800–2000	886
2000–2200	259
2200–2400	326

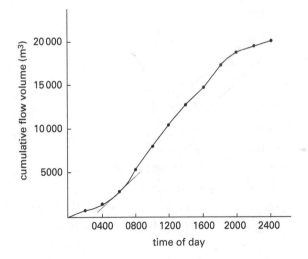

The storage capacity required to equalize the flow is most nearly

(A) 2500 m³

(B) 5500 m³

(C) 11 000 m³

(D) 20 000 m³

24. Monitoring at a wastewater treatment plant reveals the daily flow variations summarized as shown in the table.

period	average flow (m^3/h)
0000–0200	275
0200–0400	389
0400–0600	621
0600–0800	1340
0800–1000	1383
1000–1200	1312
1200–1400	1098
1400–1600	1027
1600–1800	1084
1800–2000	886
2000–2200	259
2200–2400	326

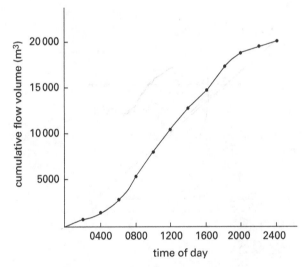

The time of day that the storage tank is at its peak level is most nearly

(A) 0500 h

(B) 1000 h

(C) 1900 h

(D) 2400 h

25. Monitoring at a wastewater treatment plant reveals the daily flow variations summarized as shown in the table.

period	average flow (m^3/h)
0000–0200	275
0200–0400	389
0400–0600	621
0600–0800	1340
0800–1000	1383
1000–1200	1312
1200–1400	1098
1400–1600	1027
1600–1800	1084
1800–2000	886
2000–2200	259
2200–2400	326

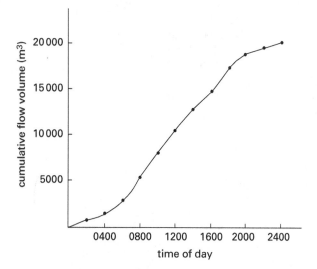

The time of day that the storage tank is at its minimum level is most nearly

(A) 0500 h

(B) 1000 h

(C) 1900 h

(D) 2400 h

26. Pilot studies evaluating BOD removal in a mixed lagoon have produced the following results.

influent BOD (mg/L)	effluent BOD (mg/L)	hydraulic resident time (d)
280	141	1
280	96	2
280	74	3
280	59	4
280	45	5

Most nearly, what is the value of the kinetic constant for lagoon design?

(A) 0.34 d^{-1}

(B) 0.69 d^{-1}

(C) 0.97 d^{-1}

(D) 1.5 d^{-1}

27. The characteristics of an activated sludge process are as shown.

> flow rate = 5000 m^3/d
> influent TSS = 40 mg/L
> effluent TSS = 20 mg/L
> mixed liquor suspended solids = 3100 mg/L
> wasted suspended solids = 15 000 mg/L
> bioreactor hydraulic residence time = 6 h
> design mean cell residence time = 10 d

The daily volume of sludge wasted is most nearly

(A) 20 m^3/d

(B) 80 m^3/d

(C) 200 m^3/d

(D) 600 m^3/d

28. The characteristics of an activated sludge process are as shown.

> flow rate = 5000 m^3/d
> influent TSS = 40 mg/L
> effluent TSS = 20 mg/L
> mixed liquor suspended solids = 3100 mg/L
> wasted suspended solids = 15 000 mg/L
> bioreactor hydraulic residence time = 6 h
> design mean cell residence time = 10 d
> daily volume of sludge wasted = 20 m^3/d

The daily mass of sludge wasted is most nearly

(A) 300 kg/d

(B) 1200 kg/d

(C) 3000 kg/d

(D) 9000 kg/d

29. A secondary clarifier accepts effluent from a bioreactor at a flow rate of 8300 m^3/d. The effluent contains total suspended solids of 1600 mg/L. The solids loading rate is most nearly

(A) 1300 kg/d

(B) 1600 kg/d

(C) 13 000 kg/d

(D) 17 000 kg/d

30. A secondary clarifier accepts effluent from a bioreactor at a flow rate of 8300 m^3/d. The effluent contains total suspended solids of 1600 mg/L. The solids flux for the suspension is 2.6 kg/m^2·h. The required surface area based on solids flux is most nearly

(A) 130 m^2

(B) 210 m^2

(C) 270 m^2

(D) 430 m^2

31. A secondary clarifier accepts effluent from a bioreactor at a flow rate of 8300 m^3/d. The particle settling velocity is 1.29 m/h. The required surface area based on particle settling velocity is most nearly

(A) 130 m^2

(B) 210 m^2

(C) 270 m^2

(D) 430 m^2

32. A secondary clarifier accepts effluent from a bioreactor at a flow rate of 8300 m^3/d. The effluent contains total suspended solids of 1600 mg/L. The solids flux for the suspension is 2.6 kg/m^2·h, and the particle settling velocity is 1.29 m/h. For a design surface area of 270 m^2, the design overflow rate is most nearly

(A) 0.8 m^3/m^2·h

(B) 1.3 m^3/m^2·h

(C) 1.6 m^3/m^2·h

(D) 2.6 m^3/m^2·h

33. The operating characteristics of an activated sludge process are given.

influent flow rate = 3500 m^3/d

hydraulic residence time = 4 h

mixed liquor volatile suspended
  solids concentration = 2600 mg/L

waste solids concentration = 12 000 mg/L

clarified solids effluent concentration = 20 mg/L

Most nearly, what is the wasted solids flow rate required to maintain the solids residence time at 3 d?

(A) 30 m^3/d

(B) 40 m^3/d

(C) 60 m^3/d

(D) 100 m^3/d

34. A 100 hp blower is operated at 20°C with an inlet pressure of 1 atm and an outlet pressure of 3.2 atm. The blower efficiency is 82%. The airflow rate produced by the blower is most nearly

(A) 1.2 lbf/sec

(B) 1.4 lbf/sec

(C) 2.1 lbf/sec

(D) 4.1 lbf/sec

35. A reverse-osmosis process is selected to treat a groundwater with high total dissolved solids (TDS). Following treatment, the groundwater is blended with other water to serve the potable water needs of a small city. The characteristics of the reverse-osmosis process are shown.

groundwater pumping rate	2700 m^3/d
groundwater TDS concentration	1120 mg/L
permeate recovery	81%
dissolved solids rejection	97%

The concentrate TDS concentration and daily volume requiring disposal is most nearly

(A) 1340 mg/L at 2600 m^3/d

(B) 1380 mg/L at 2200 m^3/d

(C) 3020 mg/L at 81 m^3/d

(D) 5720 mg/L at 510 m^3/d

Water

SOLUTIONS

1. Calculate the minimum mean cell residence time for BOD oxidation.

$$\frac{1}{\theta_c^m} = \frac{\mu_m S_o}{K_s + S_o} - k_d$$

$$= \frac{(1.17 \text{ d}^{-1})\left(168 \text{ }\frac{\text{mg}}{\text{L}}\right)}{0.60 \text{ }\frac{\text{mg}}{\text{L}} + 168 \text{ }\frac{\text{mg}}{\text{L}}} - 0.081 \text{ d}^{-1}$$

$$= 1.08 \text{ d}^{-1}$$

$$\theta_c^m = \frac{1}{1.08 \text{ d}^{-1}}$$

$$= 0.93 \text{ d}$$

Calculate the minimum mean cell residence time for nitrification.

$$\frac{1}{\theta_c^m} = \frac{\mu_m' N_o}{K_s + N_o} - k_d$$

$$= \frac{(0.54 \text{ d}^{-1})\left(36 \text{ }\frac{\text{mg}}{\text{L}}\right)}{0.57 \text{ }\frac{\text{mg}}{\text{L}} + 36 \text{ }\frac{\text{mg}}{\text{L}}} - 0.066 \text{ d}^{-1}$$

$$= 0.47 \text{ d}^{-1}$$

$$\theta_c^m = \frac{1}{0.47 \text{ d}^{-1}}$$

$$= 2.13 \text{ d}$$

Nitrification controls design, with $\theta_c^m = 2.13$ d (2.1 d).

The answer is (A).

2. Find the maximum specific growth rate using the Monod kinetic model. The half-velocity coefficient, K_s, is equal to the substrate concentration, S, when the specific growth rate is one-half the maximum growth rate, or $\mu/\mu_m = 0.5$.

K_s = half-velocity coefficient, mg/L

S = substrate concentration, mg/L

μ = specific growth rate, d^{-1}

μ_m = maximum specific growth rate, d^{-1}

$$\mu = \mu_m\left(\frac{S}{K_s + S}\right)$$

$$\frac{\mu}{\mu_m} = \frac{S}{K_s + S}$$

$$= 0.5$$

$$\mu_m = \frac{\mu}{0.5} = \frac{\dfrac{6}{\text{d}}}{0.5}$$

$$= 12 \text{ d}^{-1}$$

The answer is (D).

3. Use the equation for finding the MLVSS concentration, and solve for $1/\theta_c$.

θ = hydraulic residence time, d

θ_c = mean cell residence time, d

$\quad(\theta_c = \theta$ when reactors are operated without recycle and no solids are wasted)

Y = yield coefficient, g/g

S_0, S_e = influent and effluent BOD, mg/L

k_d = endogenous decay rate coefficient, d^{-1}

X_A = mixed liquor volatile suspended solids (MLVSS), mg/L

Activated Sludge

$$X_A = \frac{\theta_c Y(S_0 - S_e)}{\theta(1 + k_d\theta_c)}$$

$$\frac{1}{\theta_c} = \frac{Y(S_0 - S_e)}{\theta X_A} - k_d$$

The yield coefficient and the endogenous decay rate coefficient can be evaluated by plotting the equation or using linear regression.

test reactor	$1/\theta_c$ (d^{-1})	$S_0 - S_e$ (mg/L)	θX_A (mg·d/L)	$(S_0 - S_e)/$ θX_A (d^{-1})
1	0.29	267	480	0.56
2	0.53	260	243	1.1
3	0.63	254	216	1.2
4	0.77	243	169	1.4
5	1.0	231	124	1.9

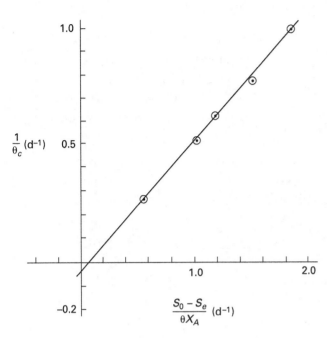

test reactor	$1/\theta_c$ (d^{-1})	$S_0 - S_e$ (mg/L)	θX_A (mg·d/L)	$(S_0 - S_e)/\theta X_A$ (d^{-1})
1	0.29	267	480	0.56
2	0.53	260	243	1.1
3	0.63	254	216	1.2
4	0.77	243	169	1.4
5	1.0	231	124	1.9

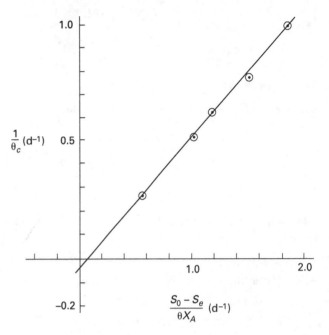

Y is the slope. From the figure,

$$Y = \frac{1.0 \text{ d}^{-1} - 0.29 \text{ d}^{-1}}{1.9 \text{ d}^{-1} - 0.56 \text{ d}^{-1}} = 0.53 \quad (0.5)$$

Using linear regression, $Y = 0.54$ (0.5).

By definition, units for Y are g/g.

$$Y = 0.5 \text{ g/g}$$

The answer is (B).

4. Use the equation for finding the MLVSS concentration, and solve for $1/\theta_c$.

θ = hydraulic residence time, d

θ_c = mean cell residence time, d

    ($\theta_c = \theta$ when reactors are operated without recycle and no solids are wasted)

Y = yield coefficient, g/g

S_0, S_e = influent and effluent BOD, mg/L

k_d = endogenous decay rate coefficient, d^{-1}

X_A = mixed liquor volatile suspended solids (MLVSS), mg/L

Activated Sludge

$$X_A = \frac{\theta_c Y (S_0 - S_e)}{\theta(1 + k_d \theta_c)}$$

$$\frac{1}{\theta_c} = \frac{Y(S_0 - S_e)}{\theta X_A} - k_d$$

The yield coefficient and the endogenous decay rate coefficient can be evaluated by plotting the equation or using linear regression.

From the figure (or through linear regression), k_d is the y-intercept.

$$Y = \frac{1.0 \text{ d}^{-1} - k_d}{1.9 \text{ d}^{-1} - 0} = 0.5$$

$$k_d = 1.0 \text{ d}^{-1} - (0.5)(1.9 \text{ d}^{-1})$$

$$= 0.05 \text{ d}^{-1}$$

Using linear regression, $k_d = 0.026$.

The answer is (B).

5. Determine the design mean cell residence time for nitrification.

θ_c^d = design mean cell residence time for nitrification, h

θ_c^m = minimum mean cell residence time, d

μ_m = maximum corrected growth rate = 0.61 d^{-1}

N_o = influent ammonia concentration

    = 112 mg/L as N

k_d = endogenous decay rate coefficient = 0.045 d^{-1}

K_s = half-saturation constant = 21.3 mg/L

SF = safety factor = 2.5

$$\frac{1}{\theta_c^m} = \frac{\mu_m N_o}{K_s + N_o} - k_d$$

$$= \frac{(0.61 \text{ d}^{-1})\left(112 \dfrac{\text{mg}}{\text{L}}\right)}{21.3 \dfrac{\text{mg}}{\text{L}} + 112 \dfrac{\text{mg}}{\text{L}}} - 0.045 \text{ d}^{-1}$$

$$= 0.47 \text{ d}^{-1}$$

$$\theta_c^m = 2.1 \text{ d}$$

$$\theta_c^d = (\text{SF})\theta_c^m = (2.5)(2.1 \text{ d})\left(24 \frac{\text{h}}{\text{d}}\right)$$

$$= 126 \text{ h} \quad (130 \text{ h})$$

The answer is (D).

6. Use the equation for finding the MLVSS concentration.

Activated Sludge

$$X_A = \frac{\theta_c Y (S_0 - S_e)}{\theta (1 + k_d \theta_c)}$$

Rearrange and substitute N_0 for S_0, N_e for S_e, and X_N for X_A.

$$\theta = \text{hydraulic residence time, h}$$
$$N_e = \text{effluent ammonia concentration}$$
$$\quad = 5 \text{ mg/L as N}$$
$$X_N = \text{nitrifying biomass concentration, mg/L}$$

$$\theta = \frac{\theta_c^d Y (N_0 - N_e)}{X_N (1 + \theta_c^d k_d)}$$

$$X_N = (0.17)\left(3200 \frac{\text{mg}}{\text{L}}\right) = 544 \text{ mg/L}$$

$$\theta_c^d = (130 \text{ h})\left(\frac{1 \text{ d}}{24 \text{ h}}\right) = 5.42 \text{ d}$$

$$\theta = \frac{(5.42 \text{ d})\left(0.2 \dfrac{\text{g}}{\text{g}}\right)\left(112 \dfrac{\text{mg}}{\text{L}} - 5 \dfrac{\text{mg}}{\text{L}}\right)\left(24 \dfrac{\text{h}}{\text{d}}\right)}{\left(544 \dfrac{\text{mg}}{\text{L}}\right)\left(1 + (5.42 \text{ d})(0.045 \text{ d}^{-1})\right)}$$

$$= 4.1 \text{ h} \quad (4.0 \text{ h})$$

The answer is (C).

7. The bioreactor volume is

$$(4.0 \text{ h})\left(10\,000 \frac{\text{m}^3}{\text{d}}\right)\left(\frac{1 \text{ d}}{24 \text{ h}}\right) = 1667 \text{ m}^3 \quad (1700 \text{ m}^3)$$

The answer is (C).

8. Calculate the daily O_2 requirement.

$$Q = \text{water flow rate} = 15\,000 \text{ m}^3/\text{d}$$
$$S_0 = \text{influent BOD}_5 = 240 \text{ mg/L}$$
$$S = \text{effluent BOD}_5 = 30 \text{ mg/L}$$
$$f = \text{ratio of BOD}_u{:}\text{BOD}_5 = 1.38$$
$$X_p = \text{volatile solids production rate}$$
$$\quad = 1400 \text{ kg/d}$$
$$\dot{m} = \text{oxygen required, kg/d}$$

$$\dot{m} = Q(S_0 - S)f - 1.42 X_p$$

$$= \left(15\,000 \frac{\text{m}^3}{\text{d}}\right)\left(240 \frac{\text{mg}}{\text{L}} - 30 \frac{\text{mg}}{\text{L}}\right)$$

$$\times \left(1000 \frac{\text{L}}{\text{m}^3}\right)\left(10^{-6} \frac{\text{kg}}{\text{mg}}\right)(1.38)$$

$$- (1.42)\left(1400 \frac{\text{kg}}{\text{d}}\right)$$

$$= 2359 \text{ kg/d} \quad (2400 \text{ kg O}_2/\text{d})$$

The answer is (B).

9. The power required at the aerator is

$$P = \frac{\dot{m}}{N} = \frac{\left(1000 \dfrac{\text{kg}}{\text{d}}\right)\left(\dfrac{1 \text{ d}}{24 \text{ h}}\right)}{0.68 \dfrac{\text{kg}}{\text{kW·h}}}$$

$$= 61 \text{ kW}$$

The answer is (B).

10. Find the biomass produced daily in the bioreactor.

$$Y_{\text{obs}} = \text{observed yield coefficient} = 0.36 \text{ g/g}$$
$$X_p = \text{biomass production rate, kg/d}$$
$$S_0 = \text{influent total BOD}_5 = 204 \text{ mg/L}$$
$$S = \text{effluent soluble BOD}_5 = (0.37)(20 \text{ mg/L})$$
$$\quad = 7.4 \text{ mg/L}$$
$$Q = \text{wastewater flow rate} = 5000 \text{ m}^3/\text{d}$$

$$X_p = Y_{\text{obs}}(S_0 - S)Q$$

$$= \left(0.36 \frac{\text{g}}{\text{g}}\right)\left(204 \frac{\text{mg}}{\text{L}} - 7.4 \frac{\text{mg}}{\text{L}}\right)\left(5000 \frac{\text{m}^3}{\text{d}}\right)$$

$$\times \left(1000 \frac{\text{L}}{\text{m}^3}\right)\left(10^{-6} \frac{\text{kg}}{\text{mg}}\right)$$

$$= 354 \text{ kg/d} \quad (350 \text{ kg/d})$$

The answer is (B).

11. The recirculated solids daily mass is

$$\left(1300 \ \frac{m^3}{d}\right)\left(15\,000 \ \frac{mg}{L}\right)\left(10^{-6} \ \frac{kg}{mg}\right)\left(1000 \ \frac{L}{m^3}\right)$$
$$= 19\,500 \ kg/d \quad (20\,000 \ kg/d)$$

The answer is (C).

12. Use the equation for complete mix-activated sludge to find the hydraulic residence time.

θ = hydraulic residence time, h
θ_c = mean cell residence time = 12 d
Y = yield coefficient = 0.23 g/g
S_0 = influent soluble BOD_5 = 235 mg/L
S_e = effluent soluble BOD_5, mg/L
$\quad = (0.3)(10 \ mg/L) = 3 \ mg/L$
X_A = mixed liquor volatile suspended solids
$\quad = 3000 \ mg/L$
k_d = endogenous decay rate coefficient = 0.05 d^{-1}

Activated Sludge

$$X_A = \frac{\theta_c Y (S_0 - S_e)}{\theta (1 + k_d \theta_c)}$$
$$\theta = \frac{\theta_c Y (S_0 - S_e)}{X_A (1 + k_d \theta_c)}$$
$$= \frac{(12 \ d)\left(0.23 \ \frac{g}{g}\right)\left(235 \ \frac{mg}{L} - 3 \ \frac{mg}{L}\right)\left(24 \ \frac{h}{d}\right)}{\left(3000 \ \frac{mg}{L}\right)\left(1 + (0.05 \ d^{-1})(12 \ d)\right)}$$
$$= 3.2 \ h$$

The answer is (A).

13. Use the equation for hydraulic residence time to find the bioreactor volume.

θ = hydraulic residence time = 3.2 h
V = bioreactor volume, m^3
Q = influent flow rate = 8000 m^3/d

Activated Sludge

$$\theta = V/Q$$
$$V = Q\theta$$
$$= \left(8000 \ \frac{m^3}{d}\right)(3.2 \ h)\left(\frac{1 \ d}{24 \ h}\right)$$
$$= 1067 \ m^3 \quad (1100 \ m^3)$$

The answer is (A).

14. Find the organic loading rate (volumetric).

VLR = organic loading rate (volumetric), kg/m^3·d
S_0 = influent soluble BOD_5 = 235 mg/L
Q_0 = influent flow rate = 8000 m^3/d
Vol = bioreactor volume = 1100 m^3

Activated Sludge

$$VLR = Q_0 S_0 / Vol$$
$$= \frac{\left(235 \ \frac{mg}{L}\right)\left(8000 \ \frac{m^3}{d}\right)}{1100 \ m^3}$$
$$\times \frac{\left(10^{-6} \ \frac{kg}{mg}\right)\left(1000 \ \frac{L}{m^3}\right)}{1100 \ m^3}$$
$$= 1.7 \ kg/m^3·d$$

The answer is (B).

15. Find the organic loading rate.

F:M = organic loading rate, d^{-1}
S_0 = influent soluble BOD_5 = 235 mg/L
θ = hydraulic residence time = 3.2 h
X_A = mixed liquor volatile suspended solids

Activated Sludge

$$F:M = Q_0 S_0 / Vol \ X_A$$
$$\theta = \frac{Vol}{Q_0}$$
$$F:M = \frac{S_0}{\theta X_A}$$
$$= \frac{\left(235 \ \frac{mg}{L}\right)\left(24 \ \frac{h}{d}\right)}{(3.2 \ h)\left(3000 \ \frac{mg}{L}\right)}$$
$$= 0.59 \ d^{-1} \quad (0.6 \ d^{-1})$$

The answer is (A).

16. The winter loading rate controls the design. Calculating the surface area gives

$$
\text{surface area} = \dfrac{\left(2300 \; \dfrac{m^3}{d}\right)\left(670 \; \dfrac{mg}{L}\right) \times \left(1000 \; \dfrac{L}{m^3}\right)\left(10^{-6} \; \dfrac{kg}{mg}\right)}{0.122 \; \dfrac{kg}{m^2 \cdot d}}
$$

$$
= 12\,631 \; m^2 \quad (12\,600 \; m^2)
$$

The answer is (B).

17. Calculating the BOD removal efficiency, E_1, gives

$$
E_1 = \dfrac{434 \; \dfrac{mg}{L} - 150 \; \dfrac{mg}{L}}{434 \; \dfrac{mg}{L}} \times 100\% = 65.4\%
$$

Calculating the BOD loading to filter, W, gives

$$
W = \left(434 \; \dfrac{mg}{L}\right)\left(1600 \; \dfrac{m^3}{day}\right)\left(\dfrac{2.204 \; lbf}{10^6 \; mg}\right)\left(1000 \; \dfrac{L}{m^3}\right)
$$

$$
= 1530 \; lbf/day
$$

Use the National Research Council equation for trickling filter performance, and solve for the volume of the filter media in 10^3 cubic feet.

$$
F = \text{recirculation factor} = 2
$$
$$
V = \text{filter volume} = 10^3 \; ft^3
$$

National Research Council (NRC) Trickling Filter Performance

$$
E_1 = \dfrac{100}{1 + 0.0561\sqrt{\dfrac{W}{VF}}}
$$

$$
V = \dfrac{W}{\left(\dfrac{100 - E_1}{0.0561 E_1}\right)^2 F}
$$

$$
= \dfrac{1530 \; \dfrac{lbf}{day}}{\left(\dfrac{100\% - 65.4\%}{(0.0561)(65.4\%)}\right)^2 (2)}
$$

$$
= 8.497 \quad (8497 \; ft^3)
$$

Calculating the cross-sectional surface area, A_s, gives

$$
A_s = \dfrac{V}{D} = \dfrac{8497 \; ft^3}{(3.7 \; m)\left(3.28 \; \dfrac{ft}{m}\right)} = 700.1 \; ft^2 \quad (700 \; ft^2)
$$

The answer is (A).

18. Taking into account the recirculation ratio, use Stokes' law to calculate the hydraulic loading rate.

$$
Q = \text{flow rate, m}^3/d
$$
$$
A_s = \text{cross-sectional surface area, m}^2
$$
$$
\text{HLR} = \text{hydraulic loading rate} = Q/A
$$
$$
R = \text{recirculation ratio}
$$

Stokes' Law

$$
\text{HLR} = Q/A
$$

$$
= \dfrac{Q + QR}{A_s}
$$

$$
= \dfrac{1600 \; \dfrac{m^3}{d} + (2)\left(1600 \; \dfrac{m^3}{d}\right)}{\left(1440 \; \dfrac{min}{d}\right)(64 \; m^2)}
$$

$$
= 0.052 \; m^3/m^2 \cdot min
$$

The answer is (B).

19. Find the organic loading rate (volumetric).

$$
Q_0 = \text{influent flow rate} = 1600 \; m^3/d
$$
$$
S_0 = \text{influent BOD concentration} = 434 \; mg/L
$$
$$
A_s = \text{cross-sectional surface area, m}^2
$$
$$
D = \text{filter depth, m}
$$
$$
\text{OLR} = \text{organic loading rate}
$$

Activated Sludge

$$
\text{OLR} = Q_0 S_0 / Vol
$$

$$
= \dfrac{Q_0 S_0}{A_s D}
$$

$$
= \dfrac{\left(1600 \; \dfrac{m^3}{d}\right)\left(434 \; \dfrac{mg}{L}\right)\left(1000 \; \dfrac{L}{m^3}\right)}{(64 \; m^2)(3.7 \; m)\left(10^6 \; \dfrac{mg}{kg}\right)}
$$

$$
= 2.9 \; kg/m^3 \cdot d
$$

The answer is (C).

20. Determine the distribution arm rotation speed.

ω = distribution arm rotation speed, rpm

HLR = hydraulic loading rate

$\quad = 0.052$ m^3/m^2·min

N = number of distribution arms = 2

DR = dosing rate = 20 cm/rev

$$\omega = \frac{\text{HLR}}{N(\text{DR})}$$

$$= \frac{\left(0.052 \ \dfrac{\text{m}^3}{\text{m}^2\text{·min}}\right)\left(100 \ \dfrac{\text{cm}}{\text{m}}\right)}{(2)\left(20 \ \dfrac{\text{cm}}{\text{rev}}\right)}$$

$$= 0.13 \text{ rpm}$$

The answer is (B).

21. Giardia Cyst control requires a 4-LOG inactivation. Direct filtration allows a 1-LOG inactivation credit for Giardia Cysts, leaving a 3-LOG inactivation requirement for chlorination. For water at 10°C and 6.5 pH with a free chlorine concentration of 2.2 mg/L, $CT = 105$. [CT Values for 3-LOG Inactivation of Giardia Cysts by Free Chlorine]

Calculate the minimum contact time for a perfect reactor.

C = free chlorine concentration, mg/L

$T = T10$, minimum contact time for
a perfect reactor, min

Secondary Drinking Water Standards

$$\text{contact time} = CT$$
$$CT = 105$$
$$T = \frac{105}{C} = \frac{105}{2.2 \ \dfrac{\text{mg}}{\text{L}}}$$
$$= 48 \text{ min}$$

For average baffling, the baffling factor BF is = 0.5. [EPA Baffling Factors]

$T_{\text{theoretical}}$ = hydraulic residence time corrected
for hydraulic efficiency, min

Secondary Drinking Water Standards

$$\text{BF} = \frac{T10}{T_{\text{theoretical}}}$$

$$T_{\text{theoretical}} = \frac{T10}{\text{BF}} = \frac{48 \text{ min}}{0.5}$$

$$= 96 \text{ min}$$

The answer is (C).

22. The precipitation reaction is

$$\text{FeCl}_3 + \text{PO}_4^{3-} \rightarrow \text{FePO}_4 \downarrow + 3\text{Cl}^-$$

Calculate the molecular weights of PO_4^{3-} and FeCl_3. [Periodic Table of Elements]

$$\text{MW}_{\text{PO}_4^{3-}} = 31 \ \frac{\text{g}}{\text{mol}} + (4)\left(16 \ \frac{\text{g}}{\text{mol}}\right)$$
$$= 95 \text{ g/mol}$$

$$\text{MW}_{\text{FeCl}_3} = 56 \ \frac{\text{g}}{\text{mol}} + (3)\left(35.5 \ \frac{\text{g}}{\text{mol}}\right)$$
$$= 162.5 \text{ g/mol}$$

The number of moles of phosphorus per liter of wastewater is

$$\frac{15 \ \dfrac{\text{mg}}{\text{L}}}{\left(95 \ \dfrac{\text{g}}{\text{mol}}\right)\left(1000 \ \dfrac{\text{mg}}{\text{g}}\right)} = 1.58 \times 10^{-4} \text{ mol PO}_4^{3-}/\text{L}$$

1 mol FeCl_3 reacts with 1 mol PO_4^{3-} to produce 1 mol FePO_4, so the number of moles of ferric chloride required per liter is

$$1.58 \times 10^{-4} \text{ mol FeCl}_3/\text{L}$$

In milligrams per liter, the dose of ferric chloride required is

$$\left(1.58 \times 10^{-4} \ \frac{\text{mol FeCl}_3}{\text{L}}\right)$$
$$\times \left(162.5 \ \frac{\text{g}}{\text{mol}}\right)\left(1000 \ \frac{\text{mg}}{\text{g}}\right) = 25.7 \text{ mg/L} \quad (30 \text{ mg/L})$$

The answer is (B).

23.

period	average flow (m^3/h)	total flow per period (m^3)	cumulative flow (m^3)
0000–0200	275	550	550
0200–0400	389	778	1328
0400–0600	621	1242	2570
0600–0800	1340	2680	5250
0800–1000	1383	2766	8016
1000–1200	1312	2624	10 640
1200–1400	1098	2196	12 836
1400–1600	1027	2054	14 890
1600–1800	1084	2168	17 058
1800–2000	886	1772	18 830
2000–2200	259	518	19 348
2200–2400	326	652	20 000
		20 000	

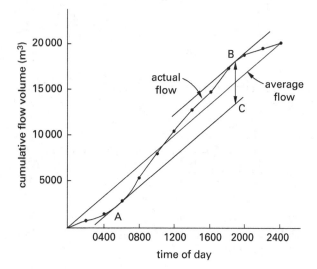

From the figure, the storage volume equals the cumulative flow volume at point B minus the cumulative flow volume at point C.

$$\text{storage volume} = 18\,000 \text{ m}^3 - 12\,500 \text{ m}^3$$
$$= 5500 \text{ m}^3$$

The answer is (B).

24.

period	average flow (m^3/h)	total flow per period (m^3)	cumulative flow (m^3)
0000–0200	275	550	550
0200–0400	389	778	1328
0400–0600	621	1242	2570
0600–0800	1340	2680	5250
0800–1000	1383	2766	8016
1000–1200	1312	2624	10 640
1200–1400	1098	2196	12 836
1400–1600	1027	2054	14 890
1600–1800	1084	2168	17 058
1800–2000	886	1772	18 830
2000–2200	259	518	19 348
2200–2400	326	652	20 000
		20 000	

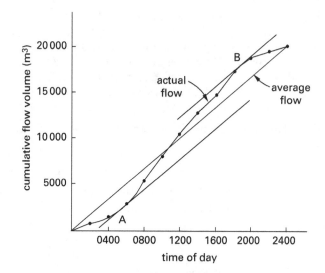

From the figure, the tank is filling whenever the slope of the actual flow line is greater than the slope of the average flow line. The tank is full at point B on the figure corresponding to 1900 h.

The answer is (C).

25.

period	average flow (m^3/h)	total flow per period (m^3)	cumulative flow (m^3)
0000–0200	275	550	550
0200–0400	389	778	1328
0400–0600	621	1242	2570
0600–0800	1340	2680	5250
0800–1000	1383	2766	8016
1000–1200	1312	2624	10 640
1200–1400	1098	2196	12 836
1400–1600	1027	2054	14 890
1600–1800	1084	2168	17 058
1800–2000	886	1772	18 830
2000–2200	259	518	19 348
2200–2400	326	652	20 000
		20 000	

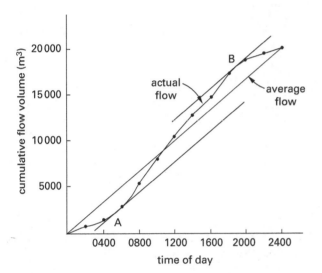

From the figure, the tank is emptying whenever the slope of the actual flow line is less than the slope of the average flow line. The tank is empty at point A on the figure corresponding to 0500 h.

The answer is (A).

26. Use the equation for BOD_5 for mixed lagoons in series, and solve for the kinetic constant in each study.

$$S^0 = \text{influent total BOD, mg/L}$$
$$S = \text{effluent total BOD, mg/L}$$
$$\theta = \text{hydraulic residence time, d}$$
$$k_p = \text{kinetic constant, d}^{-1}$$

$$\frac{S}{S^0} = \frac{1}{1 + k_p\theta}$$

$$k_p = \frac{\dfrac{S^0}{S} - 1}{\theta}$$

S^0 (mg/L)	S (mg/L)	θ (d)	k_p (d^{-1})
280	141	1	0.99
280	96	2	0.96
280	74	3	0.93
280	59	4	0.94
280	45	5	1.04
			4.86

Take the average value.

Dispersion, Mean, Median, and Mode Values

$$\overline{X} = (1/n)\sum_{i=1}^{n} X_i$$

$$\bar{k}_p = \left(\frac{1}{5}\right)(4.86 \text{ d}^{-1})$$

$$= 0.97 \text{ d}^{-1}$$

The answer is (C).

27. Find the bioreactor volume.

$$V = \left(5000\ \frac{m^3}{d}\right)(6\text{ h})\left(\frac{1\text{ d}}{24\text{ h}}\right) = 1250\text{ m}^3$$

Use the equation for solids residence time, and solve for the wasted solids flow rate.

$$X_A = \text{mixed liquor suspended solids}$$
$$= 3100\text{ mg/L}$$
$$Q_w = \text{wasted solids flow rate, m}^3/\text{d}$$
$$Q_e = \text{effluent flow rate, m}^3/\text{d}$$
$$X_w = \text{wasted suspended solids} = 15\,000\text{ mg/L}$$
$$X_e = \text{clarifier effluent TSS} = 20\text{ mg/L}$$
$$\theta_c = \text{solids residence time} = 10\text{ d}$$

Activated Sludge

$$\theta_c = \frac{V(X_A)}{Q_w X_w + Q_e X_e}$$

$$= \frac{V X_A}{Q_w X_w + (Q - Q_w)(X_e)}$$

$$Q_w = \frac{\dfrac{V X_A}{\theta_c} - Q X_e}{X_w - X_e}$$

$$= \frac{\dfrac{(1250 \text{ m}^3)\left(3100 \dfrac{\text{mg}}{\text{L}}\right)}{10 \text{ d}} - \left(5000 \dfrac{\text{m}^3}{\text{d}}\right)\left(20 \dfrac{\text{mg}}{\text{L}}\right)}{15\,000 \dfrac{\text{mg}}{\text{L}} - 20 \dfrac{\text{mg}}{\text{L}}}$$

$$= 19.2 \text{ m}^3/\text{d} \quad (20 \text{ m}^3/\text{d})$$

The answer is (A).

28. The daily sludge mass wasted is

$$\left(15\,000 \frac{\text{mg}}{\text{L}}\right)\left(20 \frac{\text{m}^3}{\text{d}}\right)$$

$$\times \left(10^{-6} \frac{\text{kg}}{\text{mg}}\right)\left(1000 \frac{\text{L}}{\text{m}^3}\right) = 300 \text{ kg/d}$$

The answer is (A).

29. The solids loading rate is

$$\left(1600 \frac{\text{mg}}{\text{L}}\right)\left(8300 \frac{\text{m}^3}{\text{d}}\right)\left(10^{-6} \frac{\text{kg}}{\text{mg}}\right)\left(1000 \frac{\text{L}}{\text{m}^3}\right)$$

$$= 13\,280 \text{ kg/d} \quad (13\,000 \text{ kg/d})$$

The answer is (C).

30. Find the surface area based on solids flux.

$$A_{\text{surface}} = \text{surface area based on solids flux, m}^2$$
$$Q = \text{flow rate} = 8300 \text{ m}^3/\text{d}$$
$$X = \text{influent TSS} = 1600 \text{ mg/L}$$
$$G = \text{solids flux} = 2.6 \text{ kg/m}^2\text{·h}$$

$$A_{\text{surface}} = \frac{QX}{G} = \frac{\left(8300 \dfrac{\text{m}^3}{\text{d}}\right)\left(1600 \dfrac{\text{mg}}{\text{L}}\right)}{\left(2.6 \dfrac{\text{kg}}{\text{m}^2\text{·h}}\right)\left(24 \dfrac{\text{h}}{\text{d}}\right)}$$
$$\quad\quad\quad\quad \times \left(10^{-6} \dfrac{\text{kg}}{\text{mg}}\right)\left(1000 \dfrac{\text{L}}{\text{m}^3}\right)$$

$$= 213 \text{ m}^2 \quad (210 \text{ m}^2)$$

The answer is (B).

31. Use the equation for critical settling velocity to find the required surface area.

$$Q = \text{flow rate} = 8300 \text{ m}^3/\text{d}$$
$$A_{\text{surface}} = \text{surface area based on settling velocity, m}^2$$
$$v_o = \text{settling velocity} = 1.29 \text{ m/h}$$

Clarifier

$$v_o = Q/A_{\text{surface}}$$

$$A_{\text{surface}} = \frac{Q}{v_o}$$

$$= \frac{\left(8300 \dfrac{\text{m}^3}{\text{d}}\right)}{\left(1.29 \dfrac{\text{m}}{\text{h}}\right)\left(24 \dfrac{\text{h}}{\text{d}}\right)}$$

$$= 268 \text{ m}^2 \quad (270 \text{ m}^2)$$

The answer is (C).

32. Use the equation for critical settling velocity to find the design overflow rate.

$$Q = \text{flow rate} = 8300 \text{ m}^3/\text{d}$$
$$A_{\text{surface}} = \text{surface area} = 270 \text{ m}^2$$
$$v_o = \text{overflow rate, m}^3/\text{m}^2\text{·h}$$

Clarifier

$$v_o = Q/A_{\text{surface}} = \frac{8300 \dfrac{\text{m}^3}{\text{d}}}{(270 \text{ m}^2)\left(24 \dfrac{\text{h}}{\text{d}}\right)}$$

$$= 1.28 \text{ m}^3/\text{m}^2\text{·h} \quad (1.3 \text{ m}^3/\text{m}^2\text{·h})$$

The answer is (B).

33. Use the equation for hydraulic residence time, and solve for the reactor volume.

$$Q = \text{influent flow rate} = 3500 \text{ m}^3/\text{d}$$

$$V = \text{reactor volume, m}^3$$

$$\theta = \text{hydraulic residence time} = 4 \text{ h}$$

Activated Sludge

$$\theta = V/Q$$

$$V = Q\theta = \left(3500 \ \frac{\text{m}^3}{\text{d}}\right)\left(\frac{4 \text{ h}}{24 \ \frac{\text{h}}{\text{d}}}\right)$$

$$= 584 \text{ m}^3$$

Use the equation for solids residence time, and solve for the wasted solids flow rate.

$$Q_e = \text{clarifier effluent flow rate, m}^3/\text{d}$$

$$Q_w = \text{wasted solids flow rate, m}^3/\text{d}$$

$$Q_e = Q - Q_w = 3500 \text{ m}^3/\text{d} - Q_w$$

$$X_A = \text{mixed liquor volatile suspended solids}$$

$$= 2600 \text{ mg/L}$$

$$X_w = \text{waste solids concentration} = 12\,000 \text{ mg/L}$$

$$X_e = \text{clarified effluent solids concentration}$$

$$= 20 \text{ mg/L}$$

$$\theta_c = \text{solids residence time} = 3 \text{ d}$$

Activated Sludge

$$\theta_c = \frac{V(X_A)}{Q_w X_w + Q_e X_e}$$

$$= \frac{VX_A}{Q_w X_w + (Q - Q_w)X_e}$$

$$Q_w = \frac{\dfrac{VX_A}{\theta_c} - QX_e}{X_w - X_e}$$

$$= \frac{\dfrac{(584 \text{ m}^3)\left(2600 \ \dfrac{\text{mg}}{\text{L}}\right)}{3 \text{ d}} - \left(3500 \ \dfrac{\text{m}^3}{\text{d}}\right)\left(20 \ \dfrac{\text{mg}}{\text{L}}\right)}{12\,000 \ \dfrac{\text{mg}}{\text{L}} - 20 \ \dfrac{\text{mg}}{\text{L}}}$$

$$= 36.5 \text{ m}^3/\text{d} \quad (40 \text{ m}^3/\text{d})$$

The answer is (B).

34. The absolute inlet temperature is

Temperature Conversions

$$T_1 = (1.8)(20°\text{C}) + 32° + 459.69°$$

$$= 527.69°\text{R}$$

The absolute inlet and outlet pressures are

$$P_1 = (1 \text{ atm})\left(14.7 \ \frac{\frac{\text{lbf}}{\text{in}^2}}{\text{atm}}\right) = 14.7 \text{ lbf/in}^2$$

$$P_2 = (3.2 \text{ atm})\left(14.7 \ \frac{\frac{\text{lbf}}{\text{in}^2}}{\text{atm}}\right) = 47.0 \text{ lbf/in}^2$$

Use the equation for blower power, and solve for the air-flow rate.

$$W = \text{weight of airflow, lbf/sec}$$

$$C = \text{blower constant} = 550 \text{ ft-lbf/sec-hp}$$

$$P_W = \text{power} = 100 \text{ hp}$$

$$R = \text{gas constant for air}$$

$$= 53.3 \text{ ft-lbf/lbf air-°R}$$

$$n = \text{constant} = 0.283 \text{ for air}$$

$$e = \text{efficiency} = 0.82$$

Blowers

$$P_W = \frac{WRT_1}{Cne}\left[\left(\frac{P_2}{P_1}\right)^{0.283} - 1\right]$$

$$W = \frac{\dfrac{P_w Cne}{RT_1}}{\left(\dfrac{P_2}{P_1}\right)^{0.2383} - 1}$$

$$= \frac{\dfrac{(100 \text{ hp})\left(550 \ \dfrac{\text{ft-lbf}}{\text{sec-hp}}\right)(0.283)(0.82)}{\left(53.3 \ \dfrac{\text{ft-lbf}}{\text{lbf-°R}}\right)(527.69°\text{R})}}{\left(\dfrac{47.0 \ \dfrac{\text{lbf}}{\text{in}^2}}{14.7 \ \dfrac{\text{lbf}}{\text{in}^2}}\right)^{0.283} - 1}$$

$$= 1.17 \text{ lbf/sec} \quad (1.2 \text{ lbf/sec})$$

The answer is (A).

35. The concentrate TDS concentration is

$$\frac{(0.97)\left(1120 \; \frac{\text{mg}}{\text{L}}\right)}{1 - 0.81} = 5717.8 \; \text{mg/L} \quad (5720 \; \text{mg/L})$$

The daily volume of concentrate is

$$\left(2700 \; \frac{\text{m}^3}{\text{d}}\right)(1 - 0.81) = 513 \; \text{m}^3/\text{d} \quad (510 \; \text{m}^3/\text{d})$$

The answer is (D).

11 Wastewater Collection Systems

Content in blue refers to the *NCEES Handbook*.

PRACTICE PROBLEMS

1. A suspended solids slurry is pumped using a positive displacement pump with a rated rotational speed of 1150 rpm. The pump suction line is 3.64 m long, and the slurry velocity in the suction line is 1.68 m/s. The fluid compressibility factor is 1.5. The acceleration head factor is 0.115. The acceleration head for the pump is most nearly

- (A) 0.92 m
- (B) 3.4 m
- (C) 55 m
- (D) 140 m

2. A fluid at a temperature of 5°C is pumped at 84% efficiency. The fluid viscosity is 980 cSt at 5°C, but it decreases to 124 cSt at 20°C. The pump efficiency exponent is 0.073. The pump efficiency at 20°C is most nearly

- (A) 81%
- (B) 86%
- (C) 96%
- (D) 98%

3. A stormwater curb inlet has a 36 in inlet length and a 4 in inlet depression. For a flow depth of 2 in, the curb inlet capacity is most nearly

- (A) 0.74 ft³/sec
- (B) 1.1 ft³/sec
- (C) 1.3 ft³/sec
- (D) 1.5 ft³/sec

4. A pump is required to pump 160 gpm against an elevation head of 32 ft. The pump inlet and associated piping has a diameter of 2 in. Head loss due to friction totals to 2.4 ft. The operating temperature is 70°F. The net positive suction head available is most nearly

- (A) −6 ft
- (B) 30 ft
- (C) 60 ft
- (D) 70 ft

SOLUTIONS

1. Calculate the acceleration head.

$$H_{ac} = \text{acceleration head, m}$$
$$L_{\text{suction}} = \text{suction line length} = 3.64 \text{ m}$$
$$v_{\text{ave}} = \text{suction line average velocity} = 1.68 \text{ m/s}$$
$$N = \text{pump rotational speed} = 1150 \text{ rpm}$$
$$K = \text{fluid compressibility factor} = 1.5$$
$$C = \text{acceleration head factor} = 0.115$$
$$g = \text{acceleration of gravity} = 9.81 \text{ m/s}^2$$

$$H_{ac} = \frac{C}{K}\left(\frac{L_{\text{suction}} v_{\text{ave}} N}{g}\right)$$

$$= \left(\frac{0.115}{1.5}\right)\left(\frac{(3.64 \text{ m})\left(1.68 \, \frac{\text{m}}{\text{s}}\right)\left(1150 \, \frac{\text{rev}}{\text{min}}\right)}{\left(9.81 \, \frac{\text{m}}{\text{s}^2}\right)\left(60 \, \frac{\text{s}}{\text{min}}\right)}\right)$$

$$= 0.92 \text{ m}$$

The answer is (A).

2. Calculate the pump efficiency at 20°C.

$$\eta = \text{fractional efficiency, \%}$$
$$n = \text{pump efficiency exponent} = 0.073$$
$$\upsilon = \text{viscosity, cSt}$$

$$\eta_{15°C} = 1 - (100\% - \eta_{5°C})\left(\frac{\upsilon_{20°C}}{\upsilon_{5°C}}\right)^n$$

$$= 1 - (100\% - 84\%)\left(\frac{124 \text{ cSt}}{980 \text{ cSt}}\right)^{0.073}$$

$$= 86.2\% \quad (86\%)$$

The answer is (B).

3. Calculate the curb inlet capacity.

$$Q_{\text{ft}^3/\text{sec}} = (0.7)\left(\begin{array}{c}\text{curb opening} \\ \text{length}\end{array}\right)\left(\begin{array}{c}\text{inlet flow} \\ \text{depth}\end{array} + \begin{array}{c}\text{curb inlet} \\ \text{depression}\end{array}\right)^{3/2}$$

$$= (0.7)\left(\frac{36 \text{ in}}{12 \, \frac{\text{in}}{\text{ft}}}\right)\left(\frac{2 \text{ in} + 4 \text{ in}}{12 \, \frac{\text{in}}{\text{ft}}}\right)^{3/2}$$

$$= 0.74 \text{ ft}^3/\text{sec}$$

The answer is (A).

4. Find the cross-sectional area of the pipe.

$$A = \text{pipe cross-sectional area, ft}^2$$
$$D = \text{pipe diameter} = 2 \text{ in}$$

$$A = \frac{\pi D^2}{4} = \frac{\pi(2 \text{ in})^2}{(4)\left(12 \, \frac{\text{in}}{\text{ft}}\right)^2} = 0.022 \text{ ft}^2$$

Use the continuity equation, and solve for the flow velocity.

$$v = \text{flow velocity}$$
$$Q = \text{flow rate} = 160 \text{ gpm}$$

Continuity Equation

$$Q = Av$$

$$v = \frac{Q}{A} = \frac{\left(160 \, \frac{\text{gal}}{\text{min}}\right)\left(0.134 \, \frac{\text{ft}^3}{\text{gal}}\right)}{(0.022 \text{ ft}^2)\left(60 \, \frac{\text{sec}}{\text{min}}\right)}$$

$$= 16.2 \text{ ft/sec}$$

Find the properties of water at 70°F. [Properties of Water (English Units)]

$$P_{\text{vapor}} = \text{water vapor pressure} = 0.36 \text{ psi}$$
$$\rho = \text{water density} = 1.936 \text{ lbf-sec}^2/\text{ft}^4$$

Find the net positive suction head available.

$$NPSH_A = \text{net positive suction head available, ft}$$
$$P_{\text{atm}} = \text{atmospheric pressure} = 14.7 \text{ lbf/in}^2$$
$$H_s = \text{elevation head} = 32 \text{ ft}$$
$$H_f = \text{friction head} = 2.4 \text{ ft}$$
$$V = \text{fluid velocity at pump inlet} = 16.2 \text{ ft/sec}$$

Centrifugal Pump Characteristics

$$NPSH_A = \frac{P_{\text{atm}}}{\rho g} \pm H_s - H_f - \frac{V^2}{2g} - \frac{P_{\text{vapor}}}{\rho g}$$

$$= \frac{\left(14.7 \ \dfrac{\text{lbf}}{\text{in}^2}\right)\left(12 \ \dfrac{\text{in}}{\text{ft}}\right)^2}{\left(1.936 \ \dfrac{\text{lbf-sec}^2}{\text{ft}^4}\right)\left(32.2 \ \dfrac{\text{ft}}{\text{sec}^2}\right)}$$

$$+ 32 \ \text{ft} - 2.4 \ \text{ft} - \frac{\left(16.2 \ \dfrac{\text{ft}}{\text{sec}}\right)^2}{(2)\left(32.2 \ \dfrac{\text{ft}}{\text{sec}^2}\right)}$$

$$- \frac{\left(0.36 \ \dfrac{\text{lbf}}{\text{in}^2}\right)\left(12 \ \dfrac{\text{in}}{\text{ft}}\right)^2}{\left(1.936 \ \dfrac{\text{lbf-sec}^2}{\text{ft}^4}\right)\left(32.2 \ \dfrac{\text{ft}}{\text{sec}^2}\right)}$$

$$= 58.6 \ \text{ft} \quad (60 \ \text{ft})$$

The answer is (C).

12 Wastewater Residuals Management

Content in blue refers to the *NCEES Handbook*.

PRACTICE PROBLEMS

1. A publicly owned treatment works generates 380 m³/d of thickened biological sludge with a biochemical oxygen demand (BOD) of 274 mg/L. The sludge is treated by anaerobic digestion to produce a stabilized sludge with a BOD of 32 mg/L. The digester wastes the equivalent of 11 kg of dry sludge solids daily. The methane gas generated by the digester has a heating value of 22 MJ/m³. The total heating value of methane at ambient conditions generated by the digester is most nearly

(A) 590 MJ/d

(B) 680 MJ/d

(C) 700 MJ/d

(D) 1680 MJ/d

2. A wastewater treatment plant produces 284 m³/d of sludge cake at 27% solids. Heating the sludge to 100°C will deactivate any pathogens and remove disposal restrictions. Before heating, the sludge temperature is the same as the ambient air at 20°C. The heat needed to heat the sludge is most nearly

(A) 3×10^6 kJ/d

(B) 7×10^7 kJ/d

(C) 1×10^8 kJ/d

(D) 3×10^8 kJ/d

3. A waste discharged at 50,000 gal/day contains hexavalent chromium (Cr(VI)) at 165 mg/L. Sodium metabisulfate, sulfuric acid, and caustic soda are used as reagents to produce the following reactions.

$$4CrO_3 + 3Na_2S_2O_5 + 3H_2SO_4$$
$$\rightarrow 3Na_2SO_4 + 2Cr_2(SO_4)_3 + 3H_2O$$
$$Cr_2(SO_4)_3 + 6NaOH$$
$$\rightarrow 2Cr(OH)_3 + 3Na_2SO_4$$

The daily dry mass of chromium sludge produced is most nearly

(A) 32 kg/d

(B) 65 kg/d

(C) 120 kg/d

(D) 31 000 kg/d

4. A sewage treatment plant processes a waste biological sludge. The sludge is dewatered to 23% solids by weight and then stabilized by raising the pH to 12.5 using lime dosed at 315 g $Ca(OH)_2$/kg dry solids. The plant wastes 18,000 gal of sludge daily at 9% solids by weight. The locally available lime contains 22% inerts. The monthly mass of lime required to stabilize the sludge is most nearly

(A) 45 000 kg/mo

(B) 58 000 kg/mo

(C) 74 000 kg/mo

(D) 190 000 kg/mo

5. A thickener receives 500 m³/d of a mixed suspension with a total suspended solids concentration of 5860 mg/L. The thickener is capable of providing a thickened solids concentration of 11 000 mg/L. The solids volume wasted through the thickener is most nearly

(A) 32 m³/d

(B) 270 m³/d

(C) 500 m³/d

(D) 2900 m³/d

SOLUTIONS

1. Calculate the daily volume of methane produced. The theoretical yield of methane is $0.35 \text{ m}^3/\text{kg}$.

V = volume of methane, m^3/d
P_x = volume of sludge wasted = 1000 kg/d
Q = flow rate of sludge generated = $380 \text{ m}^3/\text{d}$
S_o = BOD of untreated sludge = 274 mg/L
S = BOD of treated sludge = 32 mg/L
P_x = dry mass of solids wasted = 11 kg/d

$$V = \left(0.35 \frac{\text{m}^3}{\text{kg}}\right)\left|\frac{(S_o - S)Q}{1000 \frac{\text{mg} \cdot \text{m}^3}{\text{kg} \cdot \text{L}}} - 1.42 P_x\right|$$

$$= \left(0.35 \frac{\text{m}^3}{\text{kg}}\right)\left|\frac{\left(274 \frac{\text{mg}}{\text{L}} - 32 \frac{\text{mg}}{\text{L}}\right)\left(380 \frac{\text{m}^3}{\text{d}}\right)}{1000 \frac{\text{mg} \cdot \text{m}^3}{\text{kg} \cdot \text{L}}} - (1.42)\left(11 \frac{\text{kg}}{\text{d}}\right)\right|$$

$$= 26.7 \text{ m}^3/\text{d}$$

Calculate the total heating value.

q = total heating value, MJ/m^3
LHV = lower heating value = 22 MJ/m^3

$$q = (\text{LHV})V$$
$$= \left(22 \frac{\text{MJ}}{\text{m}^3}\right)\left(26.7 \frac{\text{m}^3}{\text{d}}\right)$$
$$= 587 \text{ MJ/d} \quad (590 \text{ MJ/d})$$

The answer is (A).

2. Calculate the mass of water in the sludge cake. [Specific Gravity for a Solids Slurry]

M = mass of water, kg/d
f = solids content of sludge = 27%
Q_s = daily volume of sludge cake = $284 \text{ m}^3/\text{d}$
ρ_s = water density = 1000 kg/m^3

$$Q_s = \frac{M(100)}{\rho_s(\% \text{ solids})}$$

$$M = Q_s\rho_s\left(\frac{100\% - f}{100\%}\right)$$

$$= \left(284 \frac{\text{m}^3}{\text{d}}\right)\left(1000 \frac{\text{kg}}{\text{m}^3}\right)\left(\frac{100\% - 27\%}{100\%}\right)$$

$$= 2.1 \times 10^5 \text{ kg/d}$$

Calculate the heat needed to heat the sludge from $20°C$ to $100°C$.

c = specific heat of water at $20°C$ = $4.182 \text{ kJ/kg} \cdot \text{K}$
[Thermophysical Properties of Air and Water]
Q = heat required, kJ/d
ΔT = temperature difference = $T_2 - T_1$

State Functions (properties)

$$Q = mc\Delta T = mc(T_2 - T_1)$$
$$= \left(2.1 \times 10^5 \frac{\text{kg}}{\text{d}}\right)\left(4.182 \frac{\text{kJ}}{\text{kg} \cdot \text{K}}\right)$$
$$\times \left((100°C + 273°) - (20°C + 273°)\right)$$
$$= 7 \times 10^7 \text{ kJ/d}$$

The answer is (B).

3. The constituents of interest are chromium trioxide (CrO_3), chromium sulfate ($Cr_2(SO_4)_3$), and chromium hydroxide ($Cr(OH)_3$).

Find the molecular weight of CrO_3. [Periodic Table of Elements]

$$52 \frac{\text{g}}{\text{mol}} + (3)\left(16 \frac{\text{g}}{\text{mol}}\right) = 100 \text{ g/mol}$$

The molar concentration of CrO_3 is

$$M = \frac{165 \frac{\text{mg } CrO_3}{\text{L}}}{\left(100 \frac{\text{g}}{\text{mol}}\right)\left(1000 \frac{\text{mg}}{\text{g}}\right)}$$

$$= 0.001\,65 \text{ mol } CrO_3/\text{L}$$

The molecular weight of $Cr(OH)_3$ is

$$52 \frac{\text{g}}{\text{mol}} + (3)\left(16 \frac{\text{g}}{\text{mol}} + 1 \frac{\text{g}}{\text{mol}}\right) = 103 \text{ g/mol}$$

One mole of CrO_3 will react to produce $2/4$ mol of $Cr_2(SO_4)_3$. The molar concentration of $Cr_2(SO_4)_3$ is

$$M = \left(0.001\,65 \ \frac{\text{mol } CrO_3}{\text{L}}\right)\left(\frac{\frac{2}{4} \ \text{mol } Cr_2(SO_4)_3}{1 \ \text{mol } CrO_3}\right)$$

$$= 0.000\,825 \ \text{mol } Cr_2(SO_4)_3/\text{L}$$

One mole of $Cr_2(SO_4)_3$ will react to produce 2 mol of $Cr(OH)_3$ sludge. The molar concentration of $Cr(OH)_3$ is

$$M = \left(0.000\,825 \ \frac{\text{mol } Cr_2(SO_4)_3}{\text{L}}\right)\left(\frac{2 \ \text{mol } Cr(OH)_3}{1 \ \text{mol } Cr_2(SO_4)_3}\right)$$

$$= 0.001\,65 \ \text{mol } Cr(OH)_2/\text{L}$$

The mass of sludge produced is

$$m = \left(0.001\,65 \ \frac{\text{mol } Cr(OH)_3}{\text{L}}\right)\left(103 \ \frac{\text{g}}{\text{mol}}\right)\left(50{,}000 \ \frac{\text{gal}}{\text{day}}\right)$$

$$\times \left(3.785 \ \frac{\text{L}}{\text{gal}}\right)\left(\frac{1 \ \text{kg}}{1000 \ \text{g}}\right)$$

$$= 32 \ \text{kg/d}$$

The answer is (A).

4. Use the equation for sludge volume to find the dry solids mass.

$$V = \text{sludge volume} = 18{,}000 \ \text{gal/day}$$

$$W_s = \text{dry solids mass, kg/d}$$

$$\gamma = \text{water unit mass} = 1000 \ \text{kg/m}^3$$

$$S = \text{wet sludge specific gravity} = 1.0 \ (\text{assumed})$$

$$s = \text{solids content} = 9\%$$

Specific Gravity for a Solids Slurry

$$V = \frac{W_s}{(S/100)\gamma S}$$

$$W_s = V\gamma S\left(\frac{S}{100}\right)$$

$$= \left(18{,}000 \ \frac{\text{gal}}{\text{day}}\right)\left(0.003\,785 \ \frac{\text{m}^3}{\text{gal}}\right)$$

$$\times \left(1000 \ \frac{\text{kg}}{\text{m}^3}\right)(1.0)\left(\frac{9\%}{100\%}\right)$$

$$= 6132 \ \text{kg/day}$$

The lime requirement is

$$\frac{\left(6132 \ \frac{\text{kg}}{\text{d}}\right)\left(30 \ \frac{\text{d}}{\text{mo}}\right)\left(315 \ \frac{\text{g Ca(OH)}_2}{\text{kg}}\right)}{(1-0.22)\left(1000 \ \frac{\text{g}}{\text{kg}}\right)}$$

$$= 74\,292 \ \text{kg/mo} \quad (74\,000 \ \text{kg/mo})$$

The answer is (C).

5. Find the dry solids mass.

$$Q = \text{flow rate} = 500 \ \text{m}^3/\text{d}$$

$$X = \text{TSS concentration} = 5860 \ \text{mg/L}$$

$$W_s = \text{dry solids mass, kg/d}$$

$$W_s = QX$$

$$= \left(500 \ \frac{\text{m}^3}{\text{d}}\right)\left(\frac{\left(5860 \ \frac{\text{mg}}{\text{L}}\right)\left(1000 \ \frac{\text{L}}{\text{m}^3}\right)}{10^6 \ \frac{\text{mg}}{\text{kg}}}\right)$$

$$= 2930 \ \text{kg/d}$$

Find the sludge volume.

$$V = \text{sludge volume, m}^3$$

$$\gamma = \text{water unit mass} = 1000 \ \text{kg/m}^3$$

$$S = \text{wet sludge specific gravity} = 1.0 \ (\text{assumed})$$

$$s = \text{solids content} = 11\,000 \ \text{mg}/10^6 = 0.011 \ (1.1\%)$$

Specific Gravity for a Solids Slurry

$$V = \frac{W_s}{(S/100)\gamma S}$$

$$= \frac{2930 \ \frac{\text{kg}}{\text{d}}}{\left(\frac{1.1\%}{100\%}\right)(1.0)\left(1000 \ \frac{\text{kg}}{\text{m}^3}\right)}$$

$$= 266 \ \text{m}^3/\text{d} \quad (270 \ \text{m}^3/\text{d})$$

The answer is (B).

13 Wastewater Reuse

Content in blue refers to the *NCEES Handbook.*

PRACTICE PROBLEMS

1. Reclaimed wastewater is typically required to meet public health water quality standards when the water is used for what purpose?

(A) dust suppression during road construction

(B) fighting brush and timber fires in national forests

(C) irrigation of non-food chain crops on commercial farms

(D) landscape irrigation on public or private property

SOLUTIONS

1. Reclaimed wastewater will typically be required to meet public health water quality standards only when a reasonable risk of contact with the general public is likely, such as landscape irrigation on public and private property. Dust suppression during road construction, fighting forest fires, and irrigation of non-food chain crops on commercial farmland do not present a likely risk of exposure to the general public.

The answer is (D).

14 Stormwater Pollution, Treatment, Management

Content in blue refers to the *NCEES Handbook*.

PRACTICE PROBLEMS

1. A waterway adjacent to agricultural land has a nitrogen concentration, measured as NH_4^+-N, of 5.7 mg/L. The oxygen demand created by the nitrogen is most nearly

(A) 6 mg O_2/L

(B) 13 mg O_2/L

(C) 26 mg O_2/L

(D) 31 mg O_2/L

2. A clarifier treats 150 m³/d of a water with a total suspended solids concentration of 247 mg/L. The hydraulic loading rate to the clarifier is 2.4 m/h. The overflow rate to the clarifier is most nearly

(A) 2.4 m³/m²·h

(B) 37 kg/d

(C) 62 m³/m

(D) 100 mg/m²·h

3. Stormwater runoff from a fully developed urban area is collected for treatment before it is discharged into an estuary. The stormwater enters a basin used for retention and treatment to remove suspended solids. Assume the suspended solids are typical of primary wastewater. The basin has a surface area of 9000 m² and a maximum depth of 1.2 m. The average retention time of the stormwater in the basin is 3.4 h. Is the basin likely to provide a high level of suspended solids removal?

(A) No, the overflow rate is above typical values.

(B) No, the overflow rate is below typical values.

(C) Yes, the overflow rate is above typical values.

(D) Yes, the overflow rate is below typical values.

4. Stormwater runoff from an industrialized area contains BOD at a concentration of 120 mg/L. The runoff is collected in a facultative lagoon for passive treatment. The normal interval between storm events averages 18 days. The lagoon follows a first order decay with a decay rate of 0.13 d⁻¹. What is the average BOD removal efficiency for the lagoon?

(A) 84%

(B) 88%

(C) 90%

(D) 95%

5. Stormwater runoff from a wood treatment facility is collected in an impoundment. Oil used in the wood treatment process collects on the surface of the impounded stormwater. An adjacent sod farm will use the impounded water for irrigation if the facility can eliminate the floating oil. Which of the following is NOT an appropriate method for controlling the oil at the pond discharge?

(A) Aerate the pond and discharge the water through an overflow weir.

(B) Apply chemical dispersants to emulsify the oil.

(C) Locate the discharge point near the bottom of the pond at a point distant from the inlet.

(D) Place recoverable absorbent materials near the outlet structure.

SOLUTIONS

1. The biochemical oxygen demand created by the nitrogen (NBOD) is defined by the following relationship.

$$\text{NBOD} = \frac{\text{mass of oxygen used}}{\text{mass of nitrogen oxidized}}$$

$$= \frac{(2 \text{ mol})\left((2)\left(16 \dfrac{\text{g O}_2}{\text{mol}}\right)\right)}{(1 \text{ mol})\left(14 \dfrac{\text{g N}}{\text{mol}}\right)}$$

$$= 4.57 \text{ g O}_2/\text{g N}$$

Calculate the oxygen demand created by the nitrogen. 1 g N produces 4.57 g O_2 demand.

$$\left(5.7 \frac{\text{mg N}}{\text{L}}\right)\left(4.57 \frac{\text{g O}_2}{\text{g N}}\right) = 26 \text{ mg O}_2/\text{L}$$

The answer is (C).

2. The hydraulic loading rate and the overflow rate are two terms for the same thing. [Clarifier]

The overflow rate is 2.4 m^3/m^2·h.

The answer is (A).

3. Find the volume of the basin.

$$A_{\text{surface}} = \text{surface area, } 9000 \text{ m}^2$$
$$d = \text{depth, } 1.2 \text{ m}$$
$$V = \text{volume}$$

$$V = A_{\text{surface}} d$$
$$= (9000 \text{ m}^2)(1.2 \text{ m})$$
$$= 10\,800 \text{ m}^3$$

Find the flow rate using the equation for hydraulic residence time.

$$V = \text{volume} = 10\,800 \text{ m}^3$$
$$\theta = \text{hydraulic residence time} = 3.4 \text{ h}$$
$$Q = \text{flow rate, m}^3/\text{d}$$

Clarifier

$$\theta = V/Q$$

$$Q = \frac{V}{\theta}$$

$$= \frac{(10\,800 \text{ m}^3)\left(24 \dfrac{\text{h}}{\text{d}}\right)}{3.4 \text{ h}}$$

$$= 76\,235 \text{ m}^3/\text{d}$$

Find the overflow rate using the equation for hydraulic loading rate.

$$v_o = \text{overflow rate m}^3/\text{m}^2\cdot\text{d}$$

Clarifier

$$v_o = Q/A_{\text{surface}}$$

$$= \frac{76\,235 \dfrac{\text{m}^3}{\text{d}}}{9000 \text{ m}^2}$$

$$= 8.5 \text{ m}^3/\text{m}^2\cdot\text{d}$$

Typical overflow rates for primary wastewater clarification range from 30–50 m^3/m^2·d for average flows and may approach 100 m^3/m^2·d for peak flows. An overflow rate of 8.5 m^3/m^2·d is well below typical values and suggests that the basin will provide a high level of suspended solids removal.

The answer is (D).

4. Find the final BOD concentration using the equation for first-order irreversible reaction kinetics.

$$k = \text{first-order decay rate} = 0.13 \text{ d}^{-1}$$
$$t = \text{reaction time} = 18 \text{ d}$$
$$C_{A0} = \text{initial BOD concentration} = 120 \text{ mg/L}$$
$$C_A = \text{final BOD concentration, mg/L}$$

First-Order Irreversible Reaction Kinetics

$$\ln(C_A/C_{A0}) = -kt$$

$$C_A = C_{A0}e^{-kt} = \left(120 \frac{\text{mg}}{\text{L}}\right)e^{(-0.13 \text{ 1/d})(18 \text{ d})}$$

$$= 11.6 \text{ mg/L}$$

The BOD removal efficiency is

$$\frac{120 \dfrac{\text{mg}}{\text{L}} - 11.6 \dfrac{\text{mg}}{\text{L}}}{120 \dfrac{\text{mg}}{\text{L}}} \times 100\% = 90\%$$

The answer is (C).

5. Aerating the pond and discharging the water through an overflow weir will likely not be effective in removing the oil. Aerating will momentarily mix the oil with the water, but as it moves away from the aerator toward the overflow weir, the oil will again accumulate at the surface and be discharged with the water over the weir.

Locating the discharge point near the bottom of the pond will draw oil-free water from below the surface, and locating the discharge point distant from the inlet will move oil-free water away from potential mixing that may occur there.

Using recoverable absorbent materials near the outlet structure will prevent oil from entering the structure and provide a means to remove the oil for disposal elsewhere. Even better, would be combining a near-bottom discharge point with absorbent materials.

Because the water will be used to irrigate non-food chain crops, applying chemical dispersants to emulsify the oil may also be a permissible alternative.

The answer is (A).

15 Stormwater Collection Systems

Content in blue refers to the *NCEES Handbook*.

PRACTICE PROBLEMS

1. Local public works personnel have constructed a rough earthen ditch to divert frequent floodwater away from residential property. When in use, the water depth in the ditch averages 0.8 m. The ditch is 800 m long on a slope of 1.8% with 1:1 side slopes and a 0.5 m wide bottom. The flow discharge in the earthen ditch is most nearly

(A) 1.0 m³/s

(B) 3.0 m³/s

(C) 10 m³/s

(D) 20 m³/s

2. A rectangular channel has a depth of 16 in and a top width of 24 in and is lined with finished concrete. The channel flows half full over a length of 100 ft. The maximum desired velocity is 6 ft/sec. The maximum allowable slope for the channel is most nearly

(A) 0.0003 ft/ft

(B) 0.01 ft/ft

(C) 0.02 ft/ft

(D) 0.09 ft/ft

3. The profile of a river channel is shown in the figure. The channel flow was measured at distances from the east bank at 0.6 of the total depth in shallower sections and at 0.2 and 0.8 of the total depth in deeper sections. The flow measurement data are presented in the accompanying table.

distance from east bank

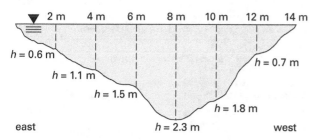

distance from east bank (m)	section depth, h (m)	flow velocity (m/s) at depth		
		$0.6h$	$0.2h$	$0.8h$
2	0.6	0.21	–	–
4	1.1	–	0.27	0.19
6	1.5	–	0.29	0.19
8	2.3	–	0.31	0.20
10	1.8	–	0.31	0.21
12	0.7	0.26	–	–

The average stream discharge is most nearly

(A) 4.0 m³/s

(B) 4.8 m³/s

(C) 5.5 m³/s

(D) 9.5 m³/s

4. In some parts of its length, the storm sewer shown in the illustration flows full during extreme rainfall events. At what manhole, if any, does the stormwater overflow through the manhole onto the street surface?

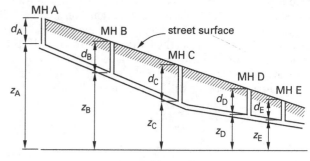

MH = manhole
d = depth below grade
z = elevation

not to scale

location	pipe centerline depth (ft)	pipe centerline elevation (ft)	water pressure (lbf/in²)
MH A	8.2	100	0
MH B	10.5	86	4.2
MH C	12.4	71	6.8
MH D	6.8	54	3.1
MH E	6.2	43	1.2

   (A)   none

   (B)   manholes A, B, and E

   (C)   manholes C and D

   (D)   manholes B, C, and D

5. Local public works personnel have constructed a rough earthen ditch to divert frequent floodwater away from residential property. When in use, the water depth in the ditch averages 0.8 m. The ditch is 800 m long on a slope of 1.8% with 1:1 side slopes and a 0.5 m wide bottom. The flow velocity in the earthen ditch is most nearly

   (A)   1.0 m/s

   (B)   3.0 m/s

   (C)   10 m/s

   (D)   20 m/s

6. Local public works personnel have constructed a rough earthen ditch to divert frequent floodwater away from residential property. The ditch has proven to be effective and the residents want the city to line the ditch with asphalt. When in use, the water depth in the ditch averages 0.8 m. The ditch is 800 m long on a slope of 1.8% with 1:1 side slopes and a 0.5 m wide bottom. The flow velocity in the earthen ditch is 3.0 m/s. If the dimensions and configuration of the ditch remain unchanged, the average depth of water in the ditch after it is lined with asphalt will be most nearly

   (A)   0.4 m

   (B)   0.6 m

   (C)   0.9 m

   (D)   1.1 m

SOLUTIONS

1. A cross section of the ditch is shown in the figure.

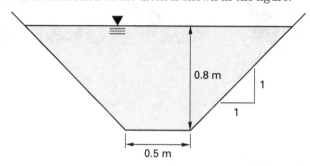

From the figure, the hydraulic radius of the ditch is

Geometric Elements of Channel Sections

$$R_H = \frac{(b + zy)y}{b + 2y\sqrt{1 + z^2}}$$

$$= \frac{(0.5 \text{ m} + (1)(0.8 \text{ m}))(0.8 \text{ m})}{0.5 \text{ m} + (2)(0.8 \text{ m})\sqrt{1 + 1^2}}$$

$$= 0.38 \text{ m}$$

Find the flow velocity.

$v =$ flow velocity, m/s

$n =$ Manning roughness coefficient

   $= 0.025$ (typical for earthen ditch)

$S =$ slope, fraction $= 1.8\%/100\% = 0.018$

$K =$ unit factor $= 1.0$ for SI units

Manning's Equation

$$v = (K/n)R_H^{2/3}S^{1/2}$$

$$= \frac{(0.38 \text{ m})^{2/3}(0.018)^{1/2}}{0.025}$$

$$= 2.8 \text{ m/s} \quad (3.0 \text{ m/s})$$

The channel cross-sectional area is

$$A = (0.8 \text{ m})^2 + (0.5 \text{ m})(0.8 \text{ m}) = 1.04 \text{ m}^2$$

Calculating the flow rate of water transported through the channel cross-sectional area gives

Continuity Equation

$$Q = Av = (1.04 \text{ m}^2)\left(3.0 \frac{\text{m}}{\text{s}}\right)$$

$$= 3.12 \text{ m}^3/\text{s} \quad (3.0 \text{ m}^3/\text{s})$$

The answer is (B).

2. Find the hydraulic radius for a rectangular channel section.

$$R_H = \text{hydraulic radius, ft}$$
$$b = \text{top and bottom width} = 24 \text{ in}$$
$$y = \text{depth of flow} = 8 \text{ in}$$

Geometric Elements of Channel Sections

$$R_H = \frac{by}{b+2y}$$

$$= \frac{(24 \text{ in})(8 \text{ in})\left(12 \dfrac{\text{in}}{\text{ft}}\right)}{\left(24 \text{ in} + (2)(8 \text{ in})\right)\left(12 \dfrac{\text{in}}{\text{ft}}\right)^2}$$

$$= 0.4 \text{ ft}$$

For finished concrete, the Manning's roughness coefficient is 0.012. Use Manning's equation and solve for the maximum slope. [Manning's Values of Roughness Coefficient, n]

$$v = \text{velocity} = 6 \text{ ft/sec}$$
$$K = \text{unit conversion} = 1.486$$
$$S = \text{slope, ft/ft}$$

Manning's Equation

$$v = (K/n)R_H^{2/3}S^{1/2}$$

$$S = \left(\frac{v}{\left(\dfrac{K}{n}\right)R_H^{2/3}}\right)^2$$

$$= \left(\frac{6 \dfrac{\text{ft}}{\text{sec}}}{\left(\dfrac{1.486}{0.012}\right)(0.4)^{2/3}}\right)^2$$

$$= 0.008 \text{ ft/ft} \quad (0.01 \text{ ft/ft})$$

An alternative method is to use the Hazen-Williams equation, which gives approximately the same answer.

The answer is (B).

3. The section flow rate is the product of the average section velocity and the section area. The flow measurement data shown includes the section flow rate for each distance from the bank. [Continuity Equation]

distance from east bank (m)	total section velocity (m/s)	section area (m²)	section flow rate (m³/s)
2	0.21	2 m × 0.6 m	0.252
4	0.23	2 m × 1.1 m	0.506
6	0.24	2 m × 1.5 m	0.720
8	0.26	2 m × 2.3 m	1.196
10	0.26	2 m × 1.8 m	0.936
12	0.26	2 m × 0.7 m	0.364
			3.97

The average stream discharge is

$$3.97 \text{ m}^3/\text{s} \quad (4.0 \text{ m}^3/\text{s})$$

The answer is (A).

4. The equation for the ground surface elevation is

$$z_g = z + d_z$$

$$d_z = \text{pipe centerline depth below grade, ft}$$
$$z = \text{pipe centerline elevation, ft}$$
$$z_g = \text{ground surface elevation, ft}$$

The equation for the water depth is

$$d_p = \frac{p}{C}$$

$$C = \text{conversion from lbf/in}^2 \text{ to ft of water,}$$
$$(0.433 \text{ lbf/in}^2)/\text{ft of water}$$
$$d_p = \text{water depth, ft}$$
$$p = \text{water pressure, lbf/in}^2$$

The equation for the elevation of the hydraulic grade line is

$$HGL = z + d_p$$

$$HGL = \text{elevation of hydraulic grade line, ft}$$

When the hydraulic grade line is greater than the ground surface elevation, water overflows through the manhole onto the street. The calculation results are summarized in the table shown.

Water

location	ground surface elevation (ft)	pipe centerline elevation (ft)	water pressure (lbf/in²)
MH A	108.2	100	0
MH B	96.5	86	4.2
MH C	83.4	71	6.8
MH D	60.8	54	3.1
MH E	49.2	43	1.2

location	water pressure depth (ft)	hydraulic grade line elevation (ft)	hydraulic grade line above ground surface
MH A	0	100	no
MH B	9.7	95.7	no
MH C	16	87.0	yes
MH D	7.2	61.2	yes
MH E	2.8	45.8	no

Water overflows through manholes C and D.

The answer is (C).

5. A cross section of the ditch is shown in the figure.

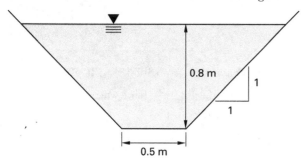

0.8 m

0.5 m

From the figure, the hydraulic radius of the ditch is

Geometric Elements of Channel Sections

$$R_H = \frac{(b + zy)y}{b + 2y\sqrt{1 + z^2}}$$

$$= \frac{(0.5 \text{ m} + (1)(0.8 \text{ m}))(0.8 \text{ m})}{0.5 \text{ m} + (2)(0.8 \text{ m})\sqrt{1 + 1^2}}$$

$$= 0.38 \text{ m}$$

Find the flow velocity.

$v = $ flow velocity, m/s

$n = $ Manning roughness coefficient

$\quad = 0.025$ (typical for earthen ditch)

$S = $ slope, fraction $= 1.8\%/100\% = 0.018$

$K = $ unit factor $= 1.0$ for SI units

Manning's Equation

$$v = (K/n)R_H^{2/3}S^{1/2}$$

$$= \frac{(0.38 \text{ m})^{2/3}(0.018)^{1/2}}{0.025}$$

$$= 2.8 \text{ m/s} \quad (3.0 \text{ m/s})$$

The answer is (B).

6. A cross section of the ditch is shown in the figure.

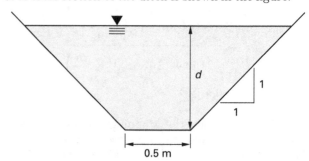

d

1
1

0.5 m

The volumetric flow rate, Q, remains unchanged before and after lining with asphalt, and the typical n for asphalt lining is 0.015. Use Manning's equation and substitute the variables for area and wetted perimeter with expressions in terms of depth, d.

$$A = \text{area} = d(0.5 \text{ m} + d)$$

$$P = \text{wetted perimeter} = 0.5 \text{ m} + (2)(d^2 + d^2)^{1/2}$$

$$= 0.5 \text{ m} + 2.83d$$

$$R_H = \text{hydraulic radius} = A/P$$

$$K = \text{unit factor} = 1.0 \text{ for SI units}$$

Manning's Equation

$$v = (K/n)R_H^{2/3}S^{1/2}$$

$$= \left(\frac{1.0}{n}\right)R_H^{2/3}S^{1/2}$$

$$\frac{Q}{A} = \frac{R_H^{2/3}S^{1/2}}{n}$$

Replace R_H with A/P.

$$= \frac{A^{2/3}S^{1/2}}{P^{2/3}n}$$

$$3.0 \frac{\text{m}}{\text{s}} = \frac{(d(0.5 \text{ m} + d))^{2/3}(0.018)^{1/2}}{(0.5 \text{ m} + 2.83d)^{2/3}(0.015)}$$

Solve for d by trial and error.

$$d = 0.34 \text{ m} \quad (0.4 \text{ m})$$

The answer is (A).

16 Potable Water Source Quality

Content in blue refers to the *NCEES Handbook*.

PRACTICE PROBLEMS

1. A chemical analysis of a water sample has produced the results shown in the table.

cation	concentration (mg/L)	anion	concentration (mg/L)
Ca^{+2}	158	SO_4^{-2}	64
Mg^{+2}	47	Cl^-	43
Na^+	26	HCO_3^-	381
K^+	19		

Is the analysis complete?

(A) No, because the cation concentration in meq/L exceeds the anion concentration in meq/L by more than 10%.

(B) Yes, because the cation concentration in meq/L exceeds the anion concentration in meq/L by more than 10%.

(C) No, because the cation concentration in mg/L exceeds the anion concentration in mg/L by more than 10%.

(D) Yes, because the cation concentration in mg/L exceeds the anion concentration in mg/L by more than 10%.

2. A chemical analysis of a water sample has produced the results shown in the table.

cation	concentration (mg/L)	anion	concentration (mg/L)
Ca^{+2}	158	SO_4^{-2}	64
Mg^{+2}	47	Cl^-	43
Na^+	26	HCO_3^-	381
K^+	19		

The total hardness is most nearly

(A) 210 mg/L

(B) 390 mg/L as $CaCO_3$

(C) 590 mg/L as $CaCO_3$

(D) 740 mg/L

3. A chemical analysis of a water sample has produced the results shown in the table.

cation	concentration (mg/L)	anion	concentration (mg/L)
Ca^{+2}	158	SO_4^{-2}	64
Mg^{+2}	47	Cl^-	43
Na^+	26	HCO_3^-	381
K^+	19		

The carbonate hardness is most nearly

(A) 210 mg/L

(B) 310 mg/L as $CaCO_3$

(C) 380 mg/L

(D) 590 mg/L as $CaCO_3$

4. A chemical analysis of a water sample has produced the results shown in the table.

cation	concentration (mg/L)	anion	concentration (mg/L)
Ca^{+2}	158	SO_4^{-2}	64
Mg^{+2}	47	Cl^-	43
Na^+	26	HCO_3^-	381
K^+	19		

The non-carbonate hardness is most nearly

(A) 210 mg/L

(B) 280 mg/L as $CaCO_3$

(C) 310 mg/L as $CaCO_3$

(D) 590 mg/L as $CaCO_3$

5. A municipal drinking water service authority is considering adding softening to its water treatment. The facility being evaluated processes a flow rate of 18 000 m³/d with the following hardness ion distribution.

$$Ca^{+2} = 347 \text{ mg/L}$$

$$Mg^{+2} = 129 \text{ mg/L}$$

$$HCO_3^- = 1256 \text{ mg/L}$$

The chemical equations used for lime-soda ash softening are

$$Ca^{2+} + 2HCO_3^- + CaO \rightarrow 2CaCO_3 \downarrow + H_2O \quad [I]$$

$$Mg^{2+} + 2HCO_3^- + 2CaO \rightarrow 2CaCO_3 \downarrow + Mg(OH)_2 \downarrow \quad [II]$$

$$Mg^{2+} + \text{other ions} + CaO + H_2O \rightarrow Ca^{2+} + Mg(OH)_2 \downarrow \quad [III]$$

$$Ca^{2+} + Na_2CO_3 \rightarrow CaCO_3 \downarrow + 2Na^+ \quad [IV]$$

Assuming 100% removal of the hardness, the annual dry sludge mass generated from lime-soda ash softening of the water is most nearly

(A) 1.6×10^5 kg/yr (dry)

(B) 2.3×10^6 kg/yr (dry)

(C) 1.8×10^7 kg/yr (dry)

(D) 2.1×10^8 kg/yr (dry)

SOLUTIONS

1. Convert concentration in mg/L to concentration in meq/L using molecular weight and valence.

$$C_{\text{meq/L}} = \frac{C_{\text{meq/L}} |\text{valence}| \frac{\text{meq}}{\text{mmol}}}{M \frac{\text{mg}}{\text{mmol}}}$$

Periodic Table of Elements

ions	C (mg/L)	valence (meq/mmol)	M (mg/mmol)	C (meq/L)
Ca^{2+}	158	2	40	7.90
Mg^{2+}	47	2	24	3.92
Na^+	26	1	23	1.13
K^+	19	1	39	0.49
SO_4^{2-}	64	2	96	1.33
Cl^-	43	1	35.5	1.21
HCO_3^-	381	1	61	6.25

Calculate the sum of the concentration of cations and anions.

$$\sum \text{cations} = 7.90 \frac{\text{meq}}{L} + 3.92 \frac{\text{meq}}{L}$$
$$+ 1.13 \frac{\text{meq}}{L} + 0.49 \frac{\text{meq}}{L}$$
$$= 13.44 \text{ meq/L}$$
$$\sum \text{anions} = 1.33 \frac{\text{meq}}{L} + 1.21 \frac{\text{meq}}{L} + 6.25 \frac{\text{meq}}{L}$$
$$= 8.79 \text{ meq/L}$$

Calculate the ratio of the cation concentration to the anion concentration.

$$\left(\frac{13.44 \frac{\text{meq}}{L} - 8.79 \frac{\text{meq}}{L}}{13.44 \frac{\text{meq}}{L}} \right) (100\%) = 35\%$$

Since 35% is greater than 10%, the analysis is deficient in anions.

The answer is (A).

2. The total hardness is

$$Ca^{+2} \frac{\text{mg}}{L} \text{ as } CaCO_3 + Mg^{+2} \frac{\text{mg}}{L} \text{ as } CaCO_3$$

$$= \left(7.90 \frac{\text{meq}}{L} + 3.92 \frac{\text{meq}}{L} \right) \left(\frac{50 \text{ mg as } CaCO_3}{1 \text{ meq}} \right)$$

$$= 591 \text{ mg/L as } CaCO_3 \quad (590 \text{ mg/L as } CaCO_3)$$

The answer is (C).

3. Carbonate hardness is the lesser of the concentration of total alkalinity and total hardness in mg/L as $CaCO_3$. All alkalinity for the water sample is from HCO_3^-.

The total alkalinity is

$$\left(6.25 \ \frac{meq}{L}\right)\left(\frac{50 \ mg \ as \ CaCO_3}{1 \ meq}\right)$$
$$= 312.5 \ mg/L \ as \ CaCO_3 \quad (310 \ mg/L \ as \ CaCO_3)$$

The total hardness is

$$Ca^{+2} \ \frac{mg}{L} \ as \ CaCO_3 + Mg^{+2} \ \frac{mg}{L} \ as \ CaCO_3$$
$$= \left(7.90 \ \frac{meq}{L} + 3.92 \ \frac{meq}{L}\right)\left(\frac{50 \ mg \ as \ CaCO_3}{1 \ meq}\right)$$
$$= 591 \ mg/L \ as \ CaCO_3 \quad (590 \ mg/L \ as \ CaCO_3)$$

310 mg/L as $CaCO_3$ is less than 590 mg/L as $CaCO_3$. Therefore, the carbonate hardness is equal to the total alkalinity.

The answer is (B).

4. Carbonate hardness is the lesser of the concentration of total alkalinity and total hardness in mg/L as $CaCO_3$.

The total alkalinity is

$$\left(6.25 \ \frac{meq}{L}\right)\left(\frac{50 \ mg \ as \ CaCO_3}{1 \ meq}\right)$$
$$= 312.5 \ mg/L \ as \ CaCO_3 \quad (310 \ mg/L \ as \ CaCO_3)$$

The total hardness is

$$Ca^{+2} \ \frac{mg}{L} \ as \ CaCO_3 + Mg^{+2} \ \frac{mg}{L} \ as \ CaCO_3$$
$$= \left(7.90 \ \frac{meq}{L} + 3.92 \ \frac{meq}{L}\right)\left(\frac{50 \ mg \ as \ CaCO_3}{1 \ meq}\right)$$
$$= 591 \ mg/L \ as \ CaCO_3 \quad (590 \ mg/L \ as \ CaCO_3)$$

The non-carbonate hardness is

non-carbonate hardness
= total hardness − carbonate hardness
$$= 590 \ \frac{mg}{L} \ as \ CaCO_3 - 310 \ \frac{mg}{L} \ as \ CaCO_3$$
$$= 280 \ mg/L \ as \ CaCO_3$$

The answer is (B).

5. The annual flow rate is

$$\left(18\,000 \ \frac{m^3}{d}\right)\left(365 \ \frac{d}{yr}\right) = 6.57 \times 10^6 \ m^3/yr$$

ion	concentration (mg/L)	mole weight (mg/mmol)	concentration (mmol/L)
Ca^{+2}	347	40	8.675
Mg^{+2}	129	24	5.375
HCO_3^-	1256	61	20.590

Find the molecular weight of $CaCO_3$. [Periodic Table of Elements]

$$40 \ \frac{mg}{mmol} + 12 \ \frac{mg}{mmol} + (3)\left(16 \ \frac{mg}{mmol}\right)$$
$$= 100 \ mg/mmol$$

The molecular weight of $Mg(OH)_2$ is

$$24 \ \frac{mg}{mmol} + (2)\left(16 \ \frac{mg}{mmol} + 1 \ \frac{mg}{mmol}\right) = 58 \ mg/mmol$$

From Eq. I, 8.675 mmol/L Ca^{+2} reacts with (2) (8.675 mmol/L) HCO_3^- and 8.675 mmol/L CaO to produce (2)(8.675 mmol/L) $CaCO_3$. Of the 20.59 mmol/L HCO_3^-, Eq. I consumes (2)(8.675 mmol/L) leaving 3.24 mmol/L for reaction in Eq. II.

From Eq. II, the remaining 3.24 mmol/L HCO_3^- reacts with (½)(3.24 mmol/L) Mg^{+2} and 3.24 mmol/L CaO to produce 3.24 mmol/L $CaCO_3$ and (½)(3.24 mmol/L) $Mg(OH)_2$. Of the 5.375 mmol/L Mg^{+2}, Eq. II consumes (½)(3.24 mmol/L) leaving 3.755 mmol/L for reaction in Eq. III.

From Eq. III, the remaining 3.755 mmol/L Mg^{+2} reacts with 3.755 mmol/L CaO to produce 3.755 mmol/L Ca^{+2}, and therefore 3.755 mmol/L CaO_3 from the reaction in Eq. IV, and 3.755 mmol/L $Mg(OH)_2$.

The total amount of $CaCO_3$ produced is

$$\frac{\left((2)\left(8.675 \ \frac{mmol}{L}\right) + 3.24 \ \frac{mmol}{L} + 3.755 \ \frac{mmol}{L}\right)}{\left(\frac{1 \ m^3}{1000 \ L}\right)\left(10^6 \ \frac{mg}{kg}\right)} \times \left(100 \ \frac{mg}{mmol}\right)\left(6.57 \times 10^6 \ \frac{m^3}{yr}\right)$$
$$= 1.6 \times 10^7 \ kg/yr \quad [dry]$$

The total amount of $Mg(OH)_2$ produced is

$$\frac{\left(\left(\frac{1}{2}\right)\left(3.24 \ \frac{mmol}{L}\right) + 3.755 \ \frac{mmol}{L}\right) \times \left(58 \ \frac{mg}{mmol}\right)\left(6.57 \times 10^6 \ \frac{m^3}{yr}\right)}{\left(\frac{1 \ m^3}{1000 \ L}\right)\left(10^6 \ \frac{mg}{kg}\right)}$$

$$= 2.0 \times 10^6 \ kg/yr \quad [dry]$$

The total amount of sludge produced is

$$1.6 \times 10^7 \ \frac{kg}{yr} \quad [dry] + 2.0 \times 10^6 \ \frac{kg}{yr} \quad [dry]$$

$$= 1.8 \times 10^7 \ kg/yr \quad [dry]$$

The answer is (C).

17 Potable Water Treatment and Management

Content in blue refers to the *NCEES Handbook*.

PRACTICE PROBLEMS

1. A plating shop produces 164 m³/d of wastewater containing cadmium at 31 mg/L as Cd^{2+}, zinc at 13 mg/L as Zn^{2+}, nickel at 21 mg/L as Ni^{2+}, and chromium at 130 mg/L as CrO_4^{2-}. Ion exchanger characteristics selected to treat the wastewater are summarized as shown.

characteristic	cation exchanger	anion exchanger
regenerant	H_2SO_4	NaOH
dosage (kg/m³)	192	76
concentration (% by weight)	5	10
regenerant density (g/cm³)	1.44	1.15
hydraulic loading rate (m³/m³·min)	0.020	0.020
resin capacity (equiv/L)	1.5	3.7

The required regeneration period for the exchanger used to recover the cations is most nearly

(A) 7 min

(B) 11 min

(C) 33 min

(D) 130 min

2. A commercial laundry is considering constructing an ion exchange process to remove hardness from the water used in laundering operations. The laundry uses 1000 m³/d of city water with Ca^{+2} at 4.5 meq/L and Mg^{+2} at 2.0 meq/L. Their goal is to reduce total hardness to 1.0 meq/L. The total bypass-water flow rate required is most nearly

(A) 0 m³/d

(B) 150 m³/d

(C) 850 m³/d

(D) 1000 m³/d

3. A commercial laundry is considering constructing an ion exchange process to remove hardness from the water used in laundering operations. The laundry uses 1000 m³/d of city water with Ca^{+2} at 4.5 meq/L and Mg^{+2} at 2.0 meq/L. Their goal is to reduce total hardness (TH) to 1.0 meq/L. The softening resin has a TH capacity of 60 kg TH/m³ of media with a recommended hydraulic loading rate of 0.4 m³/m²·min. The bypass flow rate is 154 m³/d. The total daily resin volume required to soften the water is most nearly

(A) 4.6 m³/d

(B) 5.4 m³/d

(C) 280 m³/d

(D) 330 m³/d

4. A commercial laundry is considering constructing an ion exchange process to remove hardness from the water used in laundering operations. The laundry uses 1000 m³/d of city water with Ca^{+2} at 4.5 meq/L and Mg^{+2} at 2.0 meq/L. Their goal is to reduce total hardness (TH) to 1.0 meq/L. The softening resin has a TH capacity of 60 kg TH/m³ of media with a recommended hydraulic loading rate of 0.4 m³/m²·min. The bypass flow rate is 154 m³/d. The exchanger tanks available to the laundry have a diameter of 1.0 m. Most nearly, how many exchanger tanks are required?

(A) 1

(B) 2

(C) 3

(D) 4

5. A commercial laundry is considering constructing an ion exchange process to remove hardness from the water used in laundering operations. The laundry uses two 1.0 m diameter tanks with a total daily media volume of 4.6 m^3. The resin bed depth in each exchanger tank is most nearly

(A) 1.5 m

(B) 1.9 m

(C) 2.9 m

(D) 5.8 m

6. A commercial laundry is considering constructing an ion exchange process to remove hardness from the water used in laundering operations. Regeneration can be accomplished using 100 kg NaCl/m^3 in a 15% solution at a hydraulic loading rate of 0.04 m^3/m^2·min. The daily media volume per tank is 2.3 m^3. The daily volume of NaCl regeneration solution used per tank is most nearly

(A) 0.78 m^3/d

(B) 1.5 m^3/d

(C) 1.9 m^3/d

(D) 3.8 m^3/d

7. A commercial laundry is considering constructing an ion exchange process to remove hardness from the water used in laundering operations. Regeneration can be accomplished using 100 kg NaCl/m^3 in a 15% solution at a hydraulic loading rate of 0.04 m^3/m^2·min. The exchanger tanks available to the laundry have a diameter of 1.0 m. The daily volume of regeneration solution used is 1.5 m^3. The time period required for resin regeneration in any single tank is most nearly

(A) 26 min

(B) 39 min

(C) 48 min

(D) 98 min

8. A sedimentation basin treats 2000 m^3/d at a settling time of 130 min. The volume of the settling zone is most nearly

(A) 160 m^3

(B) 170 m^3

(C) 180 m^3

(D) 200 m^3

9. A sedimentation basin has a settling zone volume of 180 m^3 and a settling zone depth of 3.0 m. The surface area of the settling zone is most nearly

(A) 52 m^2

(B) 57 m^2

(C) 60 m^2

(D) 65 m^2

10. Bench studies have determined that aluminum sulfate (alum) promotes acceptable floc formation at a dose of 23 mg/L when applied to a surface water source. Design standards for chemical flocculation require flash mixing at a minimum velocity gradient of 900/s for 120 seconds. The water demand is 19 000 m^3/d. The total tank volume required for flash mixing is most nearly

(A) 2.7 m^3

(B) 16 m^3

(C) 27 m^3

(D) 160 m^3

11. A flash mixer has a cubic volume of 27 m^3. If four tanks are desired, the dimensions of each tank are most nearly

(A) $l = 5.4$ m, $w = 1.9$ m, $d = 3.0$ m

(B) $l = 1.9$ m, $w = 1.9$ m, $d = 1.9$ m

(C) $l = 3.0$ m, $w = 3.0$ m, $d = 3.0$ m

(D) $l = 5.4$ m, $w = 5.4$ m, $d = 5.4$ m

12. Bench studies have determined that aluminum sulfate (alum) promotes acceptable floc formation at a dose of 23 mg/L when applied to a surface water source. Design standards for chemical flocculation require flash mixing at a minimum velocity gradient of 900/s for 120 seconds. The water demand is 19 000 m^3/d. The flash mixer cubic volume is 27 m^3. The water temperature is 20°C. If the motor efficiency is 85%, the total power required for flash mixing is most nearly

(A) 2.2×10^3 N·m/s

(B) 1.5×10^4 N·m/s

(C) 2.6×10^4 N·m/s

(D) 1.3×10^5 N·m/s

13. Flocculation tanks need to be sized to match three sedimentation basins 3.5 m deep and 8.0 m wide. The flow rate for both the sedimentation basins and the flocculation tanks are 4000 m³/d. The flocculation tanks will have two mixing sections with an overall average velocity gradient of 30/s. The average velocity gradient-time value has been set at 89 700. The paddles will be the wooden paddle-wheel type and will turn on a horizontal axis perpendicular to the flow, with one paddle wheel in each tank section. A common motor will turn the paddles in both sections of each tank at a constant rotational speed of 3 rpm. The water temperature is 20°C. The flocculation tank dimensions are most nearly

(A) $l = 4.0$ m, $w = 4.0$ m, $d = 4.0$ m

(B) $l = 5.2$ m, $w = 5.2$ m, $d = 5.2$ m

(C) $l = 5.0$ m, $w = 8.0$ m, $d = 3.5$ m

(D) $l = 7.0$ m, $w = 6.0$ m, $d = 3.5$ m

14. Flocculation tanks need to be sized to match three sedimentation basins 3.5 m deep and 8.0 m wide. The sedimentation basins each handle a flow of 4000 m³/d, as will the flocculation tanks. The flocculation tanks have two mixing sections with an overall average velocity gradient of 30/s. The velocity gradients for the first (inlet end) and for the second (outlet end) sections of each flocculation tank are most nearly

(A) $G_{\text{inlet}} = 15/s$, $G_{\text{outlet}} = 45/s$

(B) $G_{\text{inlet}} = 20/s$, $G_{\text{outlet}} = 40/s$

(C) $G_{\text{inlet}} = 30/s$, $G_{\text{outlet}} = 30/s$

(D) $G_{\text{inlet}} = 40/s$, $G_{\text{outlet}} = 20/s$

15. Flocculation tanks need to be sized to match three sedimentation basins 3.5 m deep and 8.0 m wide. The sedimentation basins each handle a flow of 4000 m³/d, as will the flocculation tanks. The flocculation tanks have two mixing sections with an overall average velocity gradient of 30/s. The water temperature is 20°C. The inlet velocity gradient is 40/s, and the outlet velocity gradient is 20/s. The flocculation tank dimensions are $l = 7.0$ m, $w = 6.0$ m, and $d = 3.5$ m. The power

requirements for the paddles in the first and second sections of each flocculation tank are most nearly

(A) $P_1 = 3.0$ N·m/s, $P_2 = 1.5$ N·m/s

(B) $P_1 = 66$ N·m/s, $P_2 = 66$ N·m/s

(C) $P_1 = 120$ N·m/s, $P_2 = 30$ N·m/s

(D) $P_1 = 240$ N·m/s, $P_2 = 60$ N·m/s

16. Flocculation tanks need to be sized to match three sedimentation basins 3.5 m deep and 8.0 m wide. The sedimentation basins each handle a flow of 4000 m³/d, as will the flocculation tanks. The paddles will be the wooden paddle-wheel type and will turn on a horizontal axis perpendicular to flow, with one paddle wheel in each tank section. A common motor will turn the paddles in both sections of each tank at a constant rotational speed of 3 rpm. The water temperature is 20°C. The required paddle tip velocity is most nearly

(A) 0.35 m/s

(B) 0.46 m/s

(C) 0.50 m/s

(D) 0.55 m/s

17. A treatment plant upgrade calls for the design of sedimentation basins to handle a flow rate of 30 000 m³/d. The basins are for a type I suspension with an overflow rate of 1.1 m³/m²·h, a length to width ratio of 4:1, and a maximum settling zone length of 40 m. The minimum number of tanks required to treat the flow is most nearly

(A) 1

(B) 2

(C) 3

(D) 4

18. A treatment plant upgrade calls for the design of sedimentation basins to handle a flow rate of 30 000 m³/d. The flow is evenly divided among three basins each with a 40 m length and 10 m width. The weir overflow rate is 14 m³/m·h. The required weir length for each tank is most nearly

(A) 15 m

(B) 30 m

(C) 45 m

(D) 90 m

19. A municipal drinking water service authority is considering adding softening to its water treatment. The facility being evaluated processes a flow rate of

18 000 m³/d with the following hardness ion distribution.

$$Ca^{+2} = 347 \text{ mg/L}$$

$$Mg^{+2} = 129 \text{ mg/L}$$

$$HCO_3^- = 1256 \text{ mg/L}$$

Softening chemicals are available at the following purities and costs.

CaO 67% purity at $82/1000 kg

Na_2CO_3 98% purity at $105/1000 kg

NaOH 73% purity at $215/1000 kg

The chemical equations used for lime-soda ash softening are

$$Ca^{2+} + 2HCO_3^- + CaO \rightarrow 2CaCO_3 \downarrow + H_2O \quad [I]$$

$$Mg^{2+} + 2HCO_3^- + 2CaO \rightarrow 2CaCO_3 \downarrow + Mg(OH)_2 \downarrow \quad [II]$$

$$Mg^{2+} + \text{other ions} + CaO + H_2O \rightarrow Ca^{2+} + Mg(OH)_2 \downarrow \quad [III]$$

$$Ca^{2+} + Na_2CO_3 \rightarrow CaCO_3 \downarrow + 2Na^+ \quad [IV]$$

Assuming 100% removal of the hardness, the annual cost for lime and soda ash to be used as reagents in softening of the water is most nearly

(A) $230,000/yr

(B) $450,000/yr

(C) $620,000/yr

(D) $990,000/yr

SOLUTIONS

1. The cations will be removed by the cation exchanger. In the cation exchanger, Zn, Cd, and Ni are exchanged for H^+. Size the exchanger for one day of operation. The equivalents of cations removed per liter of water treated are

$$\frac{\left(31 \frac{\text{mg Cd}}{L}\right)\left(2 \frac{\text{equiv}}{\text{mol}}\right)}{112 \frac{\text{g Cd}}{\text{mol}}} + \frac{\left(13 \frac{\text{mg Zn}}{L}\right)\left(2 \frac{\text{equiv}}{\text{mol}}\right)}{65 \frac{\text{g Zn}}{\text{mol}}}$$

$$+ \frac{\left(21 \frac{\text{mg Ni}}{L}\right)\left(2 \frac{\text{equiv}}{\text{mol}}\right)}{59 \frac{\text{g Ni}}{\text{mol}}}$$

$$= 1.67 \frac{\text{mg·equiv}}{\text{g·L}}$$

$$= 0.001\,67 \text{ equiv/L}$$

The resin volume is

$$\frac{\left(164 \frac{\text{m}^3}{\text{d}}\right)\left(0.001\,67 \frac{\text{equiv}}{L}\right)}{1.5 \frac{\text{equiv}}{\text{L resin}}} = 0.18 \text{ m}^3 \text{ resin/d}$$

The regeneration volume is

$$\frac{\left(192 \frac{\text{kg H}_2\text{SO}_4}{\text{m}^3}\right)\left(0.18 \frac{\text{m}^3}{\text{d}}\right)\left(1000 \frac{\text{g}}{\text{kg}}\right)}{\left(1.44 \frac{\text{g}}{\text{cm}^3}\right)\left(\frac{5 \text{ kg H}_2\text{SO}_4}{100 \text{ kg solution}}\right)\left(10^6 \frac{\text{cm}^3}{\text{m}^3}\right)}$$

$$= 0.48 \text{ m}^3 \text{ solution/d}$$

The regeneration period is

$$\frac{0.48 \frac{\text{m}^3 \text{ solution}}{\text{d}}}{\left(0.020 \frac{\text{m}^3 \text{ solution}}{\text{m}^3 \text{ resin·min}}\right)\left(0.18 \frac{\text{m}^3 \text{ resin}}{\text{d}}\right)}$$

$$= 133 \text{ min} \quad (130 \text{ min})$$

The answer is (D).

2. Bypass enough water to allow 1.0 meq/L total hardness (TH) in the effluent.

influent TH $= 4.5$ meq/L $+ 2.0$ meq/L $= 6.5$ meq/L

The TH concentration of the water comprising the treatment flow rate $= 0.0$ meq/L.

$$Q_B = \text{bypass flow rate}$$
$$1000 \ \text{m}^3/\text{d} - Q_B = \text{treatment flow rate}$$

$$1.0 \ \frac{\text{meq}}{\text{L}} = \frac{Q_B\left(6.5 \ \frac{\text{meq}}{\text{L}}\right) + \left(1000 \ \frac{\text{m}^3}{\text{d}} - Q_B\right)\left(0.0 \ \frac{\text{meq}}{\text{L}}\right)}{1000 \ \frac{\text{m}^3}{\text{d}}}$$

$$Q_B = 154 \ \text{m}^3/\text{d} \quad (150 \ \text{m}^3/\text{d})$$

The answer is (B).

3. The flow rate is

$$Q_{\text{treatment}} = 1000 \ \frac{\text{m}^3}{\text{d}} - 154 \ \frac{\text{m}^3}{\text{d}} = 846 \ \text{m}^3/\text{d}$$

The TH to be removed is

$$\left(6.5 \ \frac{\text{meq}}{\text{L}}\right)\left(50 \ \frac{\text{mg}}{\text{meq}}\right)\left(10^{-6} \frac{\text{kg}}{\text{mg}}\right)$$
$$\times \left(846 \ \frac{\text{m}^3}{\text{d}}\right)\left(1000 \ \frac{\text{L}}{\text{m}^3}\right) = 275 \ \text{kg TH/d}$$

The resin volume is

$$\left(275 \ \frac{\text{kg TH}}{\text{d}}\right)\left(\frac{1 \ \text{m}^3}{60 \ \text{kg TH}}\right) = 4.6 \ \text{m}^3 \ \text{resin/d}$$

The answer is (A).

4. The treatment flow rate is

$$Q_{\text{treatment}} = 1000 \ \frac{\text{m}^3}{\text{d}} - 154 \ \frac{\text{m}^3}{\text{d}} = 846 \ \text{m}^3/\text{d}$$

Size and configure the system for a one-day operating cycle. Find the total tank surface area using the equation for hydraulic loading rate.

Stokes' Law

Hydraulic loading rate $= Q/A$

$$A = \frac{Q}{\text{Hydraulic loading rate}}$$

$$= \frac{846 \ \frac{\text{m}^3}{\text{d}}}{\left(0.4 \ \frac{\text{m}^3}{\text{m}^2 \cdot \text{min}}\right)\left(1440 \ \frac{\text{min}}{\text{d}}\right)}$$

$$= 1.47 \ \text{m}^2$$

The tank diameter is 1.0 m; therefore, each tank's area is 0.785 m^2.

The total number of tanks is

$$\frac{1.47 \ \text{m}^2}{0.785 \ \text{m}^2} = 1.9 \ \text{tanks} \quad (2 \ \text{tanks})$$

The answer is (B).

5. The daily media volume per tank is

$$\frac{4.6 \ \text{m}^3}{2 \ \text{tanks}} = 2.3 \ \text{m}^3/\text{tank}$$

The tank diameter is 1.0 m; therefore, each tank's area is 0.785 m^2. The bed depth per tank is

$$\frac{2.3 \ \text{m}^2}{0.785 \ \text{m}^2} = 2.9 \ \text{m}$$

The answer is (C).

6. The regeneration salt required per tank is

$$\left(2.3 \ \frac{\text{m}^3}{\text{d}}\right)\left(100 \ \frac{\text{kg NaCl}}{\text{m}^3}\right) = 230 \ \text{kg NaCl/d}$$

Using a 15% salt solution, the daily volume of solution is

$$\frac{230 \ \frac{\text{kg NaCl}}{\text{d}}}{\left(\frac{15 \ \text{kg NaCl}}{100 \ \text{kg water}}\right)\left(1000 \ \frac{\text{kg}}{\text{m}^3}\right)} = 1.5 \ \text{m}^3/\text{d}$$

The answer is (B).

7. The tank diameter is 1.0 m; therefore, each tank's area is 0.785 m². The regeneration cycle time in any single tank is

$$\frac{\left(1.5 \ \frac{m^3}{d}\right)\left(1 \ \frac{d}{cycle}\right)}{\left(0.04 \ \frac{m^3}{m^2 \cdot min}\right)(0.785 \ m^2)} = 48 \ min$$

The answer is (C).

8. Use the equation for hydraulic residence time and solve for the volume of the settling zone.

$$V = \text{settling zone volume, m}^3$$
$$Q = \text{flow rate} = 2000 \ m^3/d$$
$$\theta = \text{settling time} = 130 \ min$$

Activated Sludge

$$\theta = \frac{V}{Q}$$
$$V = \theta Q$$

$$= \frac{(130 \ min)\left(2000 \ \frac{m^3}{d}\right)}{1440 \ \frac{min}{d}}$$

$$= 181 \ m^3 \quad (180 \ m^3)$$

The answer is (C).

9. Calculate the surface area of the settling zone.

$$A_s = \text{settling zone surface area, m}^2$$
$$V = \text{settling zone volume} = 180 \ m^3$$
$$Z_o = \text{settling zone depth, 3.0 m}$$

$$A_{surface} = \frac{V}{Z_o} = \frac{180 \ m^3}{3.0 \ m}$$

$$= 60 \ m^2$$

The answer is (C).

10. Find the total tank volume required for flash mixing.

$$V = \text{mixing volume, m}^3$$
$$Q = \text{flow rate} = 19\,000 \ m^3/d$$
$$t = \text{detention time} = 120 \ s$$

$$V = Qt$$

$$= \frac{\left(19\,000 \ \frac{m^3}{d}\right)(120 \ s)}{86\,400 \ \frac{s}{d}}$$

$$= 26.4 \ m^3 \quad (27 \ m^3)$$

The answer is (C).

11. Find the volume of each tank.

$$\frac{27 \ m^3}{4 \ tanks} = 6.75 \ m^3/tank$$

For mixing, cubic dimensions are desired. Therefore, $l = w = d$ and $V = w^3$.

$$w = (6.75 \ m^3)^{1/3}$$
$$= 1.9 \ m$$
$$l = 1.9 \ m$$
$$d = 1.9 \ m$$

The answer is (B).

12. Rearrange the equation for finding the velocity gradient, and solve for the total power required.

$$P = \text{power, N·m/s}$$
$$G = \text{velocity gradient} = 900/s$$
$$\mu = \text{dynamic viscosity} = 1.002 \times 10^{-3} \ N \cdot s/m^2$$
$$V = \text{volume} = 27 \ m^3$$

Rapid Mix and Flocculator Design

$$G = \sqrt{\frac{P}{\mu V}}$$

$$P = G^2 V \mu$$

$$= \left(\frac{900}{s}\right)^2 (27 \ m^3)\left(1.002 \times 10^{-3} \ \frac{N \cdot s}{m^2}\right)$$

$$= 21\,914 \ N \cdot m/s$$

P at 85% efficiency is

$$P \ [\text{at 85\% efficiency}] = \frac{21\,914 \ \frac{N \cdot m}{s}}{0.85}$$

$$= 25\,780 \ N \cdot m/s \quad (2.6 \times 10^4 \ N \cdot m/s)$$

The answer is (C).

13. Find the detention time.

t = detention time, s
Gt = time-velocity gradient, unitless = 89 700
G = velocity gradient = 30 s^{-1}

$$t = \frac{Gt}{G}$$

$$= \frac{89\,700}{\left(\dfrac{30}{s}\right)\left(60\ \dfrac{s}{min}\right)}$$

$$= 50\ min$$

Find the tank volume.

V = tank volume, m^3
Q = flow rate = 4000 m^3/d

$$V = Qt$$

$$= \frac{\left(4000\ \dfrac{m^3}{d}\right)(50\ min)}{1440\ \dfrac{min}{d}}$$

$$= 139\ m^3$$

To match the sedimentation basin, the flocculation tank depth should equal the sedimentation basin depth of 3.5 m and the maximum flocculation basin width should be equal to the sedimentation basin width of 8.0 m.

For the most efficient mixing with wooden paddle-wheel type paddles turning on a horizontal axis perpendicular to flow, each mixing section should have square dimensions along the cross section of the tank length. A minimum clearance of 0.3 m should be provided between the paddle tip and the tank walls and floors to avoid floc shear.

Applying these criteria, the flocculation tank dimensions are

$d = 3.5$ m
$l = 7.0$ m
$$w = \frac{139\ m^3}{(3.5\ m)(7.0\ m)}$$
$$= 5.7\ m \quad (6.0\ m) \quad [< 8.0\ m,\ \text{therefore OK}]$$

The tank cross section is shown in the figure.

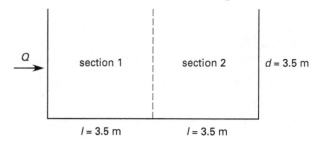

The answer is (D).

14. For good floc formation in a two-section tank, G_{inlet} at the inlet should be greater than G_{outlet} at the outlet. Let $G_{inlet} = 2G_{outlet}$.

$$\frac{G_{inlet} + G_{outlet}}{2} = \frac{2G_{outlet} + G_{outlet}}{2} = G_{ave} = 30/s$$

$$3G_{outlet} = 60/s$$
$$G_{outlet} = 20/s$$
$$2G_{outlet} = G_{inlet} = 40/s$$

The answer is (D).

15. Use the equation for finding the velocity gradient, and solve for the power requirements in the first and second sections of each tank.

P = power, N·m/s
G = velocity gradient for each section, s^{-1}
$$V = \text{volume, m}^3 = \left(\frac{7.0\ m}{2\ \text{sections}}\right)(6.0\ m)(3.5\ m)$$
$$= 73.5\ m^3$$
μ = dynamic viscosity = 1.002×10^{-3} N·s/m^2

Rapid Mix and Flocculator Design

$$G = \sqrt{\frac{P}{\mu V}}$$
$$P = G^2 V \mu$$
$$P_1 = \left(\frac{40}{s}\right)^2 (73.5\ m^3)\left(1.002 \times 10^{-3}\ \frac{N\cdot s}{m^2}\right)$$
$$= 118\ N\cdot m/s \quad (120\ N\cdot m/s)$$
$$P_2 = \left(\frac{20}{s}\right)^2 (73.5\ m^3)\left(1.002 \times 10^{-3}\ \frac{N\cdot s}{m^2}\right)$$
$$= 29.5\ N\cdot m/s \quad (30\ N\cdot m/s)$$

The answer is (C).

16. The floc tank depth matches the sedimentation basin depth of 3.5 m. Paddles rotating perpendicular to flow turn in a cross section of 3.5 m × 3.5 m. The standard clearance between the paddle tip and the sides and bottom of the floc tank is 0.3 m. Find the paddle tip velocity.

$$v = \text{paddle velocity, m/s}$$
$$v_p = \text{paddle velocity relative to water, m/s}$$
$$(\text{assume } v_p = 0.75v)$$
$$d = \text{paddle wheel diameter, m}$$
$$= 3.5 \text{ m} - 0.3 \text{ m} - 0.3 \text{ m} = 2.9 \text{ m}$$
$$\omega = \text{paddle rotational speed} = 3 \text{ rpm}$$

$$v = \pi d\omega$$
$$= \left(\frac{\pi(2.9 \text{ m})}{\text{rev}}\right)(3 \text{ rpm})\left(\frac{1 \text{ min}}{60 \text{ s}}\right)$$
$$= 0.46 \text{ m/s}$$
$$v_p = (0.75)\left(0.46 \frac{\text{m}}{\text{s}}\right) = 0.345 \text{ m/s} \quad (0.35 \text{ m/s})$$

The answer is (A).

17. Find the settling zone surface area using the equation for critical settling velocity.

$$A_{\text{surface}} = \text{settling zone surface area, m}^2$$
$$Q = \text{flow rate} = 30\,000 \text{ m}^3/\text{d}$$
$$v_o = \text{overflow rate} = 1.1 \text{ m}^3/\text{m}^2\text{·h}$$

Clarifier

$$v_o = \frac{Q}{A_{\text{surface}}}$$
$$A_{\text{surface}} = \frac{Q}{v_o}$$
$$= \frac{30\,000 \frac{\text{m}^3}{\text{d}}}{\left(1.1 \frac{\text{m}^3}{\text{m}^2\text{·h}}\right)\left(24 \frac{\text{h}}{\text{d}}\right)}$$
$$= 1136 \text{ m}^2$$

$A_{\text{surface}} = lw$, and since a $l{:}w$ of 4:1 is required, $l = 4w$. Also, $l \leq 40$ m.

For one tank,

$$A_{\text{surface}} = (4w)(w) = 4w^2 = 1136 \text{ m}^2$$
$$w = 16.9 \text{ m} \quad (17 \text{ m})$$
$$l = (4)(17 \text{ m}) = 68 \text{ m}$$

Since 68 m > 40 m, more than one tank is needed.

For two tanks,

$$A_{\text{surface}} = 4w^2 = \frac{1136 \text{ m}^2}{2} = 568 \text{ m}^2/\text{tank}$$
$$w = 11.9 \text{ m} \quad (12 \text{ m})$$
$$l = (4)(12 \text{ m}) = 48 \text{ m}$$

Since 48 m > 40 m, more than two tanks are needed.

For three tanks,

$$A_{\text{surface}} = 4w^2 = \frac{1136 \text{ m}^2}{3} = 379 \text{ m}^3/\text{tank}$$
$$w = 9.7 \text{ m} \quad (10 \text{ m})$$
$$l = (4)(10 \text{ m}) = 40 \text{ m}$$

This is OK. Use three tanks.

The answer is (C).

18. Find the required weir length.

$$Q/\text{tank} = \text{flow rate/tank} = 10\,000 \text{ m}^3/\text{d}$$
$$\text{Weir Length} = \text{weir length, m}$$
$$\text{WOR} = \text{weir overflow rate} = 14 \text{ m}^3/\text{m·h}$$

Clarifier

$$\text{WOR} = Q/\text{Weir Length}$$
$$\frac{\text{Weir Length}}{\text{tank}} = \frac{\frac{Q}{\text{tank}}}{\text{WOR}}$$
$$= \frac{10\,000 \frac{\text{m}^3}{\text{d}}}{\left(14 \frac{\text{m}^3}{\text{m·h}}\right)\left(24 \frac{\text{h}}{\text{d}}\right)}$$
$$= 29.8 \text{ m} \quad (30 \text{ m})$$

The answer is (B).

19. The annual flow rate is

$$\left(18\,000 \frac{\text{m}^3}{\text{d}}\right)\left(365 \frac{\text{d}}{\text{yr}}\right) = 6.57 \times 10^6 \text{ m}^3/\text{yr}$$

ion	concentration (mg/L)	mole weight (mg/mmol)	concentration (mmol/L)
Ca^{+2}	347	40	8.675
Mg^{+2}	129	24	5.375
HCO_3^-	1256	61	20.590

From Eq. I, 8.675 mmol/L Ca^{+2} reacts with (2) (8.675 mmol/L) HCO_3^- and 8.675 mmol/L CaO to produce (2)(8.675 mmol/L) $CaCO_3$. Of the 20.59 mmol/L HCO_3^-, Eq. I consumes (2)(8.675 mmol/L), leaving 3.24 mmol/L for reaction in Eq. II.

From Eq. II, the remaining 3.24 mmol/L HCO_3^- reacts with ($\frac{1}{2}$)(3.24 mmol/L) Mg^{+2} and 3.24 mmol/L CaO to produce 3.24 mmol/L $CaCO_3$ and ($\frac{1}{2}$)(3.24 mmol/L) $Mg(OH)_2$. Of the 5.375 mmol/L Mg^{+2}, Eq. II consumes ($\frac{1}{2}$)(3.24 mmol/L), leaving 3.755 mmol/L for reaction in Eq. III.

From Eq. III, the remaining 3.755 mmol/L Mg^{+2} reacts with 3.755 mmol/L CaO to produce 3.755 mmol/L Ca^{+2} and 3.755 mmol/L $Mg(OH)_2$.

From Eq. IV, the 3.755 mmol/L Ca^{+2} produced in Eq. III reacts with 3.755 mmol/L Na_2CO_3 to produce 3.755 mmol/L $CaCO_3$.

The molecular weight of CaO is

$$40\ \frac{mg}{mmol} + 16\ \frac{mg}{mmol} = 56\ mg/mmol$$

The molecular weight of Na_2CO_3 is

$$(2)\left(23\ \frac{mg}{mmol}\right) + 12\ \frac{mg}{mmol}$$
$$+ (3)\left(16\ \frac{mg}{mmol}\right) = 106\ mg/mmol$$

The cost of the CaO is

$$\frac{\left(8.675\ \frac{mmol}{L} + 3.24\ \frac{mmol}{L} + 3.755\ \frac{mmol}{L}\right)}{\left(\frac{67\%}{100\%}\right)\left(\frac{1\ m^3}{1000\ L}\right)\left(10^6\ \frac{mg}{kg}\right)}$$
$$= \$705,605/yr$$

The cost of the Na_2CO_3 is

$$\frac{\left(3.755\ \frac{mmol}{L}\right)\left(106\ \frac{mg}{mmol}\right)}{\left(\frac{98\%}{100\%}\right)\left(\frac{1\ m^3}{1000\ L}\right)\left(10^6\ \frac{mg}{kg}\right)} = \$280,185/yr$$

The total cost of chemical reagents is

$$\frac{\$705,605}{yr} + \frac{\$280,185}{yr} = \$985,790/yr \quad (\$990,000/yr)$$

The answer is (D).

18 Potable Water Distribution Systems

Content in blue refers to the *NCEES Handbook*.

PRACTICE PROBLEMS

1. Water flow is measured using a pitot tube. The stagnation pressure is 289 kPa and the static pressure is 265 kPa. The velocity of flow is most nearly

(A) 0.22 m/s

(B) 1.1 m/s

(C) 4.9 m/s

(D) 6.9 m/s

2. A water valve is closed suddenly, causing a compression shock wave followed by a rarefaction wave. The waves travel an out-and-back distance of 46 m. The velocity of sound in water at 10°C is 1447 m/s. The total travel time for the waves is most nearly

(A) 0.016 s

(B) 0.032 s

(C) 0.37 s

(D) 1.5 s

3. Plastic pipe is selected to convey 200 gpm of water over a distance of 280 ft. The allowable pressure is 0.02 psi/ft. Most nearly, what nominal pipe diameter should be used?

(A) 2.0 in

(B) 2.5 in

(C) 3.0 in

(D) 4.0 in

4. A pipe carrying a flow of 1 cfs divides into two parallel pipes. The pipe characteristics are as shown.

characteristic	pipe A	pipe B
flow rate (cfs)	0.5	0.5
length (ft)	38	73
friction factor	0.038	0.051
diameter (in)	4	

The diameter for pipe B is most nearly

(A) 4 in

(B) 5 in

(C) 6 in

(D) 8 in

5. A city's census records showing population growth are presented in the table.

year	population
1970	12,200
1980	18,000
1990	23,500
2000	30,000
2010	38,600

If the population growth rate is 0.021 yr^{-1}, the projected population for the city in 2050 is most nearly

(A) 48,000 people

(B) 65,000 people

(C) 100,000 people

(D) 122,000 people

6. The projected 24 h flow distribution for a small municipality is shown in the figure. The municipality currently has the capacity to pump at 3.6 m³/min on a continuous basis and is considering adding a storage tank instead of increasing pumping capacity to meet the projected demand.

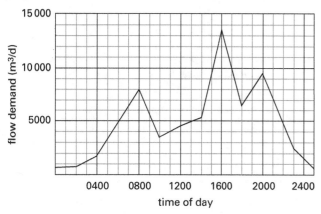

The maximum pumping rate required to meet peak demand if no storage is provided is most nearly

(A) 0.040 m³/s

(B) 0.16 m³/s

(C) 0.40 m³/s

(D) 1.6 m³/s

The required storage volume to meet a uniform 24 h pumping rate of 3.6 m³/min is most nearly

(A) 1250 m³

(B) 5200 m³

(C) 14 000 m³

(D) 34 000 m³

7. The projected 24 h flow distribution for a small municipality is shown in the figure. The municipality currently has the capacity to pump at 3.6 m³/min on a continuous basis and is considering adding a storage tank instead of increasing pumping capacity to meet the projected demand.

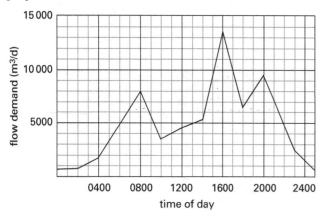

SOLUTIONS

1.

v = velocity of flow
P_0 = stagnation pressure = 289 kPa
P_s = static pressure = 265 kPa
ρ = water density = 1000 kg/m³

Pitot Tube

$$v = \sqrt{(2/\rho)(P_0 - P_s)}$$

$$= \sqrt{\left(\dfrac{2}{1000 \ \frac{\text{kg}}{\text{m}^3}}\right)(289 \text{ kPa} - 265 \text{ kPa})\left(1000 \ \dfrac{\text{Pa}}{\text{kPa}}\right)}$$
$$\times \left(\dfrac{\text{kg}}{\text{m}\cdot\text{s}^2\text{Pa}}\right)$$

$$= 6.9 \text{ m/s}$$

The answer is (D).

2. Find the total travel time.

t = total travel time
L = total distance = 46 m
a = velocity of sound in water = 1447 m/s

$$t = \frac{L}{a}$$
$$= \frac{46 \text{ m}}{1447 \ \dfrac{\text{m}}{\text{s}}}$$
$$= 0.032 \text{ s}$$

The answer is (B).

3. The Hazen-Williams coefficient for plastic pipe is 150. Use the equation for head loss and solve for the pipe diameter. [Values of Hazen-Williams Coefficient C]

D = pipe diameter, in
P = pressure loss = 0.02 psi/ft of pipe length
Q = flow = 200 gpm

Circular Pipe Head Loss Equation (Head Loss Expressed as Pressure)

$$P = \frac{4.52 Q^{1.85}}{C^{1.85} D^{4.87}}$$

$$D = \left(\frac{4.52 Q^{1.85}}{C^{1.85} P}\right)^{1/4.87}$$

$$= \left(\frac{(4.52)\left(200 \ \dfrac{\text{gal}}{\text{min}}\right)^{1.85}}{(150)^{1.85}\left(0.02 \ \dfrac{\text{lbf}}{\text{in}^2}\right)}\right)^{1/4.87}$$

$$= 3.4 \text{ in} \quad (4.0 \text{ in})$$

The standard pipe size closest to 3.4 in is 4.0 in.

The answer is (D).

4. Use the continuity equation with pipe A, and solve for the flow velocity.

Q = flow rate, cfs
A = pipe cross sectional area, ft²
v = flow velocity, ft/sec

Continuity Equation

$$Q = Av$$

$$v_{\text{pipe A}} = \frac{Q}{A} = \frac{4Q}{\pi D^2} = \frac{(4)\left(0.5 \ \dfrac{\text{ft}^3}{\text{sec}}\right)\left(12 \ \dfrac{\text{in}}{\text{ft}}\right)^2}{\pi(4 \text{ in})^2}$$

$$= 5.73 \text{ ft/sec}$$

The head loss is the same in each pipe.

f = friction factor
L = pipe length, ft
D = pipe diameter, ft

Multipath Pipeline Problems

$$h_L = f_A \frac{L_A}{D_A} \frac{v_A^2}{2g} = f_B \frac{L_B}{D_B} \frac{v_B^2}{2g}$$

Calculate the head loss for pipe A.

$$f_A \frac{L_A}{D_A} \frac{v_A^2}{2g} = (0.038)\left(\frac{(38 \text{ ft})\left(12 \ \dfrac{\text{in}}{\text{ft}}\right)}{4 \text{ in}}\right)\left(\frac{\left(5.73 \ \dfrac{\text{ft}}{\text{sec}}\right)^2}{(2)\left(32.2 \ \dfrac{\text{ft}}{\text{sec}^2}\right)}\right)$$

$$= 2.21 \text{ ft}$$

The head loss for pipe B can be expressed as

$$f_B \frac{L_B}{D_B} \frac{v_B^2}{2g} = f_B \frac{L_B}{D_B} \frac{\left(\dfrac{Q_B}{\dfrac{\pi D_B^2}{4}}\right)^2}{2g}$$

Set the head loss equation for pipe B equal to the head loss for pipe A and solve for the diameter of pipe B.

$$f_B \frac{L_B}{D_B} \frac{\left(\dfrac{Q_B}{\dfrac{\pi D_B^2}{4}}\right)^2}{2g} = 2.21 \text{ ft}$$

$$(0.051)\left(73 \ \frac{\text{ft}}{D_B}\right) \left(\frac{\left(\dfrac{0.5 \ \dfrac{\text{ft}^3}{\text{sec}}}{\dfrac{\pi D_B^2}{4}}\right)^2}{(2)\left(32.2 \ \dfrac{\text{ft}}{\text{sec}^2}\right)} \right) = 2.21 \text{ ft}$$

$$D_B = 0.403 \text{ ft}$$

$$(0.403 \text{ ft})\left(12 \ \frac{\text{in}}{\text{ft}}\right) = 4.84 \text{ in} \quad (5 \text{ in})$$

The next standard pipe size above 4.84 in is 5 in pipe.

The answer is (B).

5. The population change from decade to decade is

period	population change
1970–1980	5800
1980–1990	5500
1990–2000	6500
2000–2010	8600

The population change pattern suggests log growth. Use the equation for log growth.

P_t = population at time t

P_0 = population at time zero

k = growth rate = 0.021 yr^{-1}

Δt = elapsed time relative to time zero, yr

$$P_t = P_0 e^{k\Delta t}$$
$$= 12{,}200 e^{(0.021 \text{ yr}^{-1})(80 \text{ yr})}$$
$$= 65{,}460 \quad (65{,}000 \text{ people})$$

The answer is (B).

6. The pumping rate to meet peak demand occurs at point P on the figure.

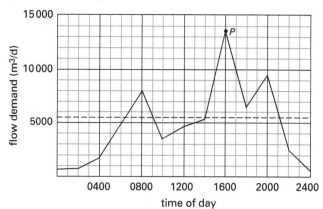

$$\left(13\,500 \ \frac{\text{m}^3}{\text{d}}\right)\left(\frac{1 \text{ d}}{86\,400 \text{ s}}\right)$$
$$= 0.156 \text{ m}^3/\text{s} \quad (0.16 \text{ m}^3/\text{s})$$

The answer is (B).

7. The average pumping rate is

$$\left(3.6 \ \frac{\text{m}^3}{\text{min}}\right)\left(1440 \ \frac{\text{min}}{\text{d}}\right) = 5184 \text{ m}^3/\text{d}$$

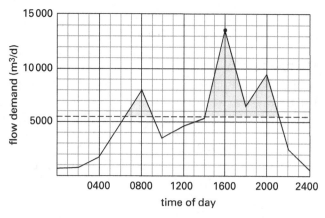

The area under the curve but above the average pumping rate line on the figure (shaded area) is the required storage volume. Integrate the shaded area by counting squares. There are approximately 30 shaded squares in the figure.

$$\text{storage volume} = (30)\left(1000 \ \frac{\text{m}^3}{\text{d}}\right)(1 \ \text{h})\left(\frac{1 \ \text{d}}{24 \ \text{h}}\right)$$
$$= 1250 \ \text{m}^3$$

The answer is (A).

19 Potable Water Residuals Management

Content in blue refers to the *NCEES Handbook*.

PRACTICE PROBLEMS

1. A waste-activated sludge with a total solids concentration of 2300 mg/L is thickened to 1.8% solids. The daily volume of waste activated sludge is 800 m³/d. The reduction in the volume of sludge from thickening is most nearly

(A) 105 m³/d

(B) 130 m³/d

(C) 670 m³/d

(D) 700 m³/d

2. Flotation is used to thicken solids generated from coagulation and precipitation processes. The flotation tank surface area is 57 m² with an air-to-solids ratio of 0.032 kg air/kg solids at a solids loading rate of 4.6 kg/m²·h. The daily air requirement is most nearly

(A) 125 kg air/d

(B) 139 kg air/d

(C) 201 kg air/d

(D) 214 kg air/d

3. A sedimentation basin produces 50,000 gal/day of chemical sludge at a concentration of 10 400 mg/L. The sludge is thickened to 9% solids and then dewatered to a produce a cake of 35% solids. Assume the specific gravity of the sludge is 1.0. The total volume reduction realized by thickening and dewatering the sludge is most nearly

(A) 5.6 m³/d

(B) 13 m³/d

(C) 34 m³/d

(D) 183 m³/d

SOLUTIONS

1. Calculate the daily mass of dry sludge solids.

$$\left(2300 \ \frac{mg}{L}\right)\left(800 \ \frac{m^3}{d}\right)\left(1000 \ \frac{L}{m^3}\right)\left(10^{-6} \ \frac{kg}{mg}\right) = 1840 \ kg/d$$

At the concentrations given, assume the specific gravity of the wet sludge is equal to 1.0, the same as that of water. Calculate the volume of wet sludge at 1.8% solids.

V = sludge volume, m³
W_s = mass of dry solids, kg
s = solids content, %
S = wet sludge specific gravity
γ = unit weight of water = 1000 kg/m³

Specific Gravity for a Solids Slurry

$$V = \frac{W_s}{(s/100)\gamma S}$$

$$= \frac{1840 \ \frac{kg}{d}}{\left(\frac{1.8}{100}\right)\left(1000 \ \frac{kg}{m^3}\right)(1.0)}$$

$$= 102 \ m^3/d$$

Calculate the daily volume reduction.

$$800 \ \frac{m^3}{d} - 102 \ \frac{m^3}{d} = 698 \ m^3/d \quad (700 \ m^3/d)$$

The answer is (D).

2. The daily mass of solids removed by flotation is

$$\left(4.6 \ \frac{kg}{m^2 \cdot h}\right)\left(24 \ \frac{h}{d}\right)(57 \ m^2) = 6292.8 \ kg \ solids/d$$

The daily mass of air required is

$$\left(\frac{0.032 \text{ kg air}}{\text{kg solids}}\right)\left(6292.8 \ \frac{\text{kg solids}}{\text{d}}\right)$$
$$= 201.37 \text{ kg air/d} \quad (201 \text{ kg air/d})$$

The answer is (C).

3. Convert gallons per day to cubic meters per day.

$$\left(50,000 \ \frac{\text{gal}}{\text{d}}\right)\left(\frac{3.785 \text{ m}^3}{1000 \text{ gal}}\right) = 189 \text{ m}^3/\text{d}$$

The daily mass of sludge produced is

$$\left(189 \ \frac{\text{m}^3}{\text{d}}\right)\left(1000 \ \frac{\text{L}}{\text{m}^3}\right)$$
$$\times \left(10\,400 \ \frac{\text{mg}}{\text{L}}\right)\left(10^{-6} \ \frac{\text{kg}}{\text{mg}}\right) = 1968.2 \text{ kg/d}$$

Find the daily volume of sludge at 65% solids.

W_s = mass of dry solids
γ = unit weight of water = 1000 kg/m^3
S = specific gravity of wet sludge
p = water content

Specific Gravity for a Solids Slurry

$$V = \frac{W_s}{[(100 - p)/100]\gamma S}$$
$$= \frac{1968.2 \ \dfrac{\text{kg}}{\text{d}}}{\left(\dfrac{100 - 65}{100}\right)\left(1000 \ \dfrac{\text{kg}}{\text{m}^3}\right)(1.0)}$$
$$= 5.62 \text{ m}^3/\text{d}$$

The volume reduction is

$$189 \ \frac{\text{m}^3}{\text{d}} - 5.62 \ \frac{\text{m}^3}{\text{d}} = 183.38 \text{ m}^3/\text{d} \quad (183 \text{ m}^3/\text{d})$$

The answer is (D).

 **Sources of Pollution**

Content in blue refers to the *NCEES Handbook.*

PRACTICE PROBLEMS

1. A design storm produces 2.3 in of runoff over an area of 24 ac. The first flush from the storm is 12% of the total runoff volume and occurs during the first 10 min period after the storm begins. The required volume of the water quality compartment in the storm sewer receiving the first flush is most nearly

(A) 0.50–0.55 ac-ft

(B) 1.1–1.2 ac-ft

(C) 4.1–4.6 ac-ft

(D) 6.0–6.6 ac-ft

2. Runoff from an urban area contains sodium (Na^+), calcium (Ca^{2+}), and magnesium (Mg^{2+}) ions at concentrations of 428 mg/L, 745 mg/L, and 312 mg/L, respectively. What is the sodium adsorption ratio (SAR) descriptor for the runoff?

(A) low sodium hazard (SAR < 9)

(B) medium sodium hazard (10 < SR < 17)

(C) high sodium hazard (18 < SAR < 25)

(D) unsuitable (SAR > 25)

SOLUTIONS

1. Calculate the volume of the first flush.

$$V_{\text{first flush}} = \left(\frac{(2.3 \text{ in})(24 \text{ ac})}{12 \ \dfrac{\text{in}}{\text{ft}}} \right) \left(\frac{12\%}{100\%} \right) = 0.55 \text{ ac-ft}$$

A water quality compartment is typically sized to retain 90% to 100% of the first flush. The required volume of the water quality compartment is 0.50–0.55 ac-ft.

The answer is (A).

2. Calculate the ion concentrations in units of meq/L. [Common Radicals in Water]

	MW (g/mol)	equivalence (eq/mol)
sodium (Na^+)	23.0	1
calcium (Ca^{2+})	40.1	2
magnesium (Mg^{2+})	24.3	2

$$\frac{\left(428 \ \dfrac{\text{mg}}{\text{L}}\right)\left(1 \ \dfrac{\text{eq}}{\text{mol}}\right)\left(1000 \ \dfrac{\text{meq}}{\text{eq}}\right)}{\left(23.0 \ \dfrac{\text{g}}{\text{mol}}\right)\left(1000 \ \dfrac{\text{mg}}{\text{g}}\right)} = 18.61 \text{ meq/L}$$

$$\frac{\left(745 \ \dfrac{\text{mg}}{\text{L}}\right)\left(2 \ \dfrac{\text{eq}}{\text{mol}}\right)\left(1000 \ \dfrac{\text{meq}}{\text{eq}}\right)}{\left(40.1 \ \dfrac{\text{g}}{\text{mol}}\right)\left(1000 \ \dfrac{\text{mg}}{\text{g}}\right)} = 37.25 \text{ meq/L}$$

$$\frac{\left(312 \ \dfrac{\text{mg}}{\text{L}}\right)\left(2 \ \dfrac{\text{eq}}{\text{mol}}\right)\left(1000 \ \dfrac{\text{meq}}{\text{eq}}\right)}{\left(24.3 \ \dfrac{\text{g}}{\text{mol}}\right)\left(1000 \ \dfrac{\text{mg}}{\text{g}}\right)} = 25.57 \text{ meq/L}$$

Calculate the sodium adsorption ratio (SAR).

$$\text{SAR} = \frac{\text{Na}^+}{\sqrt{\dfrac{\text{Ca}^{2+} + \text{Mg}^{2+}}{2}}}$$

$$= \frac{18.61\ \dfrac{\text{meq}}{\text{L}}}{\sqrt{\dfrac{37.25\ \dfrac{\text{meq}}{\text{L}} + 25.57\ \dfrac{\text{meq}}{\text{L}}}{2}}}$$

$$= 3.3$$

The sodium adsorption ratio of 3.3 is less than 9; therefore, the descriptor is "low sodium hazard."

The answer is (A).

21 Watershed Management and Planning

Content in blue refers to the *NCEES Handbook*.

PRACTICE PROBLEMS

1. A site is characterized by two adjoining parcels with the following characteristics.

parcel	area (ac)	SCS CN
A	112	80
B	94	75

The infiltration from 5.4 in of rainfall for the site is most nearly

(A) 23 ac-ft

(B) 30 ac-ft

(C) 36 ac-ft

(D) 40 ac-ft

2. The pan coefficient for evaporation from a lake is 0.65. What is the lake evaporation corresponding to a pan evaporation of 0.28 in/day?

(A) 0.18 in/day

(B) 0.23 in/day

(C) 0.43 in/day

(D) 0.93 in/day

3. A watershed occupies a 30 ha site. The overall runoff coefficient for the site is 0.18, and the average land slope is 2.1%. Because the site is upland from a residential development, the rainfall runoff from the site is collected in a catchment that discharges directly to a culvert. The overland flow distance to the catchment is 212 m. The 20 yr storm is characterized by the intensity duration curve presented in the figure.

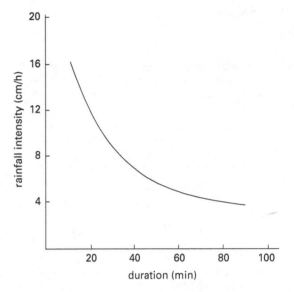

For a time of concentration of 34 min, the peak flow rate to the culvert entrance for the 20 yr storm is most nearly

(A) 0.44 m³/s

(B) 1.1 m³/s

(C) 2.1 m³/s

(D) 3.9 m³/s

4. A water balance prepared for a natural wetland covering a 359 ha area is summarized in the following table. Berm height allows a maximum water depth of 80 cm in the wetland.

source	average estimated contribution	
	October–March	April–September
inputs		
direct precipitation	100 cm	55 cm
surface inflow	0.21 m³/s	0.09 m³/s
subsurface inflow	0.0043 m³/s	–
outputs		
surface outflow	0.17 m³/s	0.02 m³/s
subsurface outflow	–	0.0073 m³/s
evapotranspiration	40 cm	121 cm

The minimum wetland storage volume is 1 500 000 m³. The maximum turnover rate during winter months is most nearly

(A) 0.32/mo

(B) 0.77/mo

(C) 3.1/mo

(D) 4.0/mo

5. A water balance prepared for a natural wetland covering a 359 ha area is summarized in the following table. Berm height allows a maximum water depth of 80 cm in the wetland.

source	average estimated contribution	
	October–March	April–September
inputs		
direct precipitation	100 cm	55 cm
surface inflow	0.21 m³/s	0.09 m³/s
subsurface inflow	0.0043 m³/s	–
outputs		
surface outflow	0.17 m³/s	0.02 m³/s
subsurface outflow	–	0.0073 m³/s
evapotranspiration	40 cm	121 cm

The minimum wetland storage volume is 1 500 000 m³. The hydraulic retention time during summer months is most nearly

(A) 0.78 mo

(B) 2.7 mo

(C) 4.0 mo

(D) 8.0 mo

6. The measured average annual pan evaporation is 0.17 in/day and the pan outer face temperature is the same as the water temperature. Using a pan coefficient of 0.7, the annual lake evaporation is most nearly

(A) 43 in/yr

(B) 53 in/yr

(C) 62 in/yr

(D) 87 in/yr

7. A partial routing schedule for a linear reservoir is shown.

time (hr)	inflow (cfs)	outflow (cfs)
0	0	0
1	10	0
2	20	2.94
3	30	10.03
4	40	18.84
5	50	28.34
6	40	38.14
7	30	42.18
8	20	37.95
9	10	30.33
10	0	21.31
11	0	11.72

The reservoir storage constant is 1.2 hr. Most nearly, what is the maximum storage volume in the reservoir?

(A) 127,000 cf

(B) 180,000 cf

(C) 182,000 cf

(D) 216,000 cf

SOLUTIONS

1. Find the total area of the site.

$$\text{total area} = 112 \text{ ac} + 94 \text{ ac} = 206 \text{ ac}$$

Calculate the composite curve number, CN.

$$CN = \left(\frac{112 \text{ ac}}{206 \text{ ac}}\right)(80) + \left(\frac{94 \text{ ac}}{206 \text{ ac}}\right)(75)$$
$$= 77.7 \quad (78)$$

Calculate the maximum basin retention, S, based on the composite CN.

NRCS (SCS) Rainfall-Runoff

$$S = \frac{1000}{CN} - 10 = \frac{1000}{78} - 10 = 2.8 \text{ in}$$

Calculate runoff, Q.

$$Q = \frac{(P - 0.2S)^2}{P + 0.8S}$$
$$= \frac{\left(5.4 \text{ in} - (0.2)(2.8 \text{ in})\right)^2}{5.4 \text{ in} + (0.8)(2.8 \text{ in})}$$
$$= 3.1 \text{ in}$$

Calculate initial abstraction, I_a.

$$I_a = 0.2S = (0.2)(2.8 \text{ in}) = 0.56 \text{ in}$$

Infiltration, F, is determined by subtracting runoff and initial abstraction from precipitation.

$$F = P - Q - I_a = 5.4 \text{ in} - 3.1 \text{ in} - 0.56 \text{ in}$$
$$= 1.74 \text{ in}$$

The volume of infiltration, V_F, is determined by multiplying the depth of infiltration by the total area.

$$V_F = \left(\frac{1.74 \text{ in}}{12 \frac{\text{in}}{\text{ft}}}\right)(206 \text{ ac}) = 30 \text{ ac-ft}$$

The answer is (B).

2. Find the evaporation rate.

E_L = lake evaporation, in/day

E_P = pan evaporation = 0.28 in/day

p_c = pan coefficient = 0.65

Pan Evaporation

$$E_L = p_c E_P = (0.65)\left(0.28 \frac{\text{in}}{\text{day}}\right)$$
$$= 0.18 \text{ in/day}$$

The answer is (A).

3. From the figure, at a time of concentration of 34 min, I is 7.5 cm/h. Find the peak flow rate.

Q = flow, m³/s

A = watershed area = 30 ha

I = rainfall intensity, cm/h

Rational Formula

$$Q = CIA$$
$$= (0.18)\left(7.5 \frac{\text{cm}}{\text{h}}\right)(30 \text{ ha})\left(\frac{1 \text{ m}}{100 \text{ cm}}\right)$$
$$\times \left(10\,000 \frac{\text{m}^2}{\text{ha}}\right)\left(\frac{1 \text{ h}}{3600 \text{ s}}\right)$$
$$= 1.125 \text{ m}^3/\text{s} \quad (1.1 \text{ m}^3/\text{s})$$

The answer is (B).

4. To estimate the maximum turnover rate, assume the minimum volume occurs from October to March.

$$\text{turnover} = \frac{\text{input}}{\text{volume}}$$

$$= \frac{(100 \text{ cm})(359 \text{ ha})\left(10\,000 \frac{\text{m}^2}{\text{ha}}\right)\left(\frac{1 \text{ m}}{100 \text{ cm}}\right)}{(6 \text{ mo})(1\,500\,000 \text{ m}^3)}$$
$$+ \frac{\left(0.21 \frac{\text{m}^3}{\text{s}} + 0.0043 \frac{\text{m}^3}{\text{s}}\right) \times \left(86\,400 \frac{\text{s}}{\text{d}}\right)\left(30 \frac{\text{d}}{\text{mo}}\right)}{(1\,500\,000 \text{ m}^3)}$$

$$= 0.77/\text{mo}$$

The answer is (B).

5. The equation for the hydraulic retention time is

$$\text{hydraulic retention time} = \frac{\text{volume}}{\text{inputs}}$$

The hydraulic retention time during the summer months is

$$
\cfrac{1\,500\,000 \text{ m}^3}{\left(\cfrac{55 \text{ cm}}{6 \text{ mo}}\right)(359 \text{ ha})\left(10\,000 \ \cfrac{\text{m}^2}{\text{ha}}\right)\left(\cfrac{1 \text{ m}}{100 \text{ cm}}\right)}
$$

$$
+\left(0.09 \ \cfrac{\text{m}^3}{\text{s}}\right)\left(86\,400 \ \cfrac{\text{s}}{\text{d}}\right)\left(30 \ \cfrac{\text{d}}{\text{mo}}\right)
$$

$$
= 2.7 \text{ mo}
$$

The answer is (B).

6. Use the equation for pan evaporation.

$E_L = $ lake evaporation, in/day
$E_P = $ pan evaporation $= 0.17$ in/day
$p_c = $ pan coefficient $= 0.7$

$$
\textit{Pan Evaporation}
$$

$$
E_L = p_c E_P
$$

$$
= (0.7)\left(0.17 \ \cfrac{\text{in}}{\text{day}}\right)\left(365 \ \cfrac{\text{days}}{\text{yr}}\right)
$$

$$
= 43.4 \text{ in/yr} \quad (43 \text{ in/yr})
$$

The answer is (A).

7. From the routing schedule table, the maximum value for O_1 occurs at 7 hr; therefore, the maximum storage volume will also occur at 7 hr.

$S = $ storage, cf
$K = $ reservoir constant $= 1.2$ hr
$O_1 = $ outflow at 7 hr, cfs

$$
S = K(O_1)
$$

$$
= \left((1.2 \text{ hr})\left(3600 \ \cfrac{\text{sec}}{\text{hr}}\right)\right)\left(42.18 \ \cfrac{\text{ft}^3}{\text{sec}}\right)
$$

$$
= 182{,}218 \text{ ft}^3 \quad (182{,}000 \text{ cf})
$$

The answer is (C).

22 Source Supply and Protection

Content in blue refers to the *NCEES Handbook*.

PRACTICE PROBLEMS

1. A fresh snow sample of 0.15 m^3 produces 2.6 L of melt water. The density of the fresh snow is most nearly

- (A) 0.062%
- (B) 1.7%
- (C) 5.8%
- (D) 17%

2. The inflow and outflow over a length of stream channel at the start and end of a 12-hour routing period are shown in the following table.

time (hr)	inflow (cfs)	outflow (cfs)
0	346	331
12	219	168

The change in storage volume in the stream channel over the routing period is most nearly

- (A) 3.7×10^5 ft^3
- (B) 1.4×10^6 ft^3
- (C) 5.2×10^6 ft^3
- (D) 2.1×10^7 ft^3

3. A raindrop falls at a velocity of 4.6 m/s. The raindrop diameter remains constant but the fall velocity increases to 5.2 m/s. The proportional increase in energy as the raindrop impacts the ground surface is most nearly

- (A) 0.782
- (B) 1.13
- (C) 1.28
- (D) 1.44

4. A reservoir has a capacity of 9.7×10^4 ac-ft and experiences a sediment load of 1.2×10^2 ac-ft/yr. The theoretical life of the reservoir is most nearly

- (A) 12 yr
- (B) 80 yr
- (C) 120 yr
- (D) 800 yr

SOLUTIONS

1. Snow density, expressed in percent, is the volume of melt water divided by the volume of unmelted snow.

$$\text{density} = \frac{V_{\text{melt}}}{V_{\text{snow}}} \times 100\%$$

$$= \frac{2.6 \text{ L}}{(0.15 \text{ m}^3)\left(1000 \frac{\text{L}}{\text{m}^3}\right)} \times 100\%$$

$$= 1.73\% \quad (1.7\%)$$

The answer is (B).

2. Find the change in storage volume.

$$\frac{I_1 + I_2}{2} - \frac{O_1 + O_2}{2} = \frac{S_2 - S_1}{\Delta t} = \frac{\Delta S}{\Delta t}$$

$$\Delta S = \left(\frac{I_1 + I_2}{2} - \frac{O_1 + O_2}{2}\right)\Delta t$$

$$= \left(\begin{array}{c} \dfrac{346 \dfrac{\text{ft}^3}{\text{sec}} + 219 \dfrac{\text{ft}^3}{\text{sec}}}{2} \\[4mm] - \dfrac{331 \dfrac{\text{ft}^3}{\text{sec}} + 168 \dfrac{\text{ft}^3}{\text{sec}}}{2} \end{array}\right)(12 \text{ hr})\left(3600 \frac{\text{sec}}{\text{hr}}\right)$$

$$= 1{,}425{,}600 \text{ ft}^3 \quad (1.4 \times 10^6 \text{ ft}^3)$$

The answer is (B).

3. The energy of a falling raindrop is proportional to the mean diameter, d, of the raindrop and the fall velocity, v, expressed as $d^3 v^2$. The impaction energy increases by a factor of

$$\frac{d_2^3 v_2^2}{d_1^3 v_1^2}$$

When $d_1 = d_2$, the impaction energy increases by a factor of

$$\frac{v_2^2}{v_1^2} = \frac{\left(5.2 \dfrac{\text{m}}{\text{s}}\right)^2}{\left(4.6 \dfrac{\text{m}}{\text{s}}\right)^2} = 1.28$$

The answer is (C).

4. Divide the reservoir capacity by the sediment load to find the theoretical life of the reservoir.

$$\frac{9.7 \times 10^4 \text{ ac-ft}}{1.2 \times 10^2 \dfrac{\text{ac-ft}}{\text{yr}}} = 808 \cdot \text{yr} \quad (800 \text{ yr})$$

The answer is (D).

Topic II Air

23 Sampling and Measurement Methods

Content in blue refers to the *NCEES Handbook*.

PRACTICE PROBLEMS

1. A series of three samplers operated over a 24 hr period are described by the data presented in the following table.

parameter	sampler 1	sampler 2	sampler 3
particle size collected, μm	unrestricted	< 10	< 2.5
clean filter mass, g	9.87	10.03	9.96
filter mass after 24 h, g	9.93	10.05	9.97
initial air flow, m^3/min	1.0	1.0	1.0
final air flow, m^3/min	0.87	0.91	0.90

The PM-2.5 concentration in the sampled air is most nearly

(A) 7.3 μg/m^3

(B) 15 μg/m^3

(C) 23 μg/m^3

(D) 45 μg/m^3

2. A series of three samplers operated over a 24 hr period are described by the data presented in the following table.

parameter	sampler 1	sampler 2	sampler 3
particle size collected, μm	unrestricted	< 10	< 2.5
clean filter mass, g	9.87	10.03	9.96
filter mass after 24 h, g	9.93	10.05	9.97
initial air flow, m^3/min	1.0	1.0	1.0
final air flow, m^3/min	0.87	0.91	0.90

The PM-10 concentration in the sampled air is most nearly

(A) 7.3 μg/m^3

(B) 15 μg/m^3

(C) 23 μg/m^3

(D) 45 μg/m^3

3. A series of three samplers operated over a 24 hr period are described by the data presented in the following table.

parameter	sampler 1	sampler 2	sampler 3
particle size collected, μm	unrestricted	< 10	< 2.5
clean filter mass, g	9.87	10.03	9.96
filter mass after 24 h, g	9.93	10.05	9.97
initial air flow, m^3/min	1.0	1.0	1.0
final air flow, m^3/min	0.87	0.91	0.90

The total particulate concentration in the sampled air is most nearly

(A) 7.3 μg/m^3

(B) 15 μg/m^3

(C) 23 μg/m^3

(D) 45 μg/m^3

4. What does the Ringlemann number signify?

(A) smoke plume dimensions

(B) black or gray smoke opacity

(C) white smoke opacity

(D) smoke plume rise rate

5. How is opacity of white smoke reported?

(A) Ringlemann number

(B) pararosaniline number

(C) percent opacity

(D) percent polarization

6. With what accuracy are opacity observations reported?

(A) 2.5% opacity or $1/8$ Ringlemann number

(B) 5% opacity or $1/4$ Ringlemann number

(C) 7.5% opacity or $3/8$ Ringlemann number

(D) 10% opacity or $1/2$ Ringlemann number

7. The maximum concentrations of air pollutants measured in a metropolitan area of the United States during a single day are shown in the table.

pollutant	duration (h)	concentration
O_3	8	0.021 ppm
NO_2	1	80 ppb
CO	8	9.5 ppm
SO_2	1	67 ppb
PM-10	24	100 $\mu g/m^3$

The pollutant with the highest air quality index (AQI) value is

(A) ozone

(B) carbon monoxide

(C) sulfur dioxide

(D) particulate

8. The maximum concentrations of air pollutants measured in a metropolitan area of the United States during a single day are shown in the table.

pollutant	duration (h)	concentration
O_3	8	0.021 ppm
NO_2	1	80 ppb
CO	8	9.5 ppm
SO_2	1	67 ppb
PM-10	24	100 $\mu g/m^3$

The air quality index (AQI) value for the day is most nearly

(A) 75

(B) 101

(C) 194

(D) 484

9. The maximum concentrations of air pollutants measured in a metropolitan area of the United States during a single day are shown in the table.

pollutant	duration (h)	concentration
O_3	8	0.021 ppm
NO_2	1	80 ppb
CO	8	9.5 ppm
SO_2	1	67 ppb
PM-10	24	100 $\mu g/m^3$

The descriptor corresponding to the air quality index (AQI) value for the day is

(A) moderate

(B) unhealthy for sensitive groups

(C) unhealthy

(D) very unhealthy

SOLUTIONS

1. PM-2.5 particles are less than 2.5 μm. Use data from sampler 3.

The average airflow is

$$(0.5)\left(1.0 \ \frac{m^3}{min} + 0.90 \ \frac{m^3}{min}\right) = 0.95 \ m^3/min$$

The concentration is

$$\frac{(9.97 \ g - 9.96 \ g)\left(10^6 \ \frac{\mu g}{g}\right)}{\left(0.95 \ \frac{m^3}{min}\right)(24 \ h)\left(60 \ \frac{min}{h}\right)} = 7.3 \ \mu g/m^3$$

The answer is (A).

2. According to USEPA, PM-10 particles are less than 10 μm. Use data from sampler 2.

The average air flow is

$$(0.5)\left(1.0 \ \frac{m^3}{min} + 0.91 \ \frac{m^3}{min}\right) = 0.955 \ m^3/min$$

The concentration is

$$\frac{(10.05 \ g - 10.03 \ g)\left(10^6 \ \frac{\mu g}{g}\right)}{\left(0.955 \ \frac{m^3}{min}\right)(24 \ h)\left(60 \ \frac{min}{h}\right)}$$
$$= 14.5 \ \mu g/m^3 \quad (15 \ \mu g/m^3)$$

The answer is (B).

3. Total particulate is measured by sampler 1.

The average air flow is

$$(0.5)\left(1.0 \ \frac{m^3}{min} + 0.87 \ \frac{m^3}{min}\right) = 0.935 \ m^3/min$$

The concentration is

$$\frac{(9.93 \ g - 9.87 \ g)\left(10^6 \ \frac{\mu g}{g}\right)}{\left(0.935 \ \frac{m^3}{min}\right)(24 \ h)\left(60 \ \frac{min}{h}\right)}$$
$$= 44.6 \ \mu g/m^3 \quad (45 \ \mu g/m^3)$$

The answer is (D).

4. The Ringlemann number is used to express the opacity of black or gray smoke.

The answer is (B).

5. The opacity of white smoke is reported by percent opacity.

The answer is (C).

6. Opacity observations are reported to 5% opacity or $\frac{1}{4}$ Ringlemann number.

The answer is (B).

7. Using the USEPA breakpoint values for the AQI, find the ACQ value for each pollutant. [Breakpoints for the AQI]

$$I_p = \text{air quality index value for}$$
$$\text{pollutant concentration } p$$
$$C_p = \text{pollutant measured concentration}$$
$$BP_{Lo} = \text{concentration breakpoint} \leq C$$
$$BP_{Hi} = \text{concentration breakpoint} \geq C$$
$$I_{Lo} = \text{air quality index value}$$
$$\text{corresponding to } BP_{Lo}$$
$$I_{Hi} = \text{air quality index value}$$
$$\text{corresponding to } BP_{Hi}$$

AQI Calculation

$$I_p = \frac{I_{Hi} - I_{Lo}}{BP_{Hi} - BP_{Lo}}(C_p - BP_{Lo}) + I_{Lo}$$

$$I_{O_3} = \left(\frac{50 - 0}{54 - 0}\right)(21 - 0) + 0 = 19$$

$$I_{NO_2} = \left(\frac{100 - 51}{100 - 54}\right)(80 - 54) + 51 = 79$$

$$I_{CO} = \left(\frac{150 - 101}{12.4 - 9.5}\right)(9.5 - 9.5) + 101 = 101$$

$$I_{SO_2} = \left(\frac{100 - 51}{75 - 36}\right)(67 - 36) + 51 = 90$$

$$I_{PM-10} = \left(\frac{100 - 51}{154 - 55}\right)(100 - 55) + 51 = 73$$

Carbon monoxide has the highest AQI value at 101.

The answer is (B).

8. Using the USEPA breakpoint values for the AQI, find the AQI value for each pollutant. [Breakpoints for the AQI]

$$I_p = \text{air quality index value for}$$
$$\text{pollutant concentration } p$$
$$C_p = \text{pollutant measured concentration}$$
$$BP_{Lo} = \text{concentration breakpoint} \leq C$$
$$BP_{Hi} = \text{concentration breakpoint} \geq C$$
$$I_{Lo} = \text{air quality index value}$$
$$\text{corresponding to } BP_{Lo}$$
$$I_{Hi} = \text{air quality index value}$$
$$\text{corresponding to } BP_{Hi}$$

AQI Calculation

$$I_p = \frac{I_{Hi} - I_{Lo}}{BP_{Hi} - BP_{Lo}}(C_p - BP_{Lo}) + I_{Lo}$$

$$I_{O_3} = \left(\frac{50 - 0}{54 - 0}\right)(21 - 0) + 0 = 19$$

$$I_{NO_2} = \left(\frac{100 - 51}{100 - 54}\right)(80 - 54) + 51 = 79$$

$$I_{CO} = \left(\frac{150 - 101}{12.4 - 9.5}\right)(9.5 - 9.5) + 101 = 101$$

$$I_{SO_2} = \left(\frac{100 - 51}{75 - 36}\right)(67 - 36) + 51 = 90$$

$$I_{PM-10} = \left(\frac{100 - 51}{154 - 55}\right)(100 - 55) + 51 = 73$$

The AQI value for the day is equal to the highest index value for any individual pollutant. The highest AQI value is for carbon monoxide at 101. Therefore, the AQI value for the day is 101.

The answer is (B).

9. Using the USEPA breakpoint values for the AQI, find the AQI value for each pollutant. [Breakpoints for the AQI]

$$I_p = \text{air quality index value for}$$
$$\text{pollutant concentration } p$$
$$C_p = \text{pollutant measured concentration}$$
$$BP_{Lo} = \text{concentration breakpoint} \leq C$$
$$BP_{Hi} = \text{concentration breakpoint} \geq C$$
$$I_{Lo} = \text{air quality index value}$$
$$\text{corresponding to } BP_{Lo}$$
$$I_{Hi} = \text{air quality index value}$$
$$\text{corresponding to } BP_{Hi}$$

AQI Calculation

$$I_p = \frac{I_{Hi} - I_{Lo}}{BP_{Hi} - BP_{Lo}}(C_p - BP_{Lo}) + I_{Lo}$$

$$I_{O_3} = \left(\frac{50 - 0}{54 - 0}\right)(21 - 0) + 0 = 19$$

$$I_{NO_2} = \left(\frac{100 - 51}{100 - 54}\right)(80 - 54) + 51 = 79$$

$$I_{CO} = \left(\frac{150 - 101}{12.4 - 9.5}\right)(9.5 - 9.5) + 101 = 101$$

$$I_{SO_2} = \left(\frac{100 - 51}{75 - 36}\right)(67 - 36) + 51 = 90$$

$$I_{PM-10} = \left(\frac{100 - 51}{154 - 55}\right)(100 - 55) + 51 = 73$$

The highest AQI value is 101, so the descriptor is "unhealthy for sensitive groups."

The answer is (B).

Content in blue refers to the *NCEES Handbook*.

PRACTICE PROBLEMS

1. What are the criteria pollutants under the Clean Air Act?

(A) ground level ozone, carbon dioxide, sulfur dioxide, small particulates, nitrogen dioxide, and radon

(B) ground level ozone, carbon monoxide, sulfur trioxide, small particulates, nitrogen dioxide, and radon

(C) ground level ozone, carbon monoxide, sulfur trioxide, small particulates, nitrogen dioxide, and lead

(D) ground level ozone, carbon monoxide, sulfur dioxide, small particulates, nitrogen dioxide, and lead

2. The maximum concentrations of air pollutants measured in a metropolitan area of the United States during a single day are shown in the table.

pollutant	duration (h)	concentration
O_3	8	0.21 ppm
NO_2	1	80 ppb
CO	8	9.5 ppm
SO_2	1	67 ppb
PM-10	24	100 $\mu g/m^3$

From the table, the pollutants that exceed the National Ambient Air Quality Standards (NAAQS) are

(A) ozone and particulate

(B) ozone and carbon monoxide

(C) carbon monoxide and particulate

(D) sulfur dioxide

3. Under the Clean Air Act, what emission controls must be applied by new sources in nonattainment areas?

(A) best available control technology (BACT)

(B) advanced pollutant control and management (APCAM)

(C) lowest achievable emission rate (LAER)

(D) comprehensive health and environment abatement technology (CHEAT)

4. What distinguishes National Ambient Air Quality Standards (NAAQS) from New Source Performance Standards (NSPS)?

(A) NAAQS define acceptable concentrations of pollutants in the air, whereas NSPS define allowable rates at which pollutants can be emitted.

(B) NSPS define acceptable concentrations of pollutants in the air, whereas NAAQS define allowable rates at which pollutants can be emitted.

(C) NAAQS define acceptable concentrations in ambient air outside of specified urban areas, whereas NSPS define air quality goals within specified urban areas.

(D) NSPS apply exclusively to limit emissions from mobile sources, whereas NAAQS are not source specific and define acceptable concentrations of pollutants in the air overall.

5. What are the six contaminants that typically constitute the major components of urban air pollution?

(A) smog, HCl, NOx, SOx, Pb, and O_3

(B) smog, NOx, SOx, CO_2, Ar, and particulate

(C) NOx, SOx, COx, Ar, O_3, and particulate

(D) NOx, SOx, CO, Pb, O_3, and particulate

6. Under the Clean Air Act, how do primary and secondary pollutants differ from primary and secondary standards?

(A) Primary and secondary standards define emission limits for primary and secondary pollutants, respectively.

(B) Primary and secondary standards are regulatory emission limits, while primary and secondary pollutants are defined by their source or by reactions in the atmosphere.

(C) Primary and secondary standards limit emissions based on technological and economic factors, while primary and secondary pollutants are defined by their health and environmental impacts.

(D) Primary and secondary standards are regulatory emission limits, while primary and secondary pollutants are defined based on technological factors required for their control.

SOLUTIONS

1. The criteria pollutants are ground level ozone, carbon monoxide, sulfur dioxide, small particulates, nitrogen dioxide, and lead.

The answer is (D).

2. Compare the actual concentrations with the National Ambient Air Quality Standards. [National Ambient Air Quality Standards]

pollutant	NAAQS maximum concentration	actual concentration
O_3	0.070 ppm	0.21 ppm
NO_2	100 ppb	80 ppb
CO	9 ppm	9.5 ppm
SO_2	75 ppb	67 ppb
PM-10	150 $\mu g/m^3$	100 $\mu g/m^3$

Ozone and carbon monoxide exceed the NAAQS.

The answer is (B).

3. The lowest achievable emission rate (LAER) must be applied to control emissions from new sources in nonattainment areas.

The answer is (C).

4. NAAQS define acceptable concentrations of pollutants in the air, whereas NSPS define allowable rates at which pollutants can be emitted.

The answer is (A).

5. The six contaminants that typically constitute the major components of urban air pollution are NOx, SOx, CO, Pb, O_3, and particulate.

The answer is (D).

6. Primary and secondary standards are regulatory emission limits, while primary and secondary pollutants are defined by their source or by reactions in the atmosphere.

The answer is (B).

25 Chemistry

Content in blue refers to the *NCEES Handbook*.

PRACTICE PROBLEMS

1. A remedial action program for contaminated soils has identified thermal desorption as a potentially feasible treatment alternative. The desorption process operates at 650°C under 2.3 atm pressure. The reactor vessel volume is 86 m³ with 30% of the volume occupied by solids. The volatilized contaminants are present in a 90% nitrogen gas environment at the percentages provided in the table.

component	concentration (% by volume)	molecular weight (g/mol)
contaminant 1	4	78
contaminant 2	2	113
contaminant 3	2	106
contaminant 4	1	92
contaminant 5	1	63
nitrogen gas	90	28

Under normal operating conditions, the extracted gases are contained and treated through air pollution control unit processes. However, under emergency shutdown conditions, the contents of the reactor vessel are directly vented to the atmosphere. The ambient air temperature is 20°C, and the ambient air pressure is 1 atm. The total mass of the contaminant gas in the reactor vessel is most nearly

(A) 0.42 kg

(B) 0.59 kg

(C) 17 kg

(D) 24 kg

2. A remedial action program for contaminated soils has identified thermal desorption as a potentially feasible treatment alternative. The desorption process operates at 650°C under 2.3 atm pressure. The reactor vessel volume is 86 m³ with 30% of the volume occupied by solids. The volatilized contaminants are present in a 90% nitrogen gas environment at the percentages provided in the table.

component	concentration (% by volume)	molecular weight (g/mol)
contaminant 1	4	78
contaminant 2	2	113
contaminant 3	2	106
contaminant 4	1	92
contaminant 5	1	63
nitrogen gas	90	28

Under normal operating conditions, the extracted gases are contained and treated through air pollution control unit processes. However, under emergency shutdown conditions, the contents of the reactor vessel are directly vented to the atmosphere. The ambient air temperature is 20°C, and the ambient air pressure is 1 atm. Most nearly, what is the total volume of the contaminant gas that would be directly emitted to the atmosphere under emergency conditions?

(A) 4.4 m³

(B) 6.0 m³

(C) 8.6 m³

(D) 26 m³

3. Air component concentrations are presented in the table. The ambient temperature and pressure are 30°C and 0.98 atm, respectively.

component	concentration (ppmv)
N_2	780 900
O_2	209 400
Ar	9350
CO_2	350
	1 000 000

The apparent molecular weight of the air is most nearly

(A) 15 g/mol

(B) 29 g/mol

(C) 36 g/mol

(D) 140 g/mol

4. The temperature and pressure of the ambient air are 30°C and 0.98 atm, respectively. The molar volume of the air is most nearly

(A) 2.7×10^{-7} m³/mol

(B) 2.5×10^{-3} m³/mol

(C) 0.025 m³/mol

(D) 250 m³/mol

5. The ambient temperature and pressure of air at a certain location are 30°C and 0.98 atm, respectively. The molecular weight of air is 29 g/mol. The density of the air is most nearly

(A) 580 g/m³

(B) 1100 g/m³

(C) 1400 g/m³

(D) 5600 g/m³

6. If air at 30°C and 0.98 atm has a density of 1100 g/m³, the specific volume of the air is most nearly

(A) 0.00018 m³/g

(B) 0.00071 m³/g

(C) 0.00091 m³/g

(D) 0.0017 m³/g

7. At a certain location the air temperature is 30°C and the atmospheric pressure is 0.98 atm. The Henry's constant for air under these conditions is 7.71×10^4 atm/mol fraction. The molecular weight of the air is

29 g/mol. The solubility of the air in water is most nearly

(A) 11 mg/L

(B) 21 mg/L

(C) 26 mg/L

(D) 100 mg/L

8. Stack emissions from a manufacturing facility contain SO_2 at 0.18 ppmv and 95°C. The ambient air temperature is 18°C, and the air pressure is 1 atm. The SO_2 concentration in µg/m³ at the stack is most nearly

(A) 15 µg/m³

(B) 190 µg/m³

(C) 380 µg/m³

(D) 540 µg/m³

9. What is the concentration of carbon dioxide in dry ambient air at 1 atm and 20°C?

(A) 6.4×10^{-7} g/cm³

(B) 4.1×10^{-7} g/cm³

(C) 6.4×10^{-5} g/cm³

(D) 3.5×10^{-4} g/cm³

10. For a partial pressure of water vapor of 0.023 atm, what is the mass of water vapor in 100 kg of dry air at 1 atm and 20°C?

(A) 1,430 g/100 kg

(B) 1,460 g/100 kg

(C) 2,300 g/100 kg

(D) 6,080 g/100 kg

SOLUTIONS

1. Assume the ideal gas law applies. [Ideal Gas Constants]

P = pressure, atm

V = volume, m^3

n = number of moles

T = temperature, K

R = specific gas constant = 8.2×10^{-5} $m^3 \cdot atm/mol \cdot K$

$$PV = nRT$$

Assume Dalton's law applies.

In a gas mixture, each gas exerts pressure independently of the others and the resulting partial pressure of each gas is proportional to the amount of that gas in the mixture.

Conditions in the reactor vessel are presented in the following table.

component	concentration (% V/V)	MW (g/mol)	partial pressure (atm)	moles	mass (kg)
contaminant 1	4	78	0.092	73	5.7
contaminant 2	2	113	0.046	36	4.1
contaminant 3	2	106	0.046	36	3.9
contaminant 4	1	92	0.023	18	1.7
contaminant 5	1	63	0.023	18	1.1
nitrogen gas	90	28	2.07		
	100		2.3		16.5

The volume is

$$V = \frac{(86 \text{ m}^3)(100\% - 30\%)}{100\%} = 60 \text{ m}^3$$

The equation for the pressure is

$$P_i = \frac{(2.3 \text{ atm})\left(\frac{\% V}{V}\right)}{100\%}$$

$$n = \frac{PV}{RT}$$

The total mass of the contaminant gas is

$$n_i = \frac{(\text{partial pressure, atm})(\text{volume, m}^3)}{\left(8.2 \times 10^{-5} \frac{\text{m}^3 \cdot \text{atm}}{\text{mol} \cdot \text{K}}\right)(923\text{K})}$$

$$m_{\text{total}} = \sum m_i = \sum n_i \text{MW}_i = 16.5 \text{ kg} \quad (17 \text{ kg})$$

The answer is (C).

2. Assume the ideal gas law applies. [Ideal Gas Constants]

P = pressure, atm

V = volume, m^3

n = number of moles

T = temperature, K

R = specific gas constant = 8.2×10^{-5} $m^3 \cdot atm/mol \cdot K$

$$PV = nRT$$

Assume Dalton's law applies.

In a gas mixture, each gas exerts pressure independently of the others and the resulting partial pressure of each gas is proportional to the amount of that gas in the mixture.

Conditions in the ambient air are presented in the following table.

component	concentration (% V/V)	MW (g/mol)	moles	volume (m³)
contaminant 1	4	78	73	1.75
contaminant 2	2	113	36	0.88
contaminant 3	2	106	36	0.88
contaminant 4	1	92	18	0.43
contaminant 5	1	63	18	0.43
nitrogen gas	90	28		—
	100			4.37

The total volume is

$$V_i = \frac{n_i RT}{P} = \frac{n_i \left(8.2 \times 10^{-5} \frac{\text{m}^3 \cdot \text{atm}}{\text{mol} \cdot \text{K}}\right)(293\text{K})}{1.0 \text{ atm}}$$

$$V_t = \sum V_i = 4.37 \text{ m}^3 \quad (4.4 \text{ m}^3)$$

The answer is (A).

3. The equation for the fractional molecular weight is

(volumetric fraction)

$\times$ (component mole weight, g/mol)

component	volumetric fraction	component mole weight (g/mol)	fractional mole weight (g/mol)
N_2	0.7809	28	21.87
O_2	0.2094	32	6.70
Ar	0.00935	40	0.37
CO_2	0.00035	44	0.015
			28.96

The apparent molecular weight is 28.96 g/mol (29 g/mol).

The answer is (B).

4. Assume air behaves as an ideal gas. [Ideal Gas Constants]

P = pressure = 0.98 atm

V = volume, m^3

n = number of moles

R = specific gas constant = 8.2×10^{-5} $\text{m}^3 \cdot \text{atm/mol} \cdot \text{K}$

T = temperature = $30°\text{C} + 273° = 303\text{K}$

$$PV = nRT$$

The molar volume is

$$\frac{V}{n} = \frac{RT}{P}$$

$$= \frac{\left(8.2 \times 10^{-5} \ \dfrac{\text{m}^3 \cdot \text{atm}}{\text{mol} \cdot \text{K}}\right)(303\text{K})}{0.98 \text{ atm}}$$

$$= 0.025 \text{ m}^3/\text{mol}$$

The answer is (C).

5. Assume the air behaves as an ideal gas. [Ideal Gas Constants]

ρ = density, g/m^3

MW = molecular weight = 29 g/mol

P = pressure = 0.98 atm

V = volume, m^3

n = number of moles

R = specific gas constant = 8.2×10^{-5} $\text{m}^3 \cdot \text{atm/mol} \cdot \text{K}$

T = temperature = $30°\text{C} + 273° = 303\text{K}$

$$PV = nRT$$

$$\frac{P}{RT} = \frac{n}{V} = \frac{\rho}{\text{MW}}$$

$$\rho = \frac{P(\text{MW})}{RT}$$

$$= \frac{(0.98 \text{ atm})\left(29 \ \dfrac{\text{g}}{\text{mol}}\right)}{\left(8.2 \times 10^{-5} \ \dfrac{\text{m}^3 \cdot \text{atm}}{\text{mol} \cdot \text{K}}\right)(303\text{K})}$$

$$= 1144 \text{ g/m}^3 \quad (1100 \text{ g/m}^3)$$

The answer is (B).

6. The specific volume, V_s, is

$$V_s = \frac{1}{\rho} = \frac{1}{1100 \ \dfrac{\text{g}}{\text{m}^3}} = 0.00091 \text{ m}^3/\text{g}$$

The answer is (C).

7. Assume Henry's law applies.

x = mole fraction

P = partial pressure = 0.98 atm

h = Henry's constant

$\qquad = \dfrac{7.71 \times 10^4 \text{ atm}}{\text{mol fraction}}$ at $30°\text{C}$

$$x = \frac{P}{h} = \frac{0.98 \text{ atm}}{\left(\dfrac{7.71 \times 10^4 \text{ atm}}{\text{mol fraction}}\right)}$$

$$= 1.3 \times 10^{-5} \text{ mol fraction}$$

$$x = \frac{\text{mol air}}{\text{mol air} + \text{mol water}}$$

$$= 1.3 \times 10^{-5}$$

Henry's Law at Constant Temperature

$$P = hx_i$$

Assume a 1 L sample of water with water density = 1000 g/L.

The molecular weight of water is [Periodic Table of Elements]

$$(2)\left(1 \ \frac{\text{g}}{\text{mol}}\right) + 16 \ \frac{\text{g}}{\text{mol}} = 18 \text{ g/mol}$$

$$\text{mol water} = \frac{\left(1000 \ \dfrac{\text{g}}{\text{L}}\right)(1 \text{ L})}{18 \ \dfrac{\text{g}}{\text{mol}}} = 55.6 \text{ mol}$$

$$1.3 \times 10^{-5} = \frac{\text{mol air}}{\text{mol air} + 55.6 \text{ mol}}$$

$$\text{mol air} = 7.2 \times 10^{-4}$$

The solubility of air in water is

$$\frac{\left(29 \ \dfrac{\text{g}}{\text{mol}}\right)(7.2 \times 10^{-4} \text{ mol})\left(1000 \ \dfrac{\text{mg}}{\text{g}}\right)}{1 \text{ L}}$$

$$= 20.9 \text{ mg/L} \quad (21 \text{ mg/L})$$

The answer is (B).

8. Find the molecular weight of SO_2. [Periodic Table of Elements]

$$MWSO_2 = 32 \, \frac{g}{mol} + (2)\left(16 \, \frac{g}{mol}\right) = 64 \, g/mol$$

Assume SO_2 behaves as an ideal gas. [Ideal Gas Constants]

m = mass SO_2, g
P = pressure = 1 atm

$$V = \frac{\text{volume } SO_2}{\text{volume emitted gas}} = \frac{0.18 \, m^3}{10^6 \, m^3}$$

T = temperature = $95°C + 273° = 368K$
R = specific gas constant = $8.2 \times 10^{-5} \, m^3 \cdot atm/mol \cdot K$
n = number of moles = $\dfrac{\text{mass } SO_2}{\text{MW } SO_2} = \dfrac{m}{MW}$

$$PV = nRT \text{ or } m = \frac{(MW)PV}{RT}$$

The mass of SO_2 is

$$m = \frac{\left(64 \, \dfrac{g}{mol}\right)(1 \, atm)\left(\dfrac{0.18 \, m^3}{10^6 \, m^3}\right)}{\left(8.2 \times 10^{-5} \, \dfrac{m^3 \cdot atm}{mol \cdot K}\right)(368K)} = 382 \, g/10^6 \, m^3$$

The SO_2 concentration is

$$\left(\frac{382 \, g}{10^6 \, m^3}\right)\left(10^6 \, \frac{\mu g}{g}\right)$$
$$= 382 \, \mu g/m^3 \quad (380 \, \mu g/m^3)$$

The answer is (C).

9. The concentration by volume of carbon dioxide (CO_2) in dry ambient air is 0.035%, or 0.00035 L of CO_2 per liter of air. [Composition of Dry Ambient Air]

Calculate the molecular weight of CO_2. [Periodic Table of Elements]

$$MW \, CO_2 = 12 \, \frac{g}{mol} + (2)\left(16 \, \frac{g}{mol}\right)$$
$$= 44 \, g/mol$$

n = moles
P = pressure = 1 atm
V = volume = 0.00035 L
T = temperature = $20°C + 273° = 293K$
R = ideal gas constant at 1 atm and 293K
 = 0.08205 atm·L/mol·K

Calculate the number of moles of CO_2.

Ideal Gas Constants

$$PV = nRT$$
$$n = \frac{PV}{RT} = \frac{(1 \, atm)(0.00035 \, L)}{\left(0.08205 \, \dfrac{atm \cdot L}{mol \cdot K}\right)(293K)}$$
$$= 1.456 \times 10^{-5} \, mol$$

Calculate the mass of CO_2.

$$(1.456 \times 10^{-5} \, mol)\left(44 \, \frac{g}{mol}\right) = 6.4 \times 10^{-4} \, g$$

The concentration in the air is

$$\left(\frac{6.4 \times 10^{-4} \, g \, CO_2}{L \, air}\right)\left(10^{-3} \, \frac{L}{cm^3}\right) = 6.4 \times 10^{-7} \, g/cm^3$$

The answer is (A).

10.

P_a = partial pressure of dry air, atm
P_v = partial pressure of water vapor = 0.023 atm
P = pressure of air-water mixture = 1 atm

Psychrometrics

$$P = P_a + P_v$$
$$P_a = P - P_v = 1 \, atm - 0.023 \, atm$$
$$= 0.977 \, atm$$

Psychrometrics

ω = specific humidity

$$\omega = 0.622 P_v/P_a$$
$$= (0.622)\left(\frac{0.023 \, atm}{0.977 \, atm}\right)$$
$$= 0.0146 \text{ or } 0.0146 \text{ g water vapor/g dry air}$$

m_a = mass of dry air
m_v = mass of water vapor

Let m_a = 100 kg.

Psychrometrics

$$\omega = m_v/m_a$$
$$m_v = \omega m_a$$
$$= \left(0.0146 \, \frac{g}{g}\right)(100 \, kg)$$
$$= 1.46 \text{ kg water vapor in 100 kg air} \quad (1{,}460 \, g/100 \, kg)$$

The answer is (B).

26 Fate and Transport

Content in blue refers to the *NCEES Handbook*.

PRACTICE PROBLEMS

1. Sulfur trioxide (SO_3) exists in the atmosphere at $458 \text{ g}/10^6 \text{ m}^3$ of air at the beginning of a rain storm. The natural rainwater pH is 5.5, and the ambient temperature and pressure are 25°C and 1 atm, respectively. Assume the rainwater is 100% efficient in scrubbing the SO_3 from the air and that each cubic meter of air is scrubbed by 0.27 m^3 of rainwater over the duration of the storm. The sulfate concentration in the rainwater at the end of the storm is most nearly

(A) 1.4 μg/L

(B) 2.0 μg/L

(C) 21 μg/L

(D) 30 μg/L

2. Sulfur trioxide (SO_3) exists in the atmosphere at $458 \text{ g}/10^6 \text{ m}^3$ of air at the beginning of a rain storm. The natural rainwater pH is 5.5, and the ambient temperature and pressure are 25°C and 1 atm, respectively. Assume the rainwater is 100% efficient in scrubbing the SO_3 from the air and that each cubic meter of air is scrubbed by 0.27 m^3 of rainwater over the duration of the storm. The pH of the rainwater at the end of the storm is most nearly

(A) 1.4

(B) 1.7

(C) 5.5

(D) 6.9

3. What are the hydrogen to carbon ratios, by mass, of methane and acetylene?

(A) methane H:C = 1:12, acetylene H:C = 1:12

(B) methane H:C = 1:3, acetylene H:C = 1:12

(C) methane H:C = 4:1, acetylene H:C = 2:2

(D) methane H:C = 1:4, acetylene H:C = 1:13

4. What is the product of the photolytic decomposition of nitrogen dioxide in the atmosphere?

(A) nitric acid aerosol and particulate

(B) nitric oxide and free oxygen radicals

(C) photochemical smog

(D) stratospheric ozone scavenging peroxyl radicals

5. What compounds are predominantly involved in reactions with the hydroxyl radical ($\cdot OH$) to form photochemical smog?

(A) carbon monoxide, aldehydes and ketones, and hydrocarbons

(B) carbon monoxide, nitrogen dioxide, and hydrocarbons

(C) carbon monoxide, nitrogen dioxide, and sulfur dioxide

(D) nitrogen dioxide, sulfur dioxide, and oxygen

6. Stack emissions from a manufacturing facility contain SO_2 at 95°C. The ambient air temperature is 18°C, and the air pressure is 1 atm. The SO_2 concentration in the air at the stack is $380 \text{ g}/10^6 \text{ m}^3$. If dispersion is neglected, the SO_2 concentration in the air at ambient temperature is most nearly

(A) 480 μg/m³

(B) 540 μg/m³

(C) 1200 μg/m³

(D) 2000 μg/m³

7. If it is raining, what is the likely fate of SO_2?

(A) The SO_2 will remain as a gas.

(B) The SO_2 will form sulfuric acid with the rainwater.

(C) The SO_2 will form sulfurous acid with the rainwater.

(D) The SO_2 will condense from a gas to a liquid.

8. Which of the following is NOT an important greenhouse gas?

(A) H_2O

(B) SO_2

(C) N_2O

(D) O_2

9. Why is CO_2 the focus of attention in efforts to control greenhouse gas formation?

(A) CO_2 accounts for over 60% of the radiative forcing caused by historical increases in greenhouse gas concentrations.

(B) CO_2 increases occur with corresponding O_2 decreases that incrementally limit respiration and initiate extinction of sensitive heterotrophic organisms.

(C) In the presence of direct sunlight, CO_2 reacts with free oxygen in the atmosphere to form carbon monoxide and ozone, both of which are greenhouse gases.

(D) Unlike other greenhouse gases, CO_2 can be easily controlled by substituting fuels such as natural gas for fuels such as coal and wood.

10. What two factors are believed to most significantly contribute to potential global warming?

(A) combustion of fossil fuels and construction of radiating surfaces such as paved areas and buildings

(B) combustion of fossil fuels and decreased biomass density on the earth's surface

(C) construction of radiating surfaces such as paved areas and buildings and decreased biomass density on the earth's surface

(D) decreased biomass density on the earth's surface and increased atmospheric water vapor from industrialization and agricultural irrigation

SOLUTIONS

1. Sulfur trioxide forms sulfuric acid when combined with water. The sulfuric acid subsequently dissociates to hydrogen ion and to sulfate.

$$SO_3 + H_2O \rightarrow H_2SO_4 \rightarrow 2H^+ + SO_4^{-2}$$

One mole of SO_3 scrubbed will yield one mole of SO_4^{-2} and two moles of H^+ in the rainwater. The SO_3 concentration in rainwater is

$$\left(\frac{458 \text{ g}}{10^6 \text{ m}^3}\right)\left(\frac{1 \text{ m}^3}{0.27 \text{ m}^3}\right)\left(10^6 \frac{\mu g}{g}\right)\left(\frac{1 \text{ m}^3}{1000 \text{ L}}\right)$$
$$= 1.7 \ \mu g/L$$

Find the molecular weight of SO_3. [Periodic Table of Elements]

$$32 \frac{g}{mol} + (3)\left(16 \frac{g}{mol}\right) = 80 \text{ g/mol}$$

The molecular weight of SO_4^{-2} is

$$32 \frac{g}{mol} + (4)\left(16 \frac{g}{mol}\right) = 96 \text{ g/mol}$$

The SO_4^{-2} concentration in the rainwater is

$$\frac{\left(1.7 \frac{\mu g}{L}\right)\left(\frac{1 \text{ mol } SO_4^{-2}}{1 \text{ mol } SO_3}\right)\left(96 \frac{\mu g}{\mu mol \ SO_4^{-2}}\right)}{80 \frac{\mu g}{\mu mol \ SO_3}}$$
$$= 2.04 \ \mu g/L \quad (2.0 \ \mu g/L)$$

The answer is (B).

2. Sulfur trioxide forms sulfuric acid when combined with water. The sulfuric acid subsequently dissociates to hydrogen ion and to sulfate.

$$SO_3 + H_2O \rightarrow H_2SO_4 \rightarrow 2H^+ + SO_4^{-2}$$

One mole of SO_3 scrubbed will yield one mole of SO_4^{-2} and two moles of H^+ in the rainwater. The SO_3 concentration in rainwater is

$$\left(\frac{458 \text{ g}}{10^6 \text{ m}^3}\right)\left(\frac{1 \text{ m}^3}{0.27 \text{ m}^3}\right)\left(10^6 \frac{\mu g}{g}\right)\left(\frac{1 \text{ m}^3}{1000 \text{ L}}\right)$$
$$= 1.7 \ \mu g/L$$

Find the molecular weight of SO_3. [Periodic Table of Elements]

$$32 \ \frac{g}{mol} + (3)\left(16 \ \frac{g}{mol}\right) = 80 \ g/mol$$

The molecular weight of SO_4^{-2} is

$$32 \ \frac{g}{mol} + (4)\left(16 \ \frac{g}{mol}\right) = 96 \ g/mol$$

The SO_4^{-2} concentration in the rainwater is

$$\frac{\left(1.7 \ \frac{\mu g}{L}\right)\left(\frac{1 \ mol \ SO_4^{-2}}{1 \ mol \ SO_3}\right)\left(96 \ \frac{\mu g}{\mu mol \ SO_4^{-2}}\right)}{80 \ \frac{\mu g}{\mu mol \ SO_3}}$$

$$= 2.04 \ \mu g/L \quad (2.0 \ \mu g/L)$$

Use the equation for pH.

Acids, Bases, and pH

$$pH = \log_{10}\left(\frac{1}{[H^+]}\right) = -\log_{10}[H^+]$$

$$[H^+] = 10^{-pH}$$

At a pH of 5.5, the H^+ concentration of natural rainwater is

$$10^{-5.5} \ \frac{mol}{L} = 3.16 \times 10^{-6} \ mol/L$$

Two moles of H^+ are added for each mole of SO_4^{-2}. The concentration of H^+ added is

$$\frac{\left(2 \ \frac{mol \ H^+}{mol \ SO_4^{-2}}\right)\left(2.0 \ \frac{\mu g}{L}\right)}{\left(96 \ \frac{\mu g}{\mu mol \ SO_4^{-2}}\right)\left(10^6 \ \frac{\mu mol}{mol}\right)} = 4.17 \times 10^{-8} \ mol/L$$

The final H^+ concentration of the rainwater is

$$3.16 \times 10^{-6} \ \frac{mol}{L} + 4.17 \times 10^{-8} \ \frac{mol}{L}$$

$$= 3.20 \times 10^{-6} \ mol/L$$

The final pH of the rainwater is

$$-\log\left(3.20 \times 10^{-6} \ \frac{mol}{L}\right) = 5.49 \quad (5.5)$$

The answer is (C).

3. Find the hydrogen to carbon ratio of methane (CH_4). [Periodic Table of Elements]

$$(1 \ mol \ C)\left(12 \ \frac{g}{mol}\right) = 12 \ g \ C$$

$$(4 \ mol \ H)\left(1 \ \frac{g}{mol}\right) = 4 \ g \ H$$

$$H:C = 4:12 = 1:3$$

Find the hydrogen to carbon ratio of acetylene (C_2H_2). [Periodic Table of Elements]

$$(2 \ mol \ C)\left(12 \ \frac{g}{mol}\right) = 24 \ g \ C$$

$$(2 \ mol \ H)\left(1 \ \frac{g}{mol}\right) = 2 \ g \ H$$

$$H:C = 2:24 = 1:12$$

The answer is (B).

4. The normal NO_2 photolytic cycle is represented by the following reaction sequence.

$$NO_2 + h\nu \rightarrow NO + O$$
$$O + O_2 + N_2 \rightarrow O_3 + N_2$$
$$O_3 + NO \rightarrow NO_2 + O_2$$

The reaction shows that the photolytic decomposition of nitrogen dioxide in the atmosphere produces nitric oxide and free oxygen radicals. Subsequent reactions produce ozone, which then reacts with nitric oxide to consume the ozone and produce nitrogen dioxide and O_2, completing the cycle.

The answer is (B).

5. Carbon monoxide, aldehydes and ketones (carbonyls), and hydrocarbons are predominantly involved in reactions with the hydroxyl radical ($\cdot OH$) to form photochemical smog. If these three groups of compounds are absent, photochemical smog will not be present.

The answer is (A).

6. The mass of SO_2 will remain constant as the temperature decreases from 95°C to 18°C, but the volume of the emitted gas will change. The ambient pressure remains constant at 1 atm. [Ideal Gas Constants]

$$T_1 = 95°C + 273° = 368K$$

$$T_2 = 18°C + 273° = 291K$$

$$V_1 = \text{volume of gas emitted at } T_1$$

$$\quad = 10^6 \text{ m}^3$$

$$V_2 = \text{volume of gas emitted after temperature}$$

$$\qquad \text{decreases to } T_2, \text{m}^3$$

$$PV_1 = nRT_1$$

$$PV_2 = nRT_2$$

$$\frac{nR}{P} = \frac{V_1}{T_1} = \frac{V_2}{T_2}$$

$$V_2 = \frac{(10^6 \text{ m}^3)(291K)}{368K} = 7.9 \times 10^5 \text{ m}^3$$

The SO_2 concentration in the air at 18°C is

$$\left(\frac{380 \text{ g}}{7.9 \times 10^5 \text{ m}^3} \right) \left(10^6 \frac{\mu g}{g} \right)$$

$$\quad = 481 \ \mu g/m^3 \quad (480 \ \mu g/m^3)$$

The answer is (A).

7. When combined with water, SO_2 will form sulfurous acid according to the following reaction.

$$SO_2 + H_2O \rightarrow H_2SO_3$$

The answer is (C).

8. Important greenhouse gases include H_2O, N_2O, and O_2, but not SO_2.

The answer is (B).

9. Carbon dioxide (CO_2) is the focus of attention in efforts to control greenhouse gas formation because it accounts for over 60% of the radiative forcing caused by historical increases in greenhouse gas concentrations.

The answer is (A).

10. The two factors believed to most significantly contribute to potential global warming are combustion of fossil fuels and decreased biomass density on the earth's surface.

The answer is (B).

27 Atmospheric Science and Meteorology

Content in blue refers to the *NCEES Handbook*.

PRACTICE PROBLEMS

1. A 60 m tall stack emits a plume at a temperature of 28°C. The air temperature at ground level is 17°C. The ambient lapse rate up to an elevation of 210 m is 0.0081°C/m. Above 210 m the ambient lapse rate is −0.0053°C/m. The maximum mixing depth is most nearly

(A) 730 m

(B) 1700 m

(C) 1900 m

(D) 2600 m

2. Hydrocarbons at a concentration of 3.1×10^9 $\mu g/m^3$ are emitted through a stack with an effective height of 110 m. The gas flow at the stack is 3.8 m^3/s, and the average wind speed 10 m above grade is 6.1 m/s. Atmospheric stability conditions are Class C. The maximum ground level concentration of hydrocarbons at 200 m crosswind of the stack is most nearly

(A) 1.6×10^3 $\mu g/m^3$

(B) 8.4×10^3 $\mu g/m^3$

(C) 1.1×10^4 $\mu g/m^3$

(D) 1.7×10^5 $\mu g/m^3$

3. An air pollutant is emitted at 4.3 kg/s through a stack with an effective height of 260 m. The wind speed at 10 m above grade is 5 m/s, and atmospheric stability conditions are Class D. The maximum ground-level concentration along the plume centerline is most nearly

(A) 6.1×10^2 $\mu g/m^3$

(B) 5.5×10^3 $\mu g/m^3$

(C) 6.9×10^4 $\mu g/m^3$

(D) 1.8×10^5 $\mu g/m^3$

4. A typical wind rose is shown in the figure.

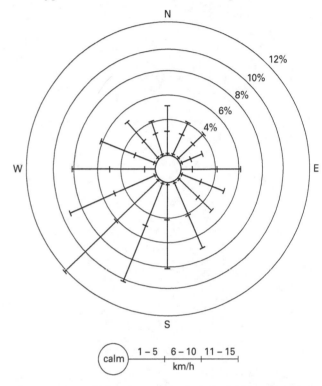

The percent of the time that the wind blows from the west is most nearly

(A) 2%

(B) 4%

(C) 6%

(D) 8%

5. The diagrams presented in the figure show common atmospheric stability patterns. The dashed line represents the dry adiabatic lapse rate and the solid line represents the ambient lapse rate.

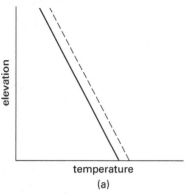

(a)

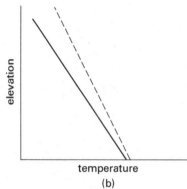

(b)

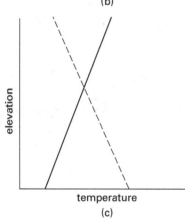

(c)

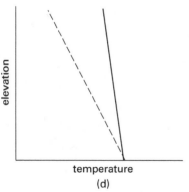

(d)

Which diagram illustrates a subadiabatic atmospheric stability?

(A)   a

(B)   b

(C)   c

(D)   d

6. A 75 m high stack emits a plume at 20°C. The ground temperature is 16°C, and the prevailing lapse rate is −10.1°C/km up to an altitude of 250 m. Above 250 m, the prevailing lapse rate is 19.8°C/km. The gas exits the stack at near-zero velocity, and wind speed is negligible. What type of plume will be formed?

(A)   coning

(B)   fanning

(C)   fumigation

(D)   looping

7. How is a dry adiabatic lapse rate affected by a water-saturated atmosphere?

(A)   A dry adiabatic lapse rate will increase in a water-saturated atmosphere.

(B)   A dry adiabatic lapse rate will decrease in a water-saturated atmosphere.

(C)   A dry adiabatic lapse rate is not directly influenced by atmospheric moisture.

(D)   A dry adiabatic lapse rate changes in a water-saturated atmosphere and that change may be an increase or a decrease.

8. A power plant burns coal with 3.8% sulfur at 11 tons/h. The SO_2 emission rate is 210 g/s. The combustion products are emitted through a stack with an effective height of 175 m. The wind speed at 10 m above the ground is 8 m/s, and atmospheric conditions are slightly unstable. The ground level SO_2 concentration 1.5 km downwind of the stack at a crosswind distance of 200 m is most nearly

(A)   42 μg/m^3

(B)   520 μg/m^3

(C)   780 μg/m^3

(D)   19 000 μg/m^3

9. The diagrams presented in the figure show common atmospheric stability patterns. The dashed line represents the dry adiabatic lapse rate and the solid line represents the ambient lapse rate.

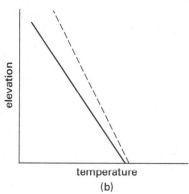

(a)

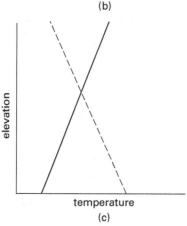

(b)

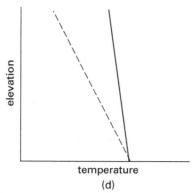

(c)

(d)

Which diagram illustrates an inversion?

(A) a

(B) b

(C) c

(D) d

10. A typical wind rose is shown in the figure.

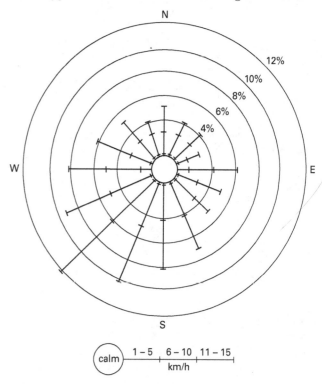

The maximum velocity for winds originating from the south-southeast is most nearly

(A) 1 to 5 km/h

(B) 6 to 10 km/h

(C) 11 to 15 km/h

(D) greater than 15 km/h

11. A power plant burns coal with 3.8% sulfur at 11 tons/h. The combustion products are emitted through a stack with an effective height of 175 m. The wind speed at 10 m above the ground is 8 m/s, and atmospheric conditions are slightly unstable. The location of the maximum ground level SO_2 concentration downwind of the power plant is most nearly

(A) 2600 m

(B) 3700 m

(C) 15 000 m

(D) 22 000 m

12. The diagrams presented in the figure show common atmospheric stability patterns. The dashed line represents the dry adiabatic lapse rate and the solid line represents the ambient lapse rate.

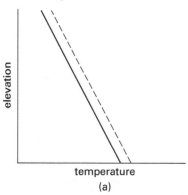

(a)

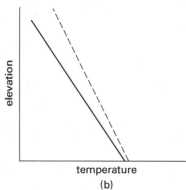

(b)

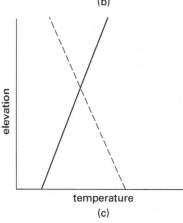

(c)

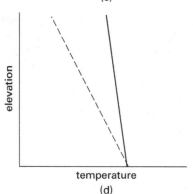

(d)

Which diagram illustrates the most stable atmospheric conditions?

(A) a

(B) b

(C) c

(D) d

13. A power plant burns coal with 3.8% sulfur at 11 tons/h. The SO_2 emission rate is 210 g/s. The combustion products are emitted through a stack with an effective height of 175 m. The wind speed at 10 m above the ground is 8 m/s, and atmospheric conditions are slightly unstable. The location of the maximum ground level SO_2 concentration is 2600 m downwind along the plume centerline. The Gaussian dispersion coefficient, σ_z, is 124 m. The maximum ground level SO_2 concentration is most nearly

(A) 92 $\mu g/m^3$

(B) 760 $\mu g/m^3$

(C) 8600 $\mu g/m^3$

(D) 130 000 $\mu g/m^3$

14. A power plant burns coal with 3.8% sulfur at 11 tons/h. The combustion products are emitted through a stack with an effective height of 175 m. The wind speed at 10 m above the ground is 8 m/s, and atmosphere conditions are slightly unstable. The SO_2 emission rate is 210 g/s. What is the ground level SO_2 concentration 1.5 km downwind of the stack along the plume centerline?

(A) 2.7 $\mu g/m^3$

(B) 78 $\mu g/m^3$

(C) 36000 $\mu g/m^3$

(D) 50000 $\mu g/m^3$

15. A 75 m high stack emits a plume at 20°C. The ground temperature is 16°C, and the prevailing lapse rate is −10.1°C/km up to an altitude of 250 m. Above 250 m, the prevailing lapse rate is 19.8°C/km. The gas exits the stack at near-zero velocity, and wind speed is negligible. What type of atmospheric stability conditions exit?

(A) subadiabatic

(B) inversion

(C) inversion over superadiabatic

(D) superadiabatic

16. A 75 m high stack emits a plume at 20°C. The ground temperature is 16°C, and the prevailing lapse rate is −10.1°C/km up to an altitude of 250 m. Above 250 m, the prevailing lapse rate is 19.8°C/km. The gas exits the stack at near-zero velocity, and wind speed is negligible. How high will the plume rise?

(A) 250 m

(B) 390 m

(C) 410 m

(D) 730 m

17. A 75 m high stack emits a plume at 20°C. The ground temperature is 16°C, and the prevailing lapse rate is −10.1°C/km up to an altitude of 250 m. Above 250 m, the prevailing lapse rate is 19.8°C/km. The gas exits the stack at near-zero velocity, and wind speed is negligible. What is the maximum mixing depth?

(A) 250 m

(B) 390 m

(C) 410 m

(D) 730 m

18. A typical wind rose is shown in the figure.

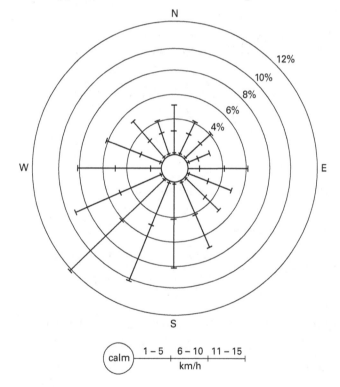

From which direction does the wind most frequently blow?

(A) northeast

(B) east-northeast

(C) southwest

(D) west-southwest

Air

SOLUTIONS

1. The plume cools at the dry adiabatic lapse rate as it rises, and will stop rising when the plume and air temperatures are equal. The dry adiabatic lapse rate is 0.98°C/100 m (−0.0098°C/m). [Selected Properties of Air]

Calculate the plume temperature at 210 m.

$$28°C + \left(-0.0098 \; \frac{°C}{m}\right)(210 \text{ m} - 60 \text{ m}) = 26.53°C$$

The air warms at the ambient lapse rate of +0.0081°C/m, up to an elevation of 210 m. Calculate the air temperature at 210 m.

$$17°C + \left(0.0081 \; \frac{°C}{m}\right)(210 \text{ m}) = 18.70°C$$

The air cools above 300 m at the ambient lapse rate of −0.0053°C/m.

Calculate the distance, z, above 210 m where the air and plume temperatures are equal.

$$26.53°C + \left(-0.0098 \; \frac{°C}{m}\right)z = 18.70°C + \left(-0.0053 \; \frac{°C}{m}\right)z$$

$$z = \frac{26.53°C - 18.70°C}{0.0098 \; \dfrac{°C}{m} - 0.0053 \; \dfrac{°C}{m}}$$

$$= 1740 \text{ m}$$

The maximum mixing depth is equal to how high the plume will rise.

$$210 \text{ m} + 1740 \text{ m} = 1950 \text{ m} \quad (1900 \text{ m})$$

The answer is (C).

2. Calculate the location of maximum ground-level concentration.

H = effective stack height, 110 m

Atmospheric Dispersion Modeling (Gaussian)

$$\sigma_z = \frac{H}{\sqrt{2}} = \frac{110 \text{ m}}{\sqrt{2}} = 78 \text{ m}$$

From a chart of standard deviations of a plume, for a standard deviation of 78 m and Class C atmospheric stability conditions, the downwind distance, $x = 1300$ m. [Vertical Standard Deviations of a Plume]

From a chart of standard deviations of a plume, for $x = 1300$ m and Class C atmospheric stability conditions, the standard deviation for the downwind distance, $s_y = 160$ m. [Horizontal Standard Deviations of a Plume]

Calculate the gas emission rate, Q.

$$Q = \left(3.8 \; \frac{\text{m}^3}{\text{s}}\right)\left(3.1 \times 10^9 \; \frac{\mu\text{g}}{\text{m}^3}\right) = 1.18 \times 10^{10} \; \mu\text{g/s}$$

Calculate the hydrocarbon concentration, C, at the monitoring station. The equation for the concentration is shown.

μ = wind speed = 6.1 m/s

z = vertical distance from ground level = 0 m

Atmospheric Dispersion Modeling (Gaussian)

$$C = \frac{Q}{2\pi\mu\sigma_y\sigma_z}\exp\left[-\frac{1}{2}\frac{y^2}{\sigma_y^2}\right]$$

$$\times\left[\exp\left[-\frac{1}{2}\frac{(z-H)^2}{\sigma_z^2}\right] + \exp\left[-\frac{1}{2}\frac{(z+H)^2}{\sigma_z^2}\right]\right]$$

For ground level concentration, the equation reduces to the following.

$$C = \frac{Q}{\pi\mu\sigma_y\sigma_z}\exp\left[-\frac{1}{2}\frac{y^2}{\sigma_y^2}\right]\exp\left[-\frac{1}{2}\frac{(H)^2}{\sigma_z^2}\right]$$

$$= \frac{1.18 \times 10^{10} \; \dfrac{\mu\text{g}}{\text{s}}}{\pi\left(6.1 \; \dfrac{\text{m}}{\text{s}}\right)(160 \text{ m})(78 \text{ m})}$$

$$\times\exp\left(-\left(\frac{1}{2}\right)\frac{(200 \text{ m})^2}{(160 \text{ m})^2}\right)\exp\left(-\left(\frac{1}{2}\right)\left(\frac{(110 \text{ m})^2}{(78 \text{ m})^2}\right)\right)$$

$$= 8.4 \times 10^3 \; \mu\text{g/m}^3$$

The answer is (B).

3. Calculate the location of maximum ground-level concentration.

H = effective stack height, 260 m

Atmospheric Dispersion Modeling (Gaussian)

$$\sigma_z = \frac{H}{\sqrt{2}} = \frac{260 \text{ m}}{\sqrt{2}} = 184 \text{ m}$$

From a chart of standard deviations of a plume, for $\sigma_z = 184$ m and Class D atmospheric stability conditions, the downwind distance, $x = 18$ km. [Vertical Standard Deviations of a Plume]

From a chart of standard deviations of a plume, for $x = 18$ km and Class D atmospheric stability conditions, the standard deviation for the downwind distance, $\sigma_y = 900$ m. [Horizontal Standard Deviations of a Plume]

Calculate the gas emission rate, Q, in units of $\mu g/s$.

$$Q = \left(4.3 \; \frac{\text{kg}}{\text{s}}\right)\left(10^9 \; \frac{\mu g}{\text{kg}}\right) = 4.3 \times 10^9 \; \mu g/s$$

Calculate the pollutant concentration.

$u =$ wind speed $= 5$ m/s

Atmospheric Dispersion Modeling (Gaussian)

$$C_{\text{max}} = \frac{Q}{\pi u \sigma_y \sigma_z}\exp\left(-\frac{1}{2}\frac{(H)^2}{\sigma_z^2}\right)$$

$$= \frac{4.3 \times 10^9 \; \frac{\mu g}{\text{s}}}{\pi\left(5\;\frac{\text{m}}{\text{s}}\right)(900\text{ m})(184\text{ m})}\exp\left(-\left(\frac{1}{2}\right)\frac{(260\text{ m})^2}{(184\text{ m})^2}\right)$$

$$= 6.1 \times 10^2 \; \mu g/\text{m}^3$$

The answer is (A).

4. From the figure, the wind blows from the west 8% of the time.

The answer is (D).

5. Diagram (d) shows the ambient lapse rate to be greater than the dry adiabatic lapse rate. This illustrates subadiabatic atmospheric stability.

The answer is (D).

6. The prevailing lapse rate of $-10.1°$C/km is less than the dry adiabatic lapse rate of $-9.8°$C/km up to a height of 250 m, indicating superadiabatic conditions. Above 250 m, the prevailing lapse rate has a positive slope and an inversion is formed. When an inversion exists over superadiabatic conditions, a fumigation plume will be formed.

The answer is (C).

7. Water vapor in air will condense, and heat will be released as the air rises. Therefore, a dry adiabatic lapse rate will increase (be less negative) in a water-saturated atmosphere.

The answer is (A).

8. From a table of atmospheric stability under various conditions, for slightly unstable stability conditions use atmospheric stability C. [Atmospheric Stability Under Various Conditions]

From a graph of horizontal standard deviations of a plume, for $x = 1500$ m and stability class C, $\sigma_y = 180$ m. [Horizontal Standard Deviations of a Plume]

From a graph of vertical standard deviations of a plume, for $x = 1500$ m and stability class C, $\sigma_z = 90$ m. [Vertical Standard Deviations of a Plume]

Use the equation for finding a steady-state concentration at a point.

Atmospheric Dispersion Modeling (Gaussian)

$$C = \frac{Q}{2\pi u \sigma_y \sigma_z}\exp\left(-\frac{1}{2}\frac{y^2}{\sigma_y^2}\right)$$

$$\times\left[\exp\left(-\frac{1}{2}\frac{(z-H)^2}{\sigma_z^2}\right) + \exp\left(-\frac{1}{2}\frac{(z+H)^2}{\sigma_z^2}\right)\right]$$

$C_{x,y} =$ concentration, g/m³, along plume centerline at distance $x = 1500$ m, and at crosswind distance $y = 200$ m. For a ground level concentration crosswind of the plume centerline, the vertical distance from the ground is zero, $z = 0$. The equation reduces to

$$C_{x,y} = \frac{Q}{\pi u \sigma_y \sigma_z}\exp\left(\left(-\frac{1}{2}\right)\frac{y^2}{\sigma_y^2}\right)\exp\left(\left(-\frac{1}{2}\right)\frac{H^2}{\sigma_z}\right)$$

$$C_{1500,200} = \frac{210\;\frac{\text{g}}{\text{s}}}{\pi\left(8\;\frac{\text{m}}{s}\right)(180\text{ m})(90\text{ m})}$$

$$\times\exp\left(\left(-\frac{1}{2}\right)\frac{(200\text{ m})^2}{(180\text{ m})^2}\right)\exp\left(\left(-\frac{1}{2}\right)\frac{(175\text{ m})^2}{(90\text{ m})^2}\right)$$

$$= 0.000042\text{ g/m}^3$$

Converting to micrograms gives

$$\left(0.000042\;\frac{\text{g}}{\text{m}^3}\right)\left(10^6\;\frac{\mu g}{\text{g}}\right) = 42\;\mu g/\text{m}^3$$

The answer is (A).

9. Diagram (c) shows the ambient lapse rate with a positive slope and the dry adiabatic lapse rate with a negative slope. This illustrates an inversion.

The answer is (C).

10. From the figure, the maximum velocity for winds originating from the south-southeast is between 11 and 15 km/h.

The answer is (C).

11. Find the Gaussian dispersion coefficient.

$\sigma_z =$ Gaussian dispersion coefficient for vertical plume concentration at downwind distance x, m

$H =$ effective stack height $= 175$ m

Atmospheric Dispersion Modeling (Gaussian)

$$\sigma_z = \frac{H}{\sqrt{2}}$$
$$= \frac{175 \text{ m}}{\sqrt{2}}$$
$$= 124 \text{ m}$$

From a table of atmospheric stability under various conditions, for slightly unstable stability conditions use atmospheric stability curve C. [Atmospheric Stability Under Various Conditions]

From a graph showing vertical standard deviations of a plume, the Gaussian dispersion for $\sigma_z = 124$ m and using stability curve C, $x = 2600$ m. [Vertical Standard Deviations of a Plume]

The answer is (A).

12. The most stable atmospheric conditions occur under an inversion as illustrated by diagram (c).

The answer is (C).

13. From a table of atmospheric stability under various conditions, for slightly unstable stability conditions use atmospheric stability C. [Atmospheric Stability Under Various Conditions]

From a graph of horizontal standard deviations of a plume, for $x = 2600$ m and using stability curve C, $\sigma_y = 270$ m. [Horizontal Standard Deviations of a Plume]

Find the maximum concentration at ground level.

$C_{x,0} =$ concentration, g/m^3, along plume
centerline at distance $x = 2600$ m

$Q =$ pollutant emission rate $= 210$ g/s

$\mu =$ average wind speed $= 8$ m/s

$\sigma_y =$ Gaussian dispersion coefficient for
horizontal plume concentration at
downwind distance $x = 270$ m

Atmospheric Dispersion Modeling (Gaussian)

$$C_{\max} = \frac{Q}{\pi \mu \sigma_y \sigma_z} \exp\left(-\frac{1}{2} \frac{(H^2)}{\sigma_z^2}\right)$$

$$C_{2600,0} = \frac{\left(210 \frac{\text{g}}{\text{s}}\right) \exp\left(-\frac{1}{2}\left(\frac{175 \text{ m}}{124 \text{ m}}\right)^2\right)}{\pi\left(8 \frac{\text{m}}{\text{s}}\right)(270 \text{ m})(124 \text{ m})}$$

$$= 0.000092 \text{ g/m}^3$$

Converting to micrograms gives

$$\left(0.000092 \frac{\text{g}}{\text{m}^3}\right)\left(10^6 \frac{\mu\text{g}}{\text{g}}\right) = 92 \ \mu\text{g/m}^3$$

The answer is (A).

14. For slightly unstable stability conditions use atmospheric stability C. [Atmospheric Stability Under Various Conditions]

For $x = 1500$ m and stability class C, $\sigma_y = 180$ m. [Horizontal Standard Deviations of a Plume]

For $x = 1500$ m and stability class C, $\sigma_z = 90$ m. [Vertical Standard Deviations of a Plume]

Find the maximum concentration at ground level.

Atmospheric Dispersion Modeling (Gaussian)

$$C_{\max} = \frac{Q}{\pi \mu \sigma_y \sigma_z} \exp\left(-\frac{1}{2} \frac{(H^2)}{\sigma_z^2}\right)$$

$$C_{1500,0} = \frac{\left(210 \frac{\text{g}}{\text{s}}\right) \exp\left(-\frac{1}{2}\left(\frac{175 \text{ m}}{90 \text{ m}}\right)^2\right)}{\pi\left(8 \frac{\text{m}}{\text{s}}\right)(180 \text{ m})(90 \text{ m})}$$

$$= 0.000078 \text{ g/m}^3$$

Converting to micrograms gives

$$\left(0.000078 \frac{\text{g}}{\text{m}^3}\right)\left(10^6 \frac{\mu\text{g}}{\text{g}}\right) = 78 \ \mu\text{g/m}^3$$

The answer is (B).

15. The prevailing lapse rate of $-10.1°$C/km is less than the dry adiabatic lapse rate of $-9.8°$C/km up to a height of 250 m, indicating superadiabatic conditions. Above 250 m, the prevailing lapse rate has a positive slope and an inversion is formed.

The answer is (C).

16. Assume the plume cools at the dry adiabatic lapse rate of $-9.8°$C/km as it rises, and the plume will stop rising when the air and plume temperatures are equal. The plume temperature at 250 m is

$$20°\text{C} + (250 \text{ m} - 75 \text{ m})\left(-9.8 \frac{°\text{C}}{\text{km}}\right)\left(\frac{1 \text{ km}}{1000 \text{ m}}\right)$$
$$= 18.3°\text{C}$$

The air temperature at 250 m is

$$16°\text{C} + (250 \text{ m})\left(-10.1 \frac{°\text{C}}{\text{km}}\right)\left(\frac{1 \text{ km}}{1000 \text{ m}}\right) = 13.5°\text{C}$$

Above 250 m, assume the air temperature will increase at 19.8°C/km and the plume temperature will continue to decrease at the dry adiabatic lapse rate.

$$18.3°C + \left(-9.8 \; \frac{°C}{km}\right)\left(\frac{1 \; km}{1000 \; m}\right)x$$
$$= 13.5°C + \left(19.8°C\right)\left(\frac{1 \; km}{1000 \; m}\right)x$$

Solve for x.

$$x = \frac{\left(18.3°C - 13.5°C\right)\left(1000 \; \frac{m}{km}\right)}{19.8 \; \frac{°C}{km} + 9.8 \; \frac{°C}{km}} = 162 \; m$$

The total height to which the plume will rise is

$$250 \; m + 162 \; m = 412 \; m \quad (410 \; m)$$

The answer is (C).

17. Assume the plume cools at the dry adiabatic lapse rate of −9.8°C/km as it rises, and the plume will stop rising when the air and plume temperatures are equal. The plume temperature at 250 m is

$$20°C + \left(250 \; m - 75 \; m\right)\left(-9.8 \; \frac{°C}{km}\right)\left(\frac{1 \; km}{1000 \; m}\right)$$
$$= 18.3°C$$

The air temperature at 250 m is

$$16°C + \left(250 \; m\right)\left(-10.1 \; \frac{°C}{km}\right)\left(\frac{1 \; km}{1000 \; m}\right) = 13.5°C$$

Above 250 m, assume the air temperature will increase at 19.8°C/km and the plume temperature will continue to decrease at the dry adiabatic lapse rate.

$$18.3°C + \left(-9.8 \; \frac{°C}{km}\right)\left(\frac{1 \; km}{1000 \; m}\right)x$$
$$= 13.5°C + \left(19.8°C\right)\left(\frac{1 \; km}{1000 \; m}\right)x$$

Solve for x.

$$x = \frac{\left(18.3°C - 13.5°C\right)\left(1000 \; \frac{m}{km}\right)}{19.8 \; \frac{°C}{km} + 9.8 \; \frac{°C}{km}} = 162 \; m$$

The total height to which the plume will rise is

$$250 \; m + 162 \; m = 412 \; m \quad (410 \; m)$$

The maximum mixing depth occurs at the elevation where the prevailing lapse rate and dry adiabatic lapse rate intersect and is equal to the height to which the plume will rise. The maximum mixing depth is 410 m.

The answer is (C).

18. From the figure, the wind most frequently blows from the southwest.

The answer is (C).

Air

 **Sources of Pollution**

Content in blue refers to the *NCEES Handbook*.

PRACTICE PROBLEMS

1. What characteristic of particulate matter presents the greatest potential human health hazard?

(A) particle size, especially where diameters exceed 10 μm

(B) potential for adsorption of toxic organics onto particle surfaces

(C) persistence which allows their re-release into the atmosphere

(D) origin from uncontrollable phenomena such as wind-borne erosion

2. What do PM-2.5 and PM-10 describe?

(A) particulates measured over 2.5 h and 10 h intervals

(B) particulate matter with diameters smaller than 2.5 μm and 10 μm

(C) total pollutant mass composited from 2.5 L and 10 L mylar sample bags

(D) preventive maintenance and monitoring of new pollution control equipment at 2.5 months and 10 months after being placed in service

3. Particulate matter with an average diameter of 10 μm is present in 20°C ambient air. The particle density is 500 kg/m^3, and the air pressure is 1 atm. The settling velocity of the particulate matter is most nearly

(A) 5.9×10^{-6} m/s

(B) 1.5×10^{-3} m/s

(C) 2.9×10^{-3} m/s

(D) 1.7×10^{-2} m/s

4. Particulate matter with an average diameter of 10 μm is present in 20°C ambient air. The particle density is 500 kg/m^3, and the air pressure is 1 atm. The settling velocity of the particulate matter is 1.5×10^{-3} m/s. What

is the Reynolds number associated with the settling particle?

(A) 1.2×10^{-13}

(B) 9.8×10^{-4}

(C) 1.2×10^{-3}

(D) 120

5. What factor does NOT directly contribute to prevent the theoretically possible continual generation of ground level ozone?

(A) cloud shading and sun position

(B) wind and temperature gradients

(C) night/day cycle

(D) rainfall and cloud cover

6. Which reaction represents a terminating reaction to the sequence of reactions leading to the accumulation of ground level ozone?

(A) $O_3 + NO \rightarrow NO_2$

(B) $HO_2 + NO \rightarrow NO_2 + OH$

(C) $OH + NO_2 \rightarrow HNO_3$

(D) $H + O_2 \rightarrow HO_2$

7. A power plant burns coal with 3.8% sulfur at 11 tons/h. The combustion products are emitted through a stack with an effective height of 175 m. The wind speed at 10 m above the ground is 8 m/s, and atmospheric conditions are neutral to slightly unstable. The SO_2 emission rate is most nearly

(A) 0.21 kg/s

(B) 0.53 kg/s

(C) 3.7 kg/s

(D) 5.6 kg/s

8. Which of the following does NOT contribute significantly to pollution from combustion?

(A) incomplete combustion

(B) combustion in air

(C) moisture in fuels and water vapor in air

(D) compounds other than C and H in fuels

9. What is the primary source of nitrogen dioxide as a pollutant in the atmosphere?

(A) Nitrogen dioxide results from fuel release and thermal formation of nitric oxide that is subsequently oxidized to nitrogen dioxide after being emitted to the atmosphere.

(B) Nitrogen dioxide is released directly to the atmosphere from fuel release and thermal formation during combustion.

(C) Nitrogen dioxide results from thermal formation of ammonia during oxygen-deficient combustion where the ammonia is oxidized to nitrogen dioxide after being emitted to the atmosphere.

(D) Nitrogen dioxide is created from atmospheric nitrogen gas during combustion of fossil fuels when excess oxygen is present at temperatures between 400°C and 600°C.

10. Which of the following are considered significant precursors to photochemical oxidants such as ozone?

(A) volatile organic compounds (VOCs) and sulfur oxides

(B) volatile organic compounds (VOCs) and nitrogen oxides

(C) volatile organic compounds (VOCs) and carbon oxides

(D) volatile organic compounds (VOCs) and sulfur and nitrogen oxides

11. The gasoline mileage of an automobile in commuter traffic is 20 mpg. The automobile owner drives 48 mi to and from work (round trip) each day from Monday through Friday. The gasoline chemical formulation is represented by C_8H_{15} with a specific weight of 0.89. In air, the ratio of N_2 to O_2 is 0.75 to 0.23 by weight. The stoichiometric air (oxygen and nitrogen) requirement for complete combustion of the gasoline is represented by

$$C_8H_{15} + 11.75O_2 + (3.73)(11.75N_2) \rightarrow$$
$$8CO_2 + (3.73)(11.75N_2) + 7.5H_2O$$

The required air:fuel ratio for complete combustion of the gasoline to occur is most nearly

(A) 3.4

(B) 7.2

(C) 11

(D) 14

12. A hazardous waste incinerator receives waste on a continuous basis at 1800 kg/h. The principal organic hazardous constituent (POHC) makes up 27% by mass of the waste mixture. Stack monitoring results show CO_2 emissions at 1600 kg/h, CO emissions at 0.13 kg/h, and POHC emissions at 0.0433 kg/h. The incinerator combustion efficiency is most nearly

(A) 88.8888%

(B) 99.9911%

(C) 99.9919%

(D) 99.9973%

13. A hazardous waste incinerator receives waste on a continuous basis at 1800 kg/h. The principal organic hazardous constituent (POHC) makes up 27% by mass of the waste mixture. Stack monitoring results show CO_2 emissions at 1600 kg/h, CO emissions at 0.13 kg/h, and POHC emissions at 0.0433 kg/h. The incinerator destruction and removal efficiency for the POHC is most nearly

(A) 88.8888%

(B) 99.9911%

(C) 99.9919%

(D) 99.9973%

14. Particulate matter with an average diameter of 10 μm is present in 23°C ambient air. The particle density is 500 kg/m^3, and the air pressure is 1 atm. The particle settling velocity is 1.5×10^{-3} m/s and the associated Reynolds number is 9.7×10^{-4}. Does Stokes' law apply?

(A) No, neither settling velocity nor particle diameter criteria are satisfied.

(B) No, settling velocity criteria are not satisfied.

(C) No, particle diameter criteria are not satisfied.

(D) Yes, both settling velocity and particle diameter criteria are satisfied.

SOLUTIONS

1. The characteristic of particulate matter that presents the greatest potential human health hazard is the potential for adsorption of toxic organics onto particle surfaces.

The answer is (B).

2. PM-2.5 and PM-10 describe particulate matter with diameters smaller than 2.5 μm and 10 μm.

The answer is (B).

3. Use Stokes' law to find the particle settling velocity. The density of air is taken from a table of thermodynamic properties of air and water. [Thermophysical Properties of Air and Water]

v_t = particle settling velocity, m/s

g = gravitational acceleration = 9.81 m/s^2

ρ_p = particle density = 500 kg/m^3

ρ_f = air density

  = 1.2042 kg/m^3 at 20°C and 1 atm

d = particle diameter = 10 μm

μ = air viscosity = 1.84 $\times$ 10^{-5} kg/m·s

Stokes' Law

$$v_t = \frac{g(\rho_p - \rho_f)d^2}{18\mu}$$

$$= \frac{\left(9.81 \ \frac{m}{s^2}\right)\left(500 \ \frac{kg}{m^3} - 1.2042 \ \frac{kg}{m^3}\right)(10 \times 10^{-6} \ m)^2}{(18)\left(1.84 \times 10^{-5} \ \frac{kg}{m \cdot s}\right)}$$

$$= 1.5 \times 10^{-3} \ m/s$$

The answer is (B).

4. Find the Reynolds number. The density of air is taken from a table of thermodynamic properties of air and water. [Thermophysical Properties of Air and Water]

v_t = particle settling velocity, m/s

g = gravitational acceleration = 9.81 m/s^2

ρ = air density

  = 1.2042 kg/m^3 at 20°C and 1 atm

d = particle diameter = 10 μm

μ = air viscosity = 1.84 $\times$ 10^{-5} kg/m·s

General Spherical

$$Re = \frac{v_t \rho d}{\mu}$$

$$= \frac{\left(1.5 \times 10^{-3} \ \frac{m}{s}\right)(10 \times 10^{-6} \ m)\left(1.2042 \ \frac{kg}{m^3}\right)}{1.84 \times 10^{-5} \ \frac{kg}{m \cdot s}}$$

$$= 9.8 \times 10^{-4}$$

The answer is (B).

5. Cloud shading and sun position, night/day cycle, and rainfall and cloud cover all directly contribute to prevent theoretically possible continual generation of ground level ozone. However, wind and temperature gradients do not.

The answer is (B).

6. $O_3 + NO \rightarrow NO_2$ is the final reaction in the normal NO_2 photolytic cycle and is not part of the chain of reactions leading to ground level ozone accumulation. $HO_2 + NO \rightarrow NO_2 + OH$ and $H + O_2 \rightarrow HO_2$ each produce a radical that is involved in subsequent reactions that contribute to ground level ozone accumulation. $OH + NO_2 \rightarrow HNO_3$ is a terminating reaction where the hydroxyl radical is removed from play by the formation of nitric acid.

The answer is (C).

7. $S + O_2 \rightarrow SO_2$

The molecular weight of SO_2 is [Periodic Table of Elements]

$$32 \ \frac{g}{mol} + (2)\left(16 \ \frac{g}{mol}\right) = 64 \ g/mol$$

$\dot{m}$ = SO_2 emission rate

$$\dot{m} = \frac{\left(11 \ \frac{tons \ coal}{h}\right)\left(0.038 \ \frac{ton \ S}{ton \ coal}\right)\left(64 \ \frac{kg \ SO_2}{kmol \ SO_2}\right)}{\left(\frac{1 \ ton}{907 \ kg}\right)\left(32 \ \frac{kg \ S}{kmol \ S}\right)\left(1 \ \frac{kmol \ S}{kmol \ SO_2}\right)\left(3600 \ \frac{s}{h}\right)}$$

$$= 0.21 \ kg/s$$

The answer is (A).

8. Incomplete combustion, combustion in air (oxygen is limited to about 20%), and compounds other than C and H in fuels all contribute significantly to pollution from combustion. Moisture existing in fuels and water vapor in air are not significant contributors.

The answer is (C).

9. The primary source of nitrogen dioxide as a pollutant is fuel release and thermal formation of nitric oxide that is subsequently oxidized to nitrogen dioxide after being emitted to the atmosphere.

The answer is (A).

10. Volatile organic compounds (VOCs) and nitrogen oxides are considered significant precursors to photochemical oxidants such as ozone.

The answer is (B).

11. Find the molecular weight of C_8H_{15}. [Periodic Table of Elements]

$$(8)\left(12 \ \frac{g}{mol}\right) + (15)\left(1 \ \frac{g}{mol}\right) = 111 \ g/mol$$

The molecular weight of O_2 is

$$(2)\left(16 \ \frac{g}{mol}\right) = 32 \ g/mol$$

The molecular weight of N_2 is

$$(2)\left(14 \ \frac{g}{mol}\right) = 28 \ g/mol$$

In air, the $N_2{:}O_2$ ratio is 0.75:0.23 by weight.

$$\frac{\left(\dfrac{0.75 \ g \ N_2}{0.23 \ g \ O_2}\right)\left(32 \ \dfrac{g \ O_2}{mol \ O_2}\right)}{\left(28 \ \dfrac{g \ N_2}{mol \ N_2}\right)} = \frac{3.73 \ mol \ N_2}{1 \ mol \ O_2}$$

The mass of C_8H_{15} is

$$(1 \ mol)\left(111 \ \frac{g}{mol}\right) = 111 \ g$$

The mass of O_2 is

$$(11.75 \ mol)\left(32 \ \frac{g}{mol}\right) = 376 \ g$$

The mass of N_2 is

$$(3.73)(11.75 \ mol)\left(28 \ \frac{g}{mol}\right) = 1227 \ g$$

The air:fuel ratio is

$$\frac{376 \ g + 1227 \ g}{111 \ g} = 14.4 \quad (14)$$

The answer is (D).

12. Find the combustion efficiency.

$$CO_2 = CO_2 \text{ emission rate}$$
$$= 1600 \ kg/h$$
$$CO = CO \text{ emission rate}$$
$$= 0.13 \ kg/h$$

Incineration

$$CE = \frac{CO_2}{CO_2 + CO} \times 100\%$$

$$= \left(\frac{1600 \ \dfrac{kg}{h}}{1600 \ \dfrac{kg}{h} + 0.13 \ \dfrac{kg}{h}}\right)(100\%)$$

$$= 99.9919\%$$

The answer is (C).

13. Find the waste mass feed rate in.

$$\dot{W}_{in} = \text{waste mass feed rate in}$$
$$\dot{W}_{out} = \text{waste mass feed rate out}$$
$$= 0.0433 \ kg/h$$

$$\dot{W}_{in} = \left(\frac{27\%}{100\%}\right)\left(1800 \ \frac{kg}{h}\right) = 486 \ kg/h$$

The destruction and removal efficiency is [Incineration]

$$DRE = \left(\frac{\dot{W}_{in} - \dot{W}_{out}}{\dot{W}_{in}}\right)(100\%)$$

$$= \left(\frac{486 \ \dfrac{kg}{h} - 0.0433 \ \dfrac{kg}{h}}{486 \ \dfrac{kg}{h}}\right)(100\%)$$

$$= 99.9911\%$$

The answer is (B).

14. Stokes' law applies for settling velocities less than 1.0 m/s and particle diameters between 0.1 and 100 μm. For this problem, the settling velocity is 1.5×10^{-3} m/s and the particle diameter is 10 μm. Therefore, Stokes' law applies.

The answer is (D).

29 Emissions Characterization, Calculations, Inventory

Content in blue refers to the *NCEES Handbook.*

PRACTICE PROBLEMS

1. Excluding in-stack filtration, what is the common EPA standard for measuring stack particulate?

(A) method 1

(B) method 5

(C) method 17

(D) method 19

2. A duct is 16 in high by 24 in wide. How far downstream from a bend should a monitoring point be located?

(A) 3 ft

(B) 7 ft

(C) 13 ft

(D) 29 ft

SOLUTIONS

1. The EPA has developed several methods for measuring particulate, but the method most commonly used for measuring stack particulate is EPA method 5.

The answer is (B).

2. Calculate the equivalent diameter, d_e, of the duct.

$$d_e = \frac{2(\text{height})(\text{width})}{\text{height} + \text{width}} = \frac{(2)(16 \text{ in})(24 \text{ in})}{16 \text{ in} + 24 \text{ in}} = 19.2 \text{ in}$$

Sampling should occur at least eight equivalent diameters downstream of a bend. Calculate the downstream distance to the sampling point.

$$\frac{(8)(19.2 \text{ in})}{12 \frac{\text{in}}{\text{ft}}} = 12.8 \text{ ft} \quad (13 \text{ ft})$$

The answer is (C).

30 Pollution Treatment, Control, Minimization, Prevention

Content in blue refers to the *NCEES Handbook*.

PRACTICE PROBLEMS

1. Which of the following is most typically associated with automobiles?

(A) photochemical smog

(B) industrial smog

(C) carbonaceous smog

(D) sulfurous smog

2. Does either acetylene or methane produce smoke when flared?

(A) methane does not because its H:C is close to 1:3, but acetylene does because its H:C is much less

(B) both will smoke since their H:C is close to 1:12

(C) neither will smoke since their H:C is greater than 1:12

(D) methane does because its H:C is greater than 1.0, but acetylene does not because its H:C is less than or equal to 1.0

3. A countercurrent wet scrubber uses lime to remove HCl from gases generated during primary metal pickling operations. The minimum gas velocity in the scrubber is 4 m/s at a gas flow rate of 10 m³/s. The minimum diameter needed for the scrubber is most nearly

(A) 0.40 m

(B) 0.71 m

(C) 1.8 m

(D) 2.5 m

4. A countercurrent wet scrubber uses lime to remove HCl from gases generated during primary metal pickling operations. Scrubber design standards and gas characteristics are shown in the table.

minimum scrubber efficiency for HCl	90%
minimum gas flow velocity	4 m/s
gas flow rate	10 m³/s
lime feed rate safety factor	5 × stoichiometric
lime slurry concentration	5%
gas HCl concentration	800 ppmv

Assume standard temperature and pressure of 25°C and 1 atm, respectively, and assume the ideal gas law applies. The daily mass of HCl removed by the scrubber is most nearly

(A) 1000 kg/d

(B) 1100 kg/d

(C) 2.8×10^5 kg/d

(D) 7.7×10^5 kg/d

5. A countercurrent wet scrubber uses lime to remove HCl from gases generated during primary metal pickling operations. The daily mass of HCl removed by the scrubber is 1000 kg. The stoichiometric requirement for lime if available as 100% CaO is most nearly

(A) 780 kg/d

(B) 890 kg/d

(C) 4.8×10^5 kg/d

(D) 19×10^5 kg/d

6. A countercurrent wet scrubber uses lime to remove HCl from gases generated during primary metal pickling operations. Scrubber design standards and gas characteristics are shown in the table.

minimum scrubber efficiency for HCl	90%
lime feed rate safety factor	5 × stoichiometric
lime slurry concentration	5%
gas HCl concentration	800 ppmv
daily mass dry lime	780 kg/d

The actual required slurry feed rate for lime if available as 89% CaO is most nearly

(A) 64 m³/d

(B) 87 m³/d

(C) 1.2 × 10⁴ m³/d

(D) 4.8 × 10⁴ m³/d

7. A baghouse is being considered for controlling particulate emissions from a metal parts sandblasting process. The airflow and baghouse parameters include the following.

airflow rate	4 m³/s
air:cloth ratio	0.01 m³/m²·s

The total area of fabric required is most nearly

(A) 260 m²

(B) 310 m²

(C) 400 m²

(D) 480 m²

8. A baghouse is being considered for controlling particulate emissions from a metal parts sandblasting process. The airflow and baghouse parameters include the following.

airflow rate	4 m³/s
air:cloth ratio	0.01 m³/m²·s
bag diameter	20 cm
bag length	2 m

The total number of bags required is most nearly

(A) 200

(B) 240

(C) 320

(D) 370

9. A cosmetics packing process produces a dust-laden air stream at a flow rate of 15 m³/s. A crossflow scrubber is being considered for removing the dust from the air stream. The gas velocity through the scrubber is 40 cm/s and the pressure at the spray nozzle is

0.05 atm. The required cross-sectional area is most nearly

(A) 0.40 m²

(B) 2.7 m²

(C) 38 m²

(D) 270 m²

10. A cosmetics packing process produces a dust-laden air stream at a flow rate of 15 m³/s. A crossflow scrubber is being considered for removing the dust from the air stream. The gas velocity through the scrubber is 40 cm/s and the pressure at the spray nozzle is 0.05 atm. The air pressure drop through the scrubber spray nozzle is most nearly

(A) 0.05 atm

(B) 0.95 atm

(C) 1.0 atm

(D) 1.05 atm

11. A farm equipment manufacturer has selected electrostatic precipitation to control atmospheric emissions from particulate-producing activities on the paint line. The airflow rate from the line is 8 m³/s at 42°C and the particulate concentration is 12 g/m³. For a particle drift velocity of 0.3 m/s, the plate area required for 90% efficiency is most nearly

(A) 21 m²

(B) 26 m²

(C) 37 m²

(D) 61 m²

12. A farm equipment manufacturer has selected electrostatic precipitation to control atmospheric emissions from particulate-producing activities on the paint line. The airflow rate from the line is 8 m³/s at 42°C and the particulate concentration is 12 g/m³. The daily particulate mass removed at 90% efficiency is most nearly

(A) 310 kg/d

(B) 630 kg/d

(C) 950 kg/d

(D) 7500 kg/d

13. A dry gas stream contains trichloroethene (TCE) at 0.0016 kg/m³ and 30 m³/s flow rate. Activated carbon adsorbers will be used to remove the TCE from the air stream. The daily mass of TCE removed is most nearly

(A) 1200 kg/d

(B) 1600 kg/d

(C) 2600 kg/d

(D) 4200 kg/d

14. A dry gas stream contains trichloroethene (TCE) at 0.0016 kg/m³ and 30 m³/s flow rate. For the temperature and moisture conditions of the gas stream, the TCE adsorption capacity of granular activated carbon (GAC) is 550 mg/g. The daily mass of TCE removed by the adsorbers is 4200 kg. The daily mass of activated carbon required to remove the TCE from the gas stream is most nearly

(A) 2200 kg/d

(B) 3000 kg/d

(C) 4700 kg/d

(D) 7700 kg/d

15. A dry gas stream contains trichloroethene (TCE) at 0.0016 kg/m³ and 30 m³/s flow rate. For the temperature and moisture conditions of the gas stream, the TCE adsorption capacity of granular activated carbon (GAC) is 550 mg/g. The gas flow capacity for each adsorber is 4 m³/s. Most nearly, how many carbon adsorbers are required to treat the gas stream?

(A) 4

(B) 6

(C) 8

(D) 10

16. A dry gas stream contains trichloroethene (TCE) at 0.0016 kg/m³ and 30 m³/s flow rate. For the temperature and moisture conditions of the gas stream, the TCE adsorption capacity of granular activated carbon (GAC) is 550 mg/g. A total of 7700 kg/d of GAC will be used among 8 adsorbers, each with a 5000 kg capacity.

Most nearly, how frequently will the adsorbers require reactivation?

(A) 2.6 d

(B) 5.2 d

(C) 11 d

(D) 14 d

17. The gasoline mileage of an automobile in commuter traffic is 20 mpg. The automobile owner drives 48 mi to and from work (round trip) each day from Monday through Friday. The gasoline chemical formulation is represented by C_8H_{15} with a specific weight of 0.89. Most nearly, how many gallons of gasoline are consumed during 1 wk of commuting to and from work?

(A) 12 gal

(B) 17 gal

(C) 24 gal

(D) 34 gal

18. The gasoline mileage of an automobile in commuter traffic is 20 mpg. The automobile owner drives 48 mi to and from work (round trip) each day from Monday through Friday. The gasoline chemical formulation is represented by C_8H_{15} with a specific weight of 0.89. What pollutant will likely increase the most if the fuel: air ratio is less than that required for complete combustion?

(A) carbon monoxide

(B) carbon dioxide

(C) hydrocarbons

(D) nitrous oxides

19. The gasoline mileage of an automobile in commuter traffic is 20 mpg. The automobile owner drives 48 mi to and from work (round trip) each day from Monday through Friday. The gasoline chemical formulation is represented by C_8H_{15} with a specific weight of 0.89. The stoichiometric air (oxygen and nitrogen) requirement for complete combustion of the gasoline is represented by

$$C_8H_{15} + 11.75O_2 + (3.73)(11.75N_2) \rightarrow$$
$$8CO_2 + (3.73)(11.75N_2) + 7.5H_2O$$

The mass of carbon dioxide that will be emitted by the automobile in one week is most nearly

(A) 2 kg/wk

(B) 13 kg/wk

(C) 130 kg/wk

(D) 820 kg/wk

20. What would be the result of injecting steam into the flame zone of an acetylene flare?

(A) It would extinguish the flame.

(B) It would trap smoke particulate in condensed steam.

(C) It would increase mixing with air to improve combustion efficiency.

(D) It would decrease flame temperature to reduce dioxin formation.

21. What measures would probably NOT contribute to overall improved urban air quality?

(A) Reduce the number of miles driven through car-pooling, increasing the use of mass transit, and improving pedestrian and bicycle access.

(B) Provide consumer incentives for the purchase of new cars to increase the ratio of new to used car sales.

(C) Increase the number of diesel-powered vehicles relative to the number of gasoline-powered vehicles through consumer incentives or manufacturer quotas.

(D) Require manufacturers to make zero emission vehicles (ZEV) available to consumers and provide incentives to promote their purchase.

22. What is the corporate average fuel economy (CAFE) efficiency of an automobile with city gasoline mileage of 23 mpg and highway mileage of 32 mpg?

(A) 26.33 mpg

(B) 27.05 mpg

(C) 27.21 mpg

(D) 27.50 mpg

23. A co-current surface condenser controls vapor losses from a tank used for acetone storage. The boiling point of acetone is 56°C and the maximum acetone outlet temperature is 30°C. The coolant is water with an inlet temperature of 15°C and an outlet temperature of 25°C.

Most nearly, what is the log mean temperature difference between the acetone and the water?

(A) 15°C

(B) 17°C

(C) 22°C

(D) 26°C

24. A co-current surface condenser controls vapor losses from a tank used for acetone storage. The coolant is water with an inlet temperature of 15°C and an outlet temperature of 25°C. The rate of heat loss through the condenser is 7.1×10^7 kJ/d and the log mean temperature difference is 22°C. The overall heat transfer coefficient for the condenser is 243 J/cm²·h·°C. Most nearly, what is the required surface area for the condenser tubes?

(A) 0.20 m²

(B) 0.50 m²

(C) 6.0 m²

(D) 56 m²

25. In a baghouse, the pressure drop through 1 cm of particulate layer is 0.012 atm at a superficial filtering velocity of 0.01 m/s and temperature of 30°C. Most nearly, what is the permeability of the particulate layer?

(A) 1.8×10^{-14} m²

(B) 1.5×10^{-12} m²

(C) 1.5×10^{-10} m²

(D) 1.6×10^{-7} m²

26. A separation cyclone used to collect particulate from an air stream is characterized as follows.

inlet height = 0.25 m
body length = 0.85 m
cone length = 1.0 m
inlet width = 0.12 m
inlet velocity = 12 m/s
particle mean density = 1400 kg/m³
air temperature = 30°C

Most nearly, what particle diameter will be collected at 50% efficiency?

(A) 1.8 μm

(B) 6 μm

(C) 42 μm

(D) 150 μm

27. A gas stream at 11 m^3/min is treated through a venturi scrubber to remove sulfates. The scrubber throat diameter is 0.15 m, and the maximum allowed pressure drop is 2.5 cm H$_2$O. Most nearly, what is the maximum water flow rate?

(A) 2.3 L/min

(B) 6.5 L/min

(C) 18 L/min

(D) 26 L/min

28. A flue gas emission rate is 1000 CFM at 15.2 psia and 139°F. Standard conditions are 14.7 psia and 80°F. Most nearly, what is the percent difference between the actual flow rate and the standard flow rate relative to actual conditions?

(A) 1.1%

(B) 4.1%

(C) 6.8%

(D) 7.3%

29. A lagooned bio-solid sludge contaminated with low concentrations of PCBs is currently being destroyed by incineration. Equipment is in place to dewater the sludge to 35% moisture and dry it to 5% moisture prior to incineration. Local community opposition to incineration has prompted the industry to investigate alternative destruction or disposal technologies. Which of the following technologies would probably require drying the sludge to 5% moisture?

(A) thermal desorption and chemical dehalogenation

(B) supercritical water oxidation

(C) chemical destruction and fixation

(D) biological oxidation

30. What USEPA test is used for evaluating vehicle emissions and fuel efficiency?

(A) acceleration/deceleration cycle—cold-start (ADCCS) test

(B) cycle vehicle (CVS-75) test

(C) dynamometer load cycle (DY-20) test

(D) power curve/emission rate (PCER) test

31. A pulse jet-felt fabric baghouse is proposed for the control of wood sawdust emissions to the atmosphere. Most nearly, what is the required fabric surface area per 1000 CFM of airflow?

(A) 12 ft^2

(B) 25 ft^2

(C) 82 ft^2

(D) 150 ft^2

SOLUTIONS

1. Photochemical smog is most typically associated with automobiles. Industrial and sulfurous smog are typically associated with industrial activities.

The answer is (A).

2. A flare will typically burn smokeless when H:C is 1:3 or greater. Therefore, methane with H:C = 1:3 will likely burn smokeless, and acetylene with H:C = 1:12 will likely burn smoky.

The answer is (A).

3. Find the scrubber cross-sectional area.

A = scrubber cross-sectional area, m^2

Q = gas flow rate = 10 m^3/s

v = gas flow velocity = 4 m/s

Continuity Equation

$$Q = Av$$

$$A = \frac{Q}{v}$$

$$= \frac{10 \; \frac{m^3}{s}}{4 \; \frac{m}{s}}$$

$$= 2.5 \; m^2$$

Find the scrubber diameter.

$$D = \left(\frac{4A}{\pi}\right)^{0.5} = \left(\frac{(4)(2.5 \; m^2)}{\pi}\right)^{0.5}$$

$$= 1.78 \; m \quad (1.8 \; m)$$

The answer is (C).

4. Find the molecular weight of HCl. [Periodic Table of Elements]

$$1 \; \frac{g}{mol} + 35 \; \frac{g}{mol} = 36 \; g/mol$$

1 mol of HCl gas at 25°C occupies 24.5 L. Find the mass density of the HCl gas. [Fundamental Constants]

$$\frac{(800 \; m^3)\left(36 \; \frac{g}{mol}\right)\left(1000 \; \frac{L}{m^3}\right)}{(10^6 \; m^3)\left(24.5 \; \frac{L}{mol}\right)} = 1.18 \; g/m^3$$

The daily mass of removed HCl is

$$\frac{\left(\frac{1.18 \; g}{m^3}\right)\left(\frac{10 \; m^3}{s}\right)\left(\frac{90\%}{100\%}\right)}{\left(\frac{1 \; d}{86\,400 \; s}\right)\left(1000 \; \frac{g}{kg}\right)}$$

$$= 918 \; kg/d \quad (1000 \; kg/d)$$

The answer is (A).

5. $2HCl + CaO \rightarrow Ca^{+2} + 2Cl^- + H_2O$

For 2 mol of HCl, 1 mol of CaO is required.

Find the molecular weight of CaO. [Periodic Table of Elements]

$$40 \; \frac{g}{mol} + 16 \; \frac{g}{mol} = 56 \; g/mol$$

The daily mass requirement of CaO is

$$\frac{\left(1000 \; \frac{kg}{d}\right)\left(56 \; \frac{kg}{kmol \; CaO}\right)}{\left(36 \; \frac{kg}{kmol \; HCl}\right)\left(\frac{2 \; kmol \; HCl}{1 \; kmol \; CaO}\right)}$$

$$= 777 \; kg/d \quad (780 \; kg/d)$$

The answer is (A).

6. Assume slurry density is the same as water density, 1000 kg/m^3. The slurry feed rate is

$$\frac{(5)\left(780 \; \frac{kg \; CaO}{d}\right)\left(\frac{1 \; m^3}{1000 \; kg}\right)}{\left(\frac{89\%}{100\%}\right)\left(\frac{5 \; kg \; CaO}{100 \; kg \; slurry}\right)} = 87 \; m^3/d$$

The answer is (B).

7. The total fabric area is

Q_g = gas (air) flow rate = 4 m^3/s

A/C = air:cloth ratio = 0.01 $m^3/m^2 \cdot s$

$$\frac{Q_g}{\frac{A}{C}} = \frac{4 \; \frac{m^3}{s}}{0.01 \; \frac{m^3}{m^2 \cdot s}} = 400 \; m^2$$

The answer is (C).

8. The total fabric area is

$$Q_g = \text{gas (air) flow rate} = 4 \text{ m}^3/\text{s}$$

$$A/C = \text{air:cloth ratio} = 0.01 \text{ m}^3/\text{m}^2\cdot\text{s}$$

$$\frac{Q_g}{\dfrac{A}{C}} = \frac{4 \dfrac{\text{m}^3}{\text{s}}}{0.01 \dfrac{\text{m}^3}{\text{m}^2\cdot\text{s}}} = 400 \text{ m}^2$$

The area/bag is

$$\pi(\text{bag diameter})(\text{bag length})$$

$$= \pi(20 \text{ cm})(2 \text{ m})\left(\frac{1 \text{ m}}{100 \text{ cm}}\right)$$

$$= 1.26 \text{ m}^2$$

The number of bags is

$$\frac{400 \text{ m}^2}{1.26 \text{ m}^2} = 317.5 \quad (320)$$

The answer is (C).

9. Find the cross-sectional area.

$$A = \text{cross-sectional area, m}^2$$

$$Q = \text{gas flow rate} = 15 \text{ m}^3/\text{s}$$

$$v = \text{gas velocity} = 40 \text{ cm/s}$$

Continuity Equation

$$Q = Av$$

$$A = \frac{Q}{v} = \frac{15 \dfrac{\text{m}^3}{\text{s}}}{\left(40 \dfrac{\text{cm}}{\text{s}}\right)\left(\dfrac{1 \text{ m}}{100 \text{ cm}}\right)}$$

$$= 37.5 \text{ m}^2 \quad (38 \text{ m}^2)$$

The answer is (C).

10. The pressure drops from 0.05 atm at the nozzle inlet to zero as a free jet at the nozzle outlet. The pressure drop is

$$0.05 \text{ atm} - 0.0 \text{ atm} = 0.05 \text{ atm}$$

The answer is (A).

11. The plate area can be found from the equation for fractional collection efficiency.

$$A = \text{collection plate area, m}^2$$

$$\eta = \text{fractional collection efficiency} = 0.90$$

$$w = \text{drift velocity} = 0.3 \text{ m/s}$$

$$Q = \text{gas stream flow rate} = 8 \text{ m}^3/\text{s}$$

Electrostatic Precipitator

$$\eta = 1 - \exp(-Aw/Q)$$

$$\exp\left(-\frac{Aw}{Q}\right) = 1 - \eta$$

$$A = \left(-\frac{Q}{w}\right)\ln(1 - \eta)$$

$$= \left(-\frac{8 \dfrac{\text{m}^3}{\text{s}}}{0.3 \dfrac{\text{m}}{\text{s}}}\right)\ln(1 - 0.90)$$

$$= 61.4 \text{ m}^2 \quad (61 \text{ m}^2)$$

The answer is (D).

12. The mass removed is

$$C_p = \text{particulate concentration} = 12 \text{ g/m}^3$$

$$Q = \text{gas stream flow rate} = 8 \text{ m}^3/\text{s}$$

$$\left(\frac{E\%}{100\%}\right)C_pQ = \frac{\left(\dfrac{90\%}{100\%}\right)\left(12 \dfrac{\text{g}}{\text{m}^3}\right)\left(8 \dfrac{\text{m}^3}{\text{s}}\right)}{\left(\dfrac{1 \text{ d}}{86\,400 \text{ s}}\right)\left(1000 \dfrac{\text{g}}{\text{kg}}\right)}$$

$$= 7465 \text{ kg/d} \quad (7500 \text{ kg/d})$$

The answer is (D).

13. The TCE mass flow rate is

$$\left(30 \dfrac{\text{m}^3}{\text{s}}\right)\left(0.0016 \dfrac{\text{kg}}{\text{m}^3}\right) = 0.048 \text{ kg/s}$$

Assume 100% adsorption efficiency.

The TCE removed is

$$\left(0.048 \dfrac{\text{kg}}{\text{s}}\right)\left(86\,400 \dfrac{\text{s}}{\text{d}}\right) = 4147 \text{ kg/d} \quad (4200 \text{ kg/d})$$

The answer is (D).

14. The mass of GAC required is

$$\frac{\left(4200 \ \frac{\text{kg TCE}}{\text{d}}\right)\left(10^6 \ \frac{\text{mg}}{\text{kg}}\right)}{\left(550 \ \frac{\text{mg TCE}}{\text{g GAC}}\right)\left(1000 \ \frac{\text{g}}{\text{kg}}\right)}$$
$$= 7636 \ \text{kg/d} \quad (7700 \ \text{kg/d})$$

The answer is (D).

15. The number of adsorbers is

$$\frac{30 \ \frac{\text{m}^3}{\text{s}}}{4 \ \frac{\text{m}^3}{\text{s}}} = 7.5 \quad (8)$$

The answer is (C).

16. The total GAC mass in 8 adsorbers is

$$(8)(5000 \ \text{kg}) = 40\,000 \ \text{kg}$$

The reactivation frequency is

$$\frac{40\,000 \ \text{kg}}{7700 \ \frac{\text{kg}}{\text{d}}} = 5.2 \ \text{d}$$

The answer is (B).

17. The amount of gasoline consumed is

$$\frac{\left(48 \ \frac{\text{mi}}{\text{d}}\right)\left(5 \ \frac{\text{d}}{\text{wk}}\right)(1 \ \text{wk})}{20 \ \frac{\text{mi}}{\text{gal}}} = 12 \ \text{gal}$$

The answer is (A).

18. As the air:fuel ratio declines below what is required for complete combustion, the most significant pollutant increase is for carbon monoxide.

The answer is (A).

19. The number of gallons of gas consumed weekly is

$$\frac{\left(48 \ \frac{\text{mi}}{\text{d}}\right)\left(5 \ \frac{\text{d}}{\text{wk}}\right)(1 \ \text{wk})}{20 \ \frac{\text{mi}}{\text{gal}}} = 12 \ \text{gal}$$

Find the molecular weight of C_8H_{15}. [Periodic Table of Elements]

$$(8)\left(12 \ \frac{\text{g}}{\text{mol}}\right) + (15)\left(1 \ \frac{\text{g}}{\text{mol}}\right) = 111 \ \text{g/mol}$$

The mass of C_8H_{15} is

$$(1 \ \text{mol})\left(111 \ \frac{\text{g}}{\text{mol}}\right) = 111 \ \text{g}$$

From the air combustion equation, 8 moles of CO_2 are produced for each mole of gasoline burned.

The molecular weight of CO_2 is

$$12 \ \frac{\text{g}}{\text{mol}} + (2)\left(16 \ \frac{\text{g}}{\text{mol}}\right) = 44 \ \text{g/mol}$$

The mass of carbon dioxide emitted is

$$\frac{\left(\begin{array}{c}\left(12 \ \frac{\text{gal}}{\text{wk}}\right)\left(3.785 \ \frac{\text{L}}{\text{gal}}\right)(0.89) \\ \times\left(1000 \ \frac{\text{g}}{\text{L}}\right)\left(44 \ \frac{\text{g}}{\text{mol CO}_2}\right)\end{array}\right)}{\left(111 \ \frac{\text{g}}{\text{mol C}_8\text{H}_{15}}\right)\left(\frac{1 \ \text{mol C}_8\text{H}_{15}}{8 \ \text{mol CO}_2}\right)\left(1000 \ \frac{\text{g}}{\text{kg}}\right)}$$
$$= 128 \ \text{kg/wk} \quad (130 \ \text{kg/wk})$$

The answer is (C).

20. Injecting steam into the flame zone of the acetylene flare will increase turbulence to promote better mixing with oxygen. This reduces the smokiness and flame length of an acetylene flare.

The answer is (C).

21. Increasing the number of diesel-powered vehicles relative to the number of gasoline-powered vehicles through consumer incentives or manufacturer quotas would not contribute to overall improved urban air quality.

The answer is (C).

22. Assume 1 mi of driving with 55% city and 45% highway as prescribed by CAFE.

$$\text{fuel economy} = \frac{1 \ \text{mi}}{\frac{0.55 \ \text{mi}}{23 \ \frac{\text{mi}}{\text{gal}}} + \frac{0.45 \ \text{mi}}{32 \ \frac{\text{mi}}{\text{gal}}}} = 26.33 \ \text{mpg}$$

The answer is (A).

23. Use the equation for log mean temperature difference for parallel flow.

T_{Ci} = inlet temperature cold fluid (water) = 15°C
T_{Co} = outlet temperature cold fluid (water) = 25°C
T_{Hi} = inlet temperature of hot fluid (acetone) = 56°C
T_{Ho} = outlet temperature of hot fluid (acetone) = 30°C

Log Mean Temperature Difference (LMTD)

$$\Delta T_{lm} = \frac{(T_{Ho} - T_{Co}) - (T_{Hi} - T_{Ci})}{\ln\left(\dfrac{T_{Ho} - T_{Co}}{T_{Hi} - T_{Ci}}\right)}$$

$$= \frac{(30°C - 25°C) - (56°C - 15°C)}{\ln\dfrac{30°C - 25°C}{56°C - 15°C}}$$

$$= 17.2°C \quad (17°C)$$

The answer is (B).

24. Use the equation for heat transfer rate in a heat exchanger, and solve for the surface area.

A = condenser tube surface area, m²
$\dot{Q}$ = rate of heat transfer = 7.1×10^7 kJ/d
U = overall heat transfer coefficient = 243 J/cm²·h·°C
T_{lm} = log mean temperature difference = 22°C
F = correction factor = 1

Heat Exchangers

$$\dot{Q} = UAF\Delta T_{lm}$$

$$A = \frac{\dot{Q}}{UF\Delta T_{lm}}$$

$$= \frac{\left(7.1 \times 10^7 \dfrac{kJ}{d}\right)\left(1000 \dfrac{J}{kJ}\right)}{\left(\left(243 \dfrac{J}{cm^2 \cdot h \cdot °C}\right) \times \left(100 \dfrac{cm}{m}\right)^2 \left(24 \dfrac{h}{d}\right)\right)(1)(22°C)}$$

$$= 55.3 \text{ m}^2 \quad (56 \text{ m}^2)$$

The answer is (D).

25. The dynamic viscosity of air at 30°C is $\mu = 1.865 \times 10^{-5}$ kg/m·s. [Thermophysical Properties of Air and Water]

Use Darcy's equation for fluid flow through fabric.

ΔP_p = pressure drop through 1 mm of particulate layer = 0.012 atm
D_p = depth of particulate layer = 1 cm
V = superficial filtering velocity = 0.01 m/s

Fabric Filtration

$$\Delta P_p = \frac{D_p \mu V}{60 K_p}$$

For the velocity units used, the 60 s/min conversion factor is not needed. Solve for the permeability.

$$K_p = \frac{D_p \mu V}{\Delta P_p}$$

$$= \frac{\left(\dfrac{1 \text{ cm}}{100 \dfrac{cm}{m}}\right)\left(1.865 \times 10^{-5} \dfrac{kg}{m \cdot s}\right) \times \left(0.01 \dfrac{m}{s}\right)}{(0.012 \text{ atm})\left(1.013 \times 10^5 \dfrac{Pa}{atm}\right) \times \left(1 \dfrac{N}{m^2 \cdot Pa}\right)\left(1 \dfrac{kg \cdot m}{s^2 \cdot N}\right)}$$

$$= 1.5 \times 10^{-12} \text{ m}^2$$

The answer is (B).

26.

H = inlet height = 0.25 m
L_b = body length = 0.85 m
L_c = cone length = 1.0 m
W = inlet width = 0.12 m
V_i = inlet velocity = 12 m/s
ρ_p = particle density = 1400 kg/m³

Find the properties of air at 30°C. [Thermophysical Properties of Air and Water]

ρ_a = air density at 30°C = 1.1644 kg/m³
μ = air dynamic viscosity at 30°C
   = 1.865×10^{-5} kg/m·s
N_e = number of effective turns

Cyclone Effective Number of Turns Approximation

$$N_e = \frac{1}{H}\left[L_b + \frac{L_c}{2}\right]$$

$$= \left(\frac{1}{0.25 \text{ m}}\right)\left(0.85 \text{ m} + \frac{1.0 \text{ m}}{2}\right)$$

$$= 5.4 \text{ turns}$$

d_{pc} = particle diameter collected with 50% efficiency, m

Cyclone 50% Collection Efficiency for Particle Diameter

$$d_{pc} = \left[\frac{9\mu W}{2\pi N_e V_i(\rho_p - \rho_g)}\right]^{0.5}$$

$$= \left(\frac{(9)\left(1.865 \times 10^{-5} \frac{\text{kg}}{\text{m·s}}\right)(0.12 \text{ m})}{(2\pi)(5.4)\left(12 \frac{\text{m}}{\text{s}}\right)\left(1400 \frac{\text{kg}}{\text{m}^3} - 1.1644 \frac{\text{kg}}{\text{m}^3}\right)}\right)^{0.5}$$

$$= 0.000006 \text{ m}$$

$$(0.000006 \text{ m})\left(10^6 \frac{\mu\text{m}}{\text{m}}\right) = 6 \text{ } \mu\text{m}$$

The answer is (B).

27. Find the cross-sectional area of the throat.

A = throat cross-sectional area, m^2
D = throat diameter = 0.15 m

$$A = \frac{\pi D^2}{4} = \frac{\pi(0.15 \text{ m})^2}{4} = 0.0177 \text{ m}^2$$

Use the continuity equation, and solve for the throat gas velocity.

Q = gas stream flow rate = 11 m^3/min
V_t = throat gas velocity, cm/s

Continuity Equation

$$Q = Av$$

$$= A V_t$$

$$V_t = \frac{Q}{A} = \frac{\left(11 \frac{\text{m}^3}{\text{min}}\right)\left(100 \frac{\text{cm}}{\text{m}}\right)}{(0.0177 \text{ m}^2)\left(60 \frac{\text{s}}{\text{min}}\right)}$$

$$= 1035 \text{ cm/s}$$

Use the equation for pressure drop across a venturi meter, and solve for the water to gas volume ratio.

L = water to gas volume ratio, L/m^3
ΔP = pressure drop across venturi = 2.5 cm

Venturi Scrubber

$$\Delta P = V_t^2 L \times 10^{-6}$$

$$L = \frac{\Delta P \times 10^6}{V_t^2} = \frac{(2.5 \text{ cm})(10^6)}{\left(1035 \frac{\text{cm}}{\text{s}}\right)^2}$$

$$= 2.33 \text{ L water/m}^3 \text{ gas}$$

The water flow rate is

$$\left(2.33 \frac{\text{L water}}{\text{m}^3 \text{ gas}}\right)\left(11 \frac{\text{m}^3 \text{ gas}}{\text{min}}\right) = 25.6 \text{ L/min} \quad (26 \text{ L/min})$$

The answer is (D).

28. Correct the flow rate for standard conditions.

Q_A = ACFM = gas flow rate at actual conditions, CFM
Q_S = SCFM = gas flow rate at standard conditions, CFM
P_A, P_S = actual and standard flue gas pressure, psia
T_A, T_S = actual and standard flue gas temperature, °R

Correcting Gas Streams for Standard Conditions

$$Q_S = Q_A\left(\frac{P_A}{P_S}\right)\left(\frac{T_S}{T_A}\right)\left(\frac{1}{1 - y_{\text{H}_2\text{O}}(g)}\right)$$

$$\times \left(\frac{12\% \text{ CO}_2 \quad \frac{\text{by volume in}}{\text{dry gas stream}}}{X\% \text{ CO}_2 \quad \frac{\text{by volume in}}{\text{dry gas stream}}}\right)$$

The standard unit for flue gas flow is CFM with other parameters expressed in English units. Because the given information is limited to pressure and temperature, other parameters are ignored and the equation reduces to

$$\text{SCFM} = \text{ACFM}\left(\frac{P_A}{P_S}\right)\left(\frac{T_S}{T_A}\right)$$

Using English units, the standard gas flow is

$$\text{SCFM} = (1000 \text{ CFM})\left(\frac{15.2 \text{ psia}}{14.7 \text{ psia}}\right)\left(\frac{80°\text{F} + 460°}{139°\text{F} + 460°}\right)$$

$$= 932 \text{ CFM}$$

The measurement difference relative to actual conditions is

$$\frac{1000 - 932}{1000} \times 100\% = 6.8\%$$

The answer is (C).

29. Thermal desorption requires heating the sludge to temperatures that will volatilize the target contaminants. High moisture in the sludge will limit the ability to heat the sludge to temperatures high enough for volatilization to occur. Therefore, for thermal desorption, drying the sludge to 5% moisture would be required for effective application of the technology. Supercritical water oxidation, chemical destruction and fixation, and biological oxidation can all be applied to sludges with relatively high moisture content.

The answer is (A).

30. The USEPA test for evaluating vehicle emissions and fuel efficiency is the cycle vehicle (CVS-75) test.

The answer is (B).

31. For a pulse jet-felt fabric application to control wood sawdust, the air-to-cloth ratio is 3.7 $\mathrm{m^3/min \cdot m^2}$. [Air-to-Cloth Ratio for Baghouses]

$$\frac{1000 \ \dfrac{\mathrm{ft^3 \ air}}{\mathrm{min}}}{\left(3.7 \ \dfrac{\mathrm{m^3 \ air}}{\mathrm{min \cdot m^2 \ fabric}}\right)\left(3.281 \ \dfrac{\mathrm{ft}}{\mathrm{m}}\right)}$$
$$= 82.4 \ \mathrm{ft^2 \ fabric} \quad (82 \ \mathrm{ft^2})$$

The answer is (C).

Topic III Solid and Hazardous Waste

Chapter

# 31 Chemistry

---

Content in blue refers to the *NCEES Handbook.*

PRACTICE PROBLEMS

1. At ambient temperature and pressure of 20°C and 1 atm, the mass of dry air required for combustion of methane is most nearly

(A) 4.8 g air/mol CH_4

(B) 64 g air/mol CH_4

(C) 120 g air/mol CH_4

(D) 280 g air/mol CH_4

SOLUTIONS

1. The stoichiometric combustion of methane is shown. [Combustion Processes]

$$CH_4 + 2O_2 \rightarrow CO_2 + 2H_2O$$

2 mol O_2 are consumed for 1 mol CH_4 flared.

V = volume, L
n = moles = 2
P = pressure = 1 atm
T = temperature = $273° + 20°C = 293K$
R = ideal gas constant at 1 atm and 293K
  $= 0.08205$ atm·L/mol·K

Ideal Gas Constants

$$PV = nRT$$

$$V_{O_2} = \frac{nRT}{P} = \frac{(2 \text{ mol})\left(0.08205 \dfrac{\text{atm·L}}{\text{mol·K}}\right) \times (293K)}{1 \text{ atm}}$$
$$= 48 \text{ L}$$

The O_2 volume in air = 20.95%. [Composition of Dry Ambient Air]

$$\frac{V_{\text{air}}}{100\%} = \frac{48 \text{ L}}{20.95\%}$$
$$V_{\text{air}} = \frac{(100\%)(48 \text{ L})}{20.95\%} = 229 \text{ L}$$

Ideal Gas Constants

$$PV = nRT$$
$$n_{\text{air}} = \frac{PV}{RT} = \frac{(1 \text{ atm})(229 \text{ L})}{\left(0.08205 \dfrac{\text{atm·L}}{\text{mol·K}}\right)(293K)}$$
$$= 9.5 \text{ mol}$$

The molecular weight of air is 28.97 g/mol. [Gases Found in Wastewater at Standard Conditions (0°C, 1 atm)]

$$m = \text{mass of air, g/mol CH}_4 \text{ flared}$$

$$m = \left(28.97 \ \frac{\text{g}}{\text{mol}}\right)(9.5 \ \text{mol})$$

$$= 275 \ \text{g air/mol CH}_4$$

$$(280 \ \text{g air/mol CH}_4)$$

The answer is (D).

32 Fate and Transport

PRACTICE PROBLEMS

1. Dioxin formation requires which of the following conditions?

(A) availability of free chlorine and temperatures between 300°C and 450°C

(B) availability of low molecular weight hydrocarbons and temperatures between 250°C and 350°C

(C) availability of nitrous oxides and temperatures between 500°C and 750°C

(D) availability of ozone and temperatures between 400°C and 600°C

2. To what classification do pesticides such as chlordane, DDT, dieldrin, and lindane belong?

(A) carbamates

(B) organochlorides

(C) organophosphates

(D) pyrethroids

3. What is the typical economic design life of a landfill gas reclamation system?

(A) <10 yr

(B) 20–25 yr

(C) 30–40 yr

(D) > 50 yr

SOLUTIONS

1. Dioxins are formed within the narrow temperature range of 300–450°C when free chlorine is present. Consequently, to avoid dioxin formation, thermal destruction methods rely on high temperatures, exceeding 850°C.

The answer is (A).

2. Pesticides such as DDT, dieldrin, and lindane are classified as organochlorides. They were early pesticides that experienced wide use beginning in the 1950s and into the 1970s. Their widespread use has been discontinued because they persist for long periods in the environment—for example, DDT has a half-life of nearly 60 years and has been found in formerly pristine locations worldwide, far removed from its points of use.

The answer is (B).

3. Landfill gas production decreases at a rate of about 3–5% annually. After about 20–25 yr, landfill gas production will have decreased to 30% of its original production level. At this production level, the cost of recovering the gas begins to exceed its economic value and the gas reclamation system will have reached the end of its economic design life.

The answer is (B).

33 Codes, Standards, Regulations, Guidelines

PRACTICE PROBLEMS

1. The Resource Conservation and Recovery Act (RCRA) defines hazardous waste according to the four criteria of ignitability, toxicity, corrosivity, and reactivity. Wastes may also be classified as hazardous if specifically listed in what subpart of 40 CFR 261?

(A) Subpart A

(B) Subpart B

(C) Subpart C

(D) Subpart D

2. The characteristics of a solid waste are as shown.

flash point	135°F
pH	8.1
lead concentration by toxicity characteristic leaching procedure (TCLP)	4.3 mg/L
toxaphene concentration by TCLP	5.8 μg/L
reactivity	nonreactive

Does the waste meet regulatory criteria for classification as a hazardous waste?

(A) Yes, the toxaphene concentration classifies the waste as hazardous waste on the basis of toxicity.

(B) Yes, the waste satisfies the criteria for classification as hazardous waste.

(C) No, the waste satisfies some, but not all, of the regulatory criteria for hazardous waste classification.

(D) No, the waste does not satisfy any of the regulatory criteria for hazardous waste classification.

3. What regulations control the transportation of hazardous materials in the United States?

(A) Title 40 CFR Parts 172–268

(B) Title 29 CFR Part 1910

(C) Title 49 CFR Parts 100–199

(D) Title 40 CFR Part 300

4. An industrial facility generates waste from four different processes. The wastes are characterized in the table shown.

characteristic	waste 1	waste 2	waste 3	waste 4
ignitability (flash point, °C)	51	112	>200	64
corrosivity (pH)	4.3	11.6	1.7	8.4
reactivity (reactive, yes/no)	no	no	no	no
toxicity—contaminant	vinyl chloride	–	–	–
concentration (mg/L)	105	–	–	–
chemical constituents	–	phenol	–	–

Which wastes are hazardous?

(A) wastes 1 and 2

(B) wastes 1 and 3

(C) wastes 2 and 3

(D) wastes 2 and 4

5. An industrial facility generates waste from four different processes. The wastes are characterized in the table shown.

characteristic	waste 1	waste 2	waste 3	waste 4
ignitability (flash point, °C)	51	112	>200	64
corrosivity (pH)	4.3	11.6	1.7	8.4
reactivity (reactive, yes/no)	no	no	no	no
toxicity—contaminant	vinyl chloride	–	–	–
concentration (mg/L)	105	–	–	–
chemical constituents	–	phenol	–	–

If the flash point of waste 4 was 60°C, would the hazardous waste characterization change?

(A) Yes, the waste would become hazardous.

(B) No, the waste would remain hazardous.

(C) Yes, the waste would become nonhazardous.

(D) No, the waste would remain nonhazardous.

6. An industrial facility generates waste from four different processes. The wastes are characterized in the table shown.

characteristic	waste 1	waste 2	waste 3	waste 4
ignitability (flash point, °C)	51	112	>200	64
corrosivity (pH)	4.3	11.6	1.7	8.4
reactivity (reactive, yes/no)	no	no	no	no
toxicity—contaminant	vinyl chloride	–	–	–
concentration (mg/L)	105	–	–	–
chemical constituents	–	phenol	–	–

If phenol was eliminated from waste 2, would the hazardous waste characterization change?

(A) Yes, the waste would become hazardous.

(B) No, the waste would remain hazardous.

(C) Yes, the waste would become nonhazardous.

(D) No, the waste would remain nonhazardous.

7. An industrial facility generates waste from four different processes. The wastes are characterized in the table shown.

characteristic	waste 1	waste 2	waste 3	waste 4
ignitability (flash point, °C)	51	112	>200	64
corrosivity (pH)	4.3	11.6	1.7	8.4
reactivity (reactive, yes/no)	no	no	no	no
toxicity—contaminant	vinyl chloride	–	–	–
concentration (mg/L)	105	–	–	–
chemical constituents	–	phenol	–	–

If waste 3 was treated by simple neutralization, would the hazardous waste characterization change?

(A) Yes, the waste would become hazardous.

(B) No, the waste would remain hazardous.

(C) Yes, the waste would become nonhazardous.

(D) No, the waste would remain nonhazardous.

8. The primary federal law and its associated regulations that address solid and hazardous waste management is the Resource Conservation & Recovery Act of 1976 (RCRA). What law amended RCRA in 1984?

(A) Comprehensive Environmental Response, Compensation, and Liability Act (CERCLA)

(B) Hazardous & Solid Waste Amendments (HSWA)

(C) Superfund Amendments and Reauthorization Act (SARA)

(D) Land Disposal Program Flexibility Act (LDPFA)

9. An industrial facility generates waste from four different processes. The wastes are characterized in the table shown.

characteristic	waste 1	waste 2	waste 3	waste 4
ignitability (flash point, °C)	51	112	>200	64
corrosivity (pH)	4.3	11.6	1.7	8.4
reactivity (reactive, yes/no)	no	no	no	no
toxicity—contaminant	vinyl chloride	–	–	–
concentration (mg/L)	105	–	–	–
chemical constituents	–	phenol	–	–

If vinyl chloride were eliminated from waste 1, would the hazardous waste characterization change?

(A) Yes, the waste would become hazardous.

(B) No, the waste would remain hazardous.

(C) Yes, the waste would become nonhazardous.

(D) No, the waste would remain nonhazardous.

10. What is the primary focus of the Emergency Planning and Community Right-to-Know Act (EPCRA) of 1986?

(A) emergency planning and notification

(B) the reporting responsibilities of facilities that use and store hazardous chemicals

(C) both (A) and (B)

(D) neither (A) nor (B)

SOLUTIONS

1. In RCRA, a hazardous waste is any solid waste that is either ignitable, toxic, corrosive, or reactive (40 CFR 261.20 to 40 CFR 261.24) or is listed in 40 CFR 261 Subpart D (Sections 261.30 to 261.33). Wastes are listed in Subpart D by specific and nonspecific source and by chemical name under the following hazard codes: ignitable waste, corrosive waste, reactive waste, toxicity characteristic waste, acute hazardous waste, and toxic waste.

The answer is (D).

2. A solid waste is a hazardous waste if it exhibits ignitability, toxic, corrosive, or reactive characteristics as defined by Title 40 of the Code of Federal Regulations (40 CFR), Secs. 261.20 to 261.24, or is listed in 40 CFR, Part 261, Subpart D, Secs. 261.30 to 261.33.

The general criteria for hazardous waste are

characteristic of ignitability (40 CFR, Sec. 261.21)	flash point less than 60°C or 140°F
characteristic for corrosivity (40 CFR, Sec. 261.22)	pH < 2 or pH > 12.5
characteristic of reactivity (40 CFR, Sec. 261.23)	reacts violently with water
characteristic of toxicity (40 CFR, Sec. 261.24)	listed in Table 1 of 40 CFR, Sec. 261.24

Assume that the waste is not listed in Subpart D because no information is given regarding the listing.

The waste meets the criteria for ignitability. This is sufficient by itself to classify the waste as hazardous waste.

The answer is (B).

3. Although other regulations, such as Title 40 CFR Part 263, address transportation of hazardous wastes and materials, these regulations reference the U.S. Department of Transportation regulations in Title 49 Parts 100 to 199. The Title 49 regulations are known as the Hazardous Materials Regulations and provide for the comprehensive regulation of hazardous waste, hazardous substance, and hazardous materials transportation in the United States.

The answer is (C).

4. Waste 1 is hazardous because of ignitability (flash point less than 60°C) according to Title 40 CFR 261.21 and toxicity (vinyl chloride concentration greater than 0.2 mg/L) according to Title 40 CFR 24. Waste 3 is hazardous because of corrosivity (pH less than 2) according to Title 40 CFR 261.22. [Hazardous Waste Characteristics]

The answer is (B).

Solid & Hazard. Waste

5. No, the waste would remain nonhazardous. The characteristic of ignitability defined in Title 40 CFR 261.21 requires the flash point to be less than, not less than or equal to, 60°C for the waste to be hazardous. [Hazardous Waste Characteristics]

The answer is (D).

6. No, the waste would remain nonhazardous. Waste 2 is not hazardous with phenol included (phenol is not listed in Title 40 CRF 261.33). Removing phenol from the waste would not change its characterization as nonhazardous waste. [Hazardous Waste Characteristics]

The answer is (D).

7. Yes, the waste would become nonhazardous. Simple neutralization would increase the pH to above 2. According to Title 40 CFR 261.22, a waste is hazardous by corrosivity if its pH is less than or equal to 2 or greater than or equal to 12.5. [Hazardous Waste Characteristics]

The answer is (C).

8. The federal law and its associated regulations that address solid and hazardous waste management is the Resource Conservation & Recovery Act of 1976 (RCRA) as amended in 1984 by the Hazardous & Solid Waste Amendments (HSWA).

The Comprehensive Environmental Response, Compensation, and Liability Act of 1980 (CERCLA or Superfund) was amended in 1986 by the Superfund Amendments and Reauthorization Act (SARA). CERCLA addresses emergency response and investigation and remediation of abandoned hazardous waste sites and associated issues.

The Land Disposal Program Flexibility Act of 1996 amended RCRA and provides regulatory flexibility for land disposal of certain wastes.

The answer is (B).

9. No, the waste would remain hazardous. If vinyl chloride were removed from the waste, the waste would still be hazardous because of the ignitability characteristic. [Hazardous Waste Characteristics]

The answer is (B).

10. The primary focus of the Emergency Planning and Community Right-to-Know Act (EPCRA) of 1986 is emergency planning and notification, as well as the reporting responsibilities of facilities that use or store hazardous chemicals, which includes SDSs and chemical release and inventory forms.

The answer is (C).

 **Risk Assessment**

Content in blue refers to the *NCEES Handbook*.

PRACTICE PROBLEMS

There are no problems in this book corresponding to Chap. 34 of the *PE Environmental Review*.

35 Sampling and Measurement Methods

Content in blue refers to the *NCEES Handbook*.

PRACTICE PROBLEMS

1. A city is considering alternatives for the management of solid waste generated by its residents. The solid waste characteristics for the city are as follows.

component	% mass	component discarded % moisture	component discarded dry density (kg/m³)	component discarded dry energy (kJ/kg)
paper	44	6	85	16 750
garden	17	60	105	6500
food	11	70	290	4650
cardboard	9	5	50	16 300
wood	7	20	240	18 600
plastic	7	2	65	32 600
miscellaneous inert materials	5	8	480	7000

The moisture content of the bulk waste as discarded is most nearly

(A) 23 kg/100 kg

(B) 68 kg/100 kg

(C) 76 kg/100 kg

(D) 88 kg/100 kg

2. A city is considering alternatives for the management of solid waste generated by its residents. The solid waste characteristics for the city are as follows.

component	% mass	component discarded % moisture	component discarded dry density (kg/m³)	component discarded dry energy (kJ/kg)
paper	44	6	85	16 750
garden	17	60	105	6500
food	11	70	290	4650
cardboard	9	5	50	16 300
wood	7	20	240	18 600
plastic	7	2	65	32 600
miscellaneous inert materials	5	8	480	7000

The bulk density of the waste when dry is most nearly

(A) 26 kg/m³

(B) 68 kg/m³

(C) 88 kg/m³

(D) 120 kg/m³

3. A city is considering alternatives for the management of solid waste generated by its residents. The solid waste characteristics for the city are as follows.

component	% mass	component discarded % moisture	component discarded dry density (kg/m³)	component discarded dry energy (kJ/kg)
paper	44	6	85	16 750
garden	17	60	105	6500
food	11	70	290	4650
cardboard	9	5	50	16 300
wood	7	20	240	18 600
plastic	7	2	65	32 600
miscellaneous inert materials	5	8	480	7000

The per capita solid waste generation rate for the 82,000 residents of the city is 2.7 kg/d. The daily energy content available from the waste when dry is most nearly

(A) 2.2×10^3 kJ/d

(B) 2.6×10^5 kJ/d

(C) 2.8×10^9 kJ/d

(D) 3.2×10^{11} kJ/d

4. The residents of a city with a population of 25,000 generate solid waste at a rate of 1.85 kg/d per capita. The annual total mass of solid waste generated by the city's population is most nearly

(A) 1.2×10^2 kg/yr

(B) 4.6×10^4 kg/yr

(C) 4.9×10^6 kg/yr

(D) 1.7×10^7 kg/yr

5. A city is considering alternatives for the management of the 221 400 kg of solid waste generated daily by its residents. The bulk solid waste moisture content is 23 kg/100 kg and the waste has a daily energy value of 2.8×10^9 kJ. Is the energy content of the waste adequate to dry the waste without having to add energy from an external source, and why?

(A) Yes; the energy content of the waste is greater than the energy required to dry it.

(B) No; the energy content of the waste is less than the energy required to dry it.

(C) Yes; the energy content of the waste is less than the energy required to dry it.

(D) No; the energy content of the waste is greater than the energy required to dry it.

SOLUTIONS

1. Calculate the moisture content based on a 100 kg sample of the waste.

component	discard mass (kg)	discarded moisture (%)	dry mass (kg)
paper	44	6	41.0
garden	17	60	6.8
food	11	70	3.3
cardboard	9	5	8.6
wood	7	20	5.6
plastic	7	2	6.9
miscellaneous	5	8	4.6
	100		76.8

The equation for the dry mass is

$$\text{dry mass, kg} = (\text{discarded mass, kg})\left(\frac{1 - \% \text{ moisture}}{100\%}\right)$$

The moisture content is

$$\frac{100 \text{ kg}}{100 \text{ kg}} - \frac{76.8 \text{ kg}}{100 \text{ kg}} = 23.2 \text{ kg/100 kg} \quad (23 \text{ kg/100 kg})$$

The answer is (A).

2. Calculate the density based on a 100 kg sample of the waste.

component	discard mass (kg)	discarded moisture (%)	dry mass (kg)	discarded dry density (kg/m³)	discarded dry volume (m³)
paper	44	6	41.0	85	0.48
garden	17	60	6.8	105	0.065
food	11	70	3.3	290	0.011
cardboard	9	5	8.6	50	0.17
wood	7	20	5.6	240	0.023
plastic	7	2	6.9	65	0.11
miscellaneous	5	8	4.6	480	0.010
	100		76.8		0.87

The equation for the dry mass is

$$\text{dry mass, kg} = (\text{discarded mass, kg})\left(\frac{1 - \% \text{ moisture}}{100\%}\right)$$

The equation for the discarded dry volume is

$$\frac{\text{discarded dry}}{\text{volume m}^3} = \frac{\text{dry mass, kg}}{\text{discarded dry density, kg/m}^3}$$

The dry waste bulk density is

$$\frac{76.8 \text{ kg}}{0.87 \text{ m}^3} = 88 \text{ kg/m}^3$$

The answer is (C).

3.

component	discard mass (kg)	discarded moisture (%)	dry mass (kg)	unit dry energy (kJ/kg)	total dry energy (kJ)
paper	44	6	41.0	16 750	686 750
garden	17	60	6.8	6500	44 200
food	11	70	3.3	4650	15 345
cardboard	9	5	8.6	16 300	140 180
wood	7	20	5.6	18 600	104 160
plastic	7	2	6.9	32 600	224 940
miscellaneous	5	8	4.6	7000	32200
	100		76.8		1 247 775

The equation for the dry mass is

$$\text{dry mass, kg} = (\text{discarded mass, kg})\left(\frac{1 - \% \text{ moisture}}{100\%}\right)$$

The equation for the total dry energy is

$$\begin{pmatrix} \text{total dry} \\ \text{energy, kJ} \end{pmatrix} = \begin{pmatrix} \text{dry mass,} \\ \text{kg} \end{pmatrix}\begin{pmatrix} \text{unit dry energy,} \\ \text{kJ/kg} \end{pmatrix}$$

The total waste generated daily is

$$\left(82{,}000 \text{ people}\right)\left(2.7 \frac{\text{kg}}{\text{person·d}}\right) = 221\,400 \text{ kg/d}$$

The dry waste energy content is

$$\left(\frac{1\,247\,775 \text{ kJ}}{100 \text{ kg}}\right)\left(221\,400 \frac{\text{kg}}{\text{d}}\right)$$
$$= 2.8 \times 10^9 \text{ kJ/d}$$

The answer is (C).

4. The annual solid waste generated is

$$(25{,}000 \text{ people})\left(1.85 \frac{\text{kg}}{\text{person·d}}\right)\left(365 \frac{\text{d}}{\text{yr}}\right)$$
$$= 1.7 \times 10^7 \text{ kg/yr}$$

The answer is (D).

5. Find the mass of the water in the solid waste. [Thermophysical Properties of Air and Water]

$$q = \text{heat (energy) added}$$
$$\text{to raise water to}$$
$$\text{vaporization temperature, kJ}$$
$$T_s = \text{vaporization temperature}$$
$$\text{of water at 1 atm}$$
$$= 100°C$$
$$T_{sat} = \text{ambient temperature,}$$
$$\text{assume} = 20°C$$
$$c = \text{specific heat of water}$$
$$= 4.184 \text{ kJ/kg·°C}$$
$$m = \text{mass of water, kg}$$

$$m = \left(221\,400 \frac{\text{kg}}{\text{d}}\right)\left(\frac{23 \text{ kg water}}{100 \text{ kg waste}}\right) = 50\,922 \text{ kg /d}$$

Use the equation for evaporation at a solid-liquid interface.

Boiling

$$q'' = h(T_s - T_{sat})$$
$$h = mc$$
$$q'' = mc(T_s - T_{sat})$$

$$q'' = mc(T_2 - T_1)$$
$$= \left(50\,922 \frac{\text{kg}}{\text{d}}\right)\left(4.184 \frac{\text{kJ}}{\text{kg·°C}}\right)$$
$$\times (100°C - 20°C)$$
$$= 1.7 \times 10^7 \text{ kJ/d}$$

Heating Value of Waste

$$q_v = \text{heat of vaporization}$$
$$= \left(2420 \frac{\text{kJ}}{\text{kg}}\right)\left(50\,922 \frac{\text{kg}}{\text{d}}\right)$$
$$= 1.28 \times 10^8 \text{ kJ/d}$$
$$q_T = \text{total heat required for boiling}$$
$$= q'' + q_v$$
$$= 1.7 \times 10^7 \frac{\text{kJ}}{\text{d}} + 1.28 \times 10^8 \frac{\text{kJ}}{\text{d}}$$
$$= 1.45 \times 10^8 \text{ kJ/d}$$

The energy content of the waste at 2.8×10^9 kJ/d is greater than the energy required at 1.45×10^8 kJ/d to dry the waste. The energy content of the waste is adequate to dry the waste without adding energy from an external source.

The answer is (A).

Solid & Hazard. Waste

36 Minimization, Reduction, Recycling

PRACTICE PROBLEMS

1. Which of the following is an example of direct recycling?

(A) donating goods either for sale or by distribution for use by others

(B) reusing discarded materials for applications that are different than their original use

(C) separating recyclable materials into separate containers for curbside pickup

(D) taking recyclable materials to a central facility and placing them in containers designated for each type of material

2. Which of the following statements about recycling is true?

(A) Overall recycling rates in the U.S. are consistently greater than 75%.

(B) Recycling is generally an important revenue source for municipalities.

(C) The highest recycling rates are for materials that are easy to separate and handle.

(D) The most successful recycling programs require high technology and dedicated infrastructure.

3. A municipality of 75,000 people generates solid waste at 1.9 kg/person·d with the proportion of individual waste components listed in the table.

component	dry percentage by weight (%)	density (kg/m^3)
paper and paper products	36	140
yard waste	18	120
food waste	9.0	300
ferrous metals	5.1	160
non-ferrous metals	4.3	240
plastics	7.6	130
glass	6.9	350
wood	3.9	220
textiles	2.1	60
rubber	3.2	130
miscellaneous inert materials	3.9	480

Regional markets make recycling potentially feasible for paper and paper products, all metals, glass, and plastics. Because local regulations allow city authorities to assess fines for failure to separate and process (i.e., clean, crush, bundle) recyclable wastes for curbside collection, participation in the recycling program is expected to be 92%. Most nearly, what percentage of the total waste volume can potentially be recycled?

(A) 41%

(B) 54%

(C) 59%

(D) 64%

SOLUTIONS

1. Donating goods for either resale or distribution to people in need is direct recycling. Examples of direct recycling include clothing, automobiles, electronic devices, eyeglasses, and other personal and household items being donated to donation programs, restaurants donating unsold food items to food banks at end-of-day, or grocery stores doing the same with expired goods. These activities are often seen as charitable giving, but they are also an important part of recycling.

The answer is (A).

2. Materials that are easy to separate and handle or present few convenient alternatives for disposal are recycled at the highest rates. Nevertheless, most municipalities in the United States have organized recycling programs. The motivation for this is not financial, as recycling typically costs more to manage than the revenue it generates. Using data compiled in 2014, of all municipal solid waste generated, 52.6% was landfilled, leaving less than 50% being diverted to recycling and energy recovery. Nearly all recycling activities are low technology, involving communities who are motivated by environmental conservation. Higher technology activities are difficult to justify until markets value recycled materials and create a financial incentive for investment.

The answer is (C).

3. Use a 100 kg sample so that dry mass can be conveniently expressed in kilograms.

component	dry mass (kg)	density (kg/m^3)	volume (m^3)	recyclables volume (m^3)
paper and paper products	36	140	0.257	0.257
yard waste	18	120	0.150	–
food waste	9.0	300	0.030	–
ferrous metals	5.1	160	0.032	0.032
non-ferrous metals	4.3	240	0.018	0.018
plastics	7.6	130	0.058	0.058
glass	6.9	350	0.020	0.020
wood	3.9	220	0.018	–
textiles	2.1	60	0.035	–
rubber	3.2	130	0.025	–
miscellaneous	3.9	480	0.008	–
inert materials			0.651	0.385

The percentage of the total waste volume that can potentially be recycled is

$$\frac{(0.385 \text{ m}^3)(0.92)}{0.651 \text{ m}^3} \times 100\% = 54\%$$

The answer is (B).

Mass and Energy Balance

Content in blue refers to the *NCEES Handbook*.

PRACTICE PROBLEMS

1. The characteristics of an activated sludge process are shown.

flow rate = 5000 m³/d
soluble BOD: total BOD = 0.37
influent TSS = 40 mg/L
effluent TSS = 20 mg/L
mixed liquor suspended solids = 3100 mg/L
wasted suspended solids = 15 000 mg/L
biomass production rate based on total solids = 438 kg/d

The recirculated solids flow rate is most nearly

(A) 130 m³/d

(B) 850 m³/d

(C) 1300 m³/d

(D) 1700 m³/d

2. A water balance prepared for a natural wetland covering a 359 ha area is summarized in the following table. Berm height allows a maximum water depth of 80 cm in the wetland.

source	average estimated contribution	
	October–March	April–September
inputs		
direct precipitation	100 cm	55 cm
surface inflow	0.21 m³/s	0.09 m³/s
subsurface inflow	0.0043 m³/s	–
outputs		
surface outflow	0.17 m³/s	0.02 m³/s
subsurface outflow	–	0.0073 m³/s
evapotranspiration	40 cm	121 cm

The wetland minimum storage volume is most nearly

(A) 980 000 m³

(B) 1 400 000 m³

(C) 1 500 000 m³

(D) 1 900 000 m³

Solid & Hazard. Waste

SOLUTIONS

1. To determine the recirculated solids flow rate, Q_R, perform a mass balance around the bioreactor as depicted in the figure.

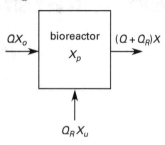

X_o = influent TSS = 40 mg/L

X_p = biomass production rate based on total solids

$\quad$ = 438 kg/d

$$\text{inputs} = \text{outputs}$$
$$QX_o + X_p + Q_R X_u = QX + Q_R X$$
$$Q_R = \frac{Q(X_o - X) + X_p}{(X - X_u)}$$
$$Q_R = \frac{Q(X_o - X) + X_p}{(X - X_u)}$$

$$= \frac{\begin{pmatrix}5000 \ \frac{m^3}{d}\end{pmatrix}\begin{pmatrix}40 \ \frac{mg}{L} - 3100 \ \frac{mg}{L}\end{pmatrix}}{\begin{pmatrix}3100 \ \frac{mg}{L} - 15\,000 \ \frac{mg}{L}\end{pmatrix}}$$
$$\times \begin{pmatrix}10^{-6} \ \frac{kg}{mg}\end{pmatrix}\begin{pmatrix}1000 \ \frac{L}{m^3}\end{pmatrix} + 438 \ \frac{kg}{d}$$
$$\times \begin{pmatrix}10^{-6} \ \frac{kg}{mg}\end{pmatrix}\begin{pmatrix}1000 \ \frac{L}{m^3}\end{pmatrix}$$

$$= 1253 \ \text{m}^3/\text{d} \quad (1300 \ \text{m}^3/\text{d})$$

The answer is (C).

2. The change in volume is

$$\Delta V = \text{inputs} - \text{outputs}$$

The change in volume from October to March is

$$\left(\frac{(100 \ \text{cm} - 40 \ \text{cm})(359 \ \text{ha})\begin{pmatrix}10\,000 \ \frac{m^2}{ha}\end{pmatrix}\begin{pmatrix}\frac{1 \ m}{100 \ cm}\end{pmatrix}}{6 \ \text{mo}}\right)$$
$$+ \begin{pmatrix}0.21 \ \frac{m^3}{s} + 0.0043 \ \frac{m^3}{s} - 0.17 \ \frac{m^3}{s}\end{pmatrix}$$
$$\times \begin{pmatrix}86\,400 \ \frac{s}{d}\end{pmatrix}\begin{pmatrix}30 \ \frac{d}{mo}\end{pmatrix}$$
$$= 473\,826 \ \text{m}^3/\text{mo} \quad [\text{increase}]$$

The change in volume from April to September is

$$\left(\frac{(55 \ \text{cm} - 121 \ \text{cm})(359 \ \text{ha})\begin{pmatrix}10\,000 \ \frac{m^2}{ha}\end{pmatrix}\begin{pmatrix}\frac{1 \ m}{100 \ cm}\end{pmatrix}}{6 \ \text{mo}}\right)$$
$$+ \begin{pmatrix}0.09 \ \frac{m^3}{s} - 0.02 \ \frac{m^3}{s} - 0.0073 \ \frac{m^3}{s}\end{pmatrix}$$
$$\times \begin{pmatrix}86\,400 \ \frac{s}{d}\end{pmatrix}\begin{pmatrix}30 \ \frac{d}{mo}\end{pmatrix}$$
$$= -232\,382 \ \text{m}^3/\text{mo} \quad [\text{decrease}]$$

From October to March, the wetland gains water at a faster rate than it loses water from April to September. Therefore, at some point during the October to March period the wetland will be at its maximum depth of 80 cm, corresponding to its maximum storage volume.

The maximum storage volume is

$$(80 \ \text{cm})(359 \ \text{ha})\begin{pmatrix}10\,000 \ \frac{m^2}{ha}\end{pmatrix}\begin{pmatrix}\frac{1 \ m}{100 \ cm}\end{pmatrix}$$
$$= 2\,872\,000 \ \text{m}^3 \quad (2\,900\,000 \ \text{m}^3)$$

The minimum storage volume is

$$2\,900\,000 \ \text{m}^3 - \begin{pmatrix}232\,382 \ \frac{m^3}{mo}\end{pmatrix}(6 \ \text{mo})$$
$$= 1\,505\,708 \ \text{m}^3 \quad (1\,500\,000 \ \text{m}^3)$$

The answer is (C).

38 Hydrology, Hydrogeology, Geology

PRACTICE PROBLEMS

1. The following figure presents a schematic of a landfill and cap, and the accompanying table summarizes the cap and drainage layer material, fill thicknesses, and hydraulic conductivities. The maximum desired leachate head is 30 cm and the leachate collection laterals are 150 m long and are placed on the bottom of the drainage layer at 25 m intervals. The overall hydraulic conductivity of the layers comprising the landfill and cap is 5.3×10^{-5} cm/s. Assume a self-cleaning velocity on the laterals of 0.6 m/s.

material	thickness (cm)	hydraulic conductivity (cm/s)
top soil	30	10^{-3}
drainage sub-layer	15	10^{-2}
clay cap layer	60	10^{-6}
waste fill	3200	10^{-3}
drainage layer	30	10^{-1}

The required spacing of the leachate collection laterals in the drainage layer is most nearly

(A) 6.0 m

(B) 26 m

(C) 36 m

(D) 43 m

2. The unit downward percolating leachate to each lateral in a landfill drainage layer is 6.8×10^{-6} m^3/m·s. The laterals are 150 m long and spaced at 26 m. The leachate flow rate to each lateral in the drainage layer is most nearly

(A) 6.8×10^{-6} m^3/s

(B) 1.8×10^{-4} m^3/s

(C) 1.0×10^{-3} m^3/s

(D) 2.7×10^{-2} m^3/s

3. The flow rate in each lateral in a landfill drainage layer is 0.001 m^3/s. The laterals are placed at a slope to maintain a minimum self-cleaning velocity of 0.6 m/s. The required diameter of the laterals is most nearly

(A) 2.3 cm

(B) 4.7 cm

(C) 40 cm

(D) 82 cm

4. Rainfall on a closed solid waste landfill with a 17 ac surface area results in the following.

total precipitation = 2.4 in
total runoff = 1.6 in
total evapotranspiration = 0.31 in
total infiltration = 0.23 in

Most nearly, what is the total change in the volume of water held in storage within the landfill cover?

- (A) 0.37 ac-ft
- (B) 1.0 ac-ft
- (C) 1.9 ac-ft
- (D) 4.4 ac-ft

5. Rainfall on a closed solid waste landfill with a 17 ac surface area causes a 0.37 ac ft increase in the volume of water retained in the landfill cover. The cover is currently unsodded. The operators plan to place sod on the cover to increase evapotranspiration from 0.31 in to 0.56 in. Most nearly, what is the reduction in the volume of water retained in the landfill cover?

- (A) 0.017 ac-ft
- (B) 0.25 ac-ft
- (C) 0.35 ac-ft
- (D) 0.62 ac-ft

SOLUTIONS

1. Find the downward percolating leachate flow to each lateral.

$q = $ unit downward percolating leachate
$\quad$ flow to each lateral, $\text{m}^3/\text{m·s}$

$K = $ overall hydraulic conductivity
$\quad = 5.3 \times 10^{-5}$ cm/s

$L_s = $ lateral spacing, m

$N = $ number of laterals draining an area $= 2$

$$q = \frac{KL_s}{N}$$

$$= \frac{\left(\begin{array}{c}\left(5.3 \times 10^{-5} \dfrac{\text{cm}}{\text{s}}\right)L_s \\ \times \left(\dfrac{1 \text{ m}^3}{10^6 \text{ cm}^3}\right)\left(10^4 \dfrac{\text{cm}^2}{\text{m}^2}\right)\end{array}\right)}{(2)}$$

$$= (2.6 \times 10^{-7} \text{ m}^3/\text{m}^2\text{·s}) L_s$$

Find the lateral spacing.

$K_d = $ hydraulic conductivity of the drainage layer
$\quad = 10^{-1}$ cm /s

$d_o = $ maximum desired leachate head $= 30$ cm

$d_c = $ distance of leachate lateral above the liner $= 0$

$$L_s = \frac{2K_d(d_o^2 - d_c^2)}{q}$$

$$= \frac{(2)\left(10^{-1} \dfrac{\text{cm}}{\text{s}}\right)}{\times\left((30)^2 \text{ cm}^2 - (0)^2 \text{ cm}^2\right)\left(\dfrac{1 \text{ m}^3}{10^6 \text{ cm}^3}\right)}{\left(2.6 \times 10^{-7} \dfrac{\text{m}^3}{\text{m}^2\text{·s}}\right)L_s}$$

$$L_s^2 = 692 \text{ m}^2$$

$$L_s = 26.3 \text{ m} \quad (26 \text{ m})$$

The answer is (B).

2. The flow rate to each lateral, Q, is

$$Q = \left(6.8 \times 10^{-6} \ \frac{m^3}{m \cdot s}\right)(150 \ m)$$

$$= 0.0010 \ m^3/s \quad (1.0 \times 10^{-3} \ m^3/s)$$

The answer is (C).

3. Find the cross-sectional area.

A_x = lateral cross-sectional area, m^2

v_c = self-cleaning velocity in the lateral

$\quad = 0.6 \ m/s$

Continuity Equation

$$Q = Av$$

$$A_x = \frac{Q}{v_c}$$

$$= \frac{0.001 \ \frac{m^3}{s}}{0.6 \ \frac{m}{s}}$$

$$= 0.0017 \ m^2$$

$$A_x = \frac{\pi D^2}{4}$$

The lateral diameter is

$$D = \left(\frac{4A_x}{\pi}\right)^{1/2} = \left(\frac{(4)(0.0017 \ m^2)}{\pi}\right)^{1/2}$$

$$= 0.0465 \ m \quad (4.7 \ cm)$$

The answer is (B).

4.

ΔS_{LC} = change in water held in storage in
a unit volume of landfill cover, in

P = precipitation per unit area = 2.4 in

R = runoff per unit area = 1.6 in

ET = evapotranspiration per unit area = 0.31 in

PER_{sw} = water percolating through a unit area
of landfill cover into compacted solid waste

$\quad = 0.23 \ in$

Soil Landfill Cover Water Balance

$$\Delta S_{LC} = P - R - ET - \text{PER}_{sw}$$

$$= 2.4 \ in - 1.6 \ in - 0.31 \ in - 0.23 \ in$$

$$= 0.26 \ in$$

The total change in storage volume is

$$\left(\frac{0.26 \ in}{12 \ \frac{in}{ft}}\right)(17 \ ac) = 0.37 \ ac\text{-}ft$$

The answer is (A).

5. This problem could be solved using the equation for soil landfill cover water balance, but for this application all parameters in the equation except evapotranspiration are constant. Therefore, a direct calculation using evapotranspiration differences is possible.

The reduction in landfill cover storage volume is

$$\left(\frac{0.56 \ in - 0.31 \ in}{12 \ \frac{in}{ft}}\right)(17 \ ac) = 0.35 \ ac\text{-}ft$$

The answer is (C).

Solid & Hazard. Waste

39 Solid Waste Storage, Collection, Transportation

PRACTICE PROBLEMS

1. The results of a time study and route analysis for curbside residential waste collection in a planned community are as follows.

population	600
per-capita solid waste generation rate	0.9 kg/person·d
number of residences	285
typical waste as-discarded density	140 kg/m³

The daily total as-discarded volume of waste requiring collection is most nearly

(A) 1.4 m³/d

(B) 3.9 m³/d

(C) 7.7 m³/d

(D) 29 m³/d

2. The results of a time study and route analysis for curbside residential waste collection in a planned community are as follows.

number of residences	285
average driving time between residences	18 s
average pick-up/load time at each residence	45 s
travel time from truck yard to route start	38 min
travel time from route end to landfill	63 min
time to unload at landfill	20 min
travel time from landfill to truck yard	45 min
truck compacted waste capacity	12 m³
truck compaction ratio	2.6:1

The total time available to a single crew in one 8 h day for curbside collection is most nearly

(A) 120 min

(B) 230 min

(C) 310 min

(D) 480 min

3. The results of a time study and route analysis for curbside residential waste collection in a planned community are as follows.

population	600
per-capita solid waste generation rate	0.9 kg/person·d
number of residences	285
available collection time	310 min
average driving time between residences	18 s
average pick-up/load time at each residence	45 s
travel time from truck yard to route start	38 min
travel time from route end to landfill	63 min
time to unload at landfill	20 min
travel time from landfill to truck yard	45 min
truck compacted waste capacity	12 m³
truck compaction ratio	2.6:1
typical waste as-discarded density	140 kg/m³

The total compacted waste volume one crew can collect during one 8 h work day if collection occurs once weekly is most nearly

(A) 7.2 m³

(B) 11 m³

(C) 15 m³

(D) 27 m³

4. The results of a time study and route analysis for curbside residential waste collection in a planned community are as follows.

population	600
per-capita solid waste generation rate	0.9 kg/person·d
number of residences	285
available collection time	310 min
average driving time between residences	18 s
average pick-up/load time at each residence	45 s
travel time from truck yard to route start	38 min
travel time from route end to landfill	63 min
time to unload at landfill	20 min
travel time from landfill to truck yard	45 min
truck compacted waste capacity	12 m³
truck compaction ratio	2.6:1
typical waste as-discarded density	140 kg/m³
collection frequency	1/wk

Most nearly, how many days should be scheduled for one crew to collect all the waste generated by the community?

(A) 1 d

(B) 2 d

(C) 3 d

(D) 4 d

5. A city of 50,000 people generates municipal solid waste at a rate of 2 kg/person·d with an as-discarded density of 120 kg/m³. The waste can be compacted to 575 kg/m³. The total compacted volume of solid waste collected from the city each week is most nearly

(A) 830 m³/wk

(B) 1200 m³/wk

(C) 5800 m³/wk

(D) 21 000 m³/wk

6. Truck crews of three men each work 8 h/d, 5 d/wk collecting solid waste, with collections occurring at each stop once weekly. The city has evaluated options for replacing its fleet of collection trucks and has selected trucks with a 23 m³ compacted capacity. The weekly compacted volume collected is 1200 m³, and a single truck can complete two trips in a day. The number of trucks required to meet the weekly collection schedule is most nearly

(A) 4

(B) 5

(C) 6

(D) 7

7. A city plans to implement a curbside recycling program for newspapers, plastic containers, aluminum cans, steel cans, and glass. Residents will be required to segregate the waste, bundle newspapers, and crush aluminum and steel cans and plastic containers. No compaction will occur on the truck, and the waste will remain segregated. The total (recyclable and nonrecyclable) solid waste generation rate is 1.3 kg/person·d. On average, there are 3.6 residents per stop, and collection will occur once weekly. Most nearly, what truck capacity is needed to complete collection on a route of 100 stops without having to unload until the end of the day?

component	percentage of total discarded by weight (%)	average curbside density (kg/m³)
newspaper	16	200
aluminum	1.3	240
steel	1.9	405
plastic	5.7	265
glass	4.1	180

(A) 1 m³

(B) 2 m³

(C) 4 m³

(D) 5 m³

SOLUTIONS

1. The total daily as-discarded waste volume requiring collection is

$$\frac{(600 \text{ people})\left(0.9 \dfrac{\text{kg}}{\text{person}\cdot\text{d}}\right)}{140 \dfrac{\text{kg}}{\text{m}^3}} = 3.9 \text{ m}^3/\text{d}$$

The answer is (B).

2. The available collection time is the total time minus the noncollection task time.

$$(8 \text{ h})\left(60 \dfrac{\text{min}}{\text{h}}\right) - \left(\begin{matrix} 38 \text{ min} + 63 \text{ min} \\ +20 \text{ min} \\ +45 \text{ min} \end{matrix}\right)$$
$$= 314 \text{ min} \quad (310 \text{ min})$$

The answer is (C).

3. The possible stops per day are

$$\frac{\text{total time}}{\dfrac{\text{time}}{\text{stop}}} + 1 = \frac{(310 \text{ min})\left(60 \dfrac{\text{s}}{\text{min}}\right)}{18 \text{ s} + 45 \text{ s}} + 1 = 296$$

The number of residences is 285.

Enough time is available to collect at all residences in a single day since 285 is less than 296.

Since waste can be collected during a single 8 h day and collection occurs once weekly, the total compacted volume collected is

$$\frac{\left(0.9 \dfrac{\text{kg}}{\text{person}\cdot\text{d}}\right)(600 \text{ people})\left(7 \dfrac{\text{d}}{\text{wk}}\right)}{\left(140 \dfrac{\text{kg}}{\text{m}^3}\right)\left(\dfrac{2.6 \text{ m}^3}{1 \text{ m}^3}\right)}$$
$$= 10.4 \text{ m}^3/\text{wk} \quad (11 \text{ m}^3/\text{wk})$$

Check truck capacity $= 12 \text{ m}^3 > 11 \text{ m}^3$. OK.

The answer is (B).

4. The possible stops per day are

$$\frac{\text{total time}}{\dfrac{\text{time}}{\text{stop}}} + 1 = \frac{(310 \text{ min})\left(60 \dfrac{\text{s}}{\text{min}}\right)}{18 \text{ s} + 45 \text{ s}} + 1 = 296$$

The number of residences is 285.

Enough time is available to collect at all residences in a single day since 285 is less than 296.

Since waste can be collected during a single 8 h day and collection occurs once weekly, the total compacted volume collected is

$$\frac{\left(0.9 \dfrac{\text{kg}}{\text{person}\cdot\text{d}}\right)(600 \text{ people})\left(7 \dfrac{\text{d}}{\text{wk}}\right)}{\left(140 \dfrac{\text{kg}}{\text{m}^3}\right)\left(\dfrac{2.6 \text{ m}^3}{1 \text{ m}^3}\right)} = 10.4 \text{ m}^3/\text{wk}$$

The time is available to collect all waste in one day, and the truck capacity of 12 m^3 is adequate since it is greater than the daily compacted volume collected of 10.4 m^3.

The answer is (A).

5. The weekly compacted volume collected is

$$\frac{(50{,}000 \text{ people})\left(2 \dfrac{\text{kg}}{\text{person}\cdot\text{d}}\right)\left(7 \dfrac{\text{d}}{\text{wk}}\right)}{575 \dfrac{\text{kg}}{\text{m}^3}}$$
$$= 1217 \text{ m}^3/\text{wk} \quad (1200 \text{ m}^3/\text{wk})$$

The answer is (B).

6.

$$\frac{\left(1200 \dfrac{\text{m}^3}{\text{wk}}\right)\left(\dfrac{1 \text{ wk}}{5 \text{ d}}\right)\left(\dfrac{1 \text{ d}}{2 \text{ trips}}\right)}{23 \dfrac{\text{m}^3}{\text{truck}\cdot\text{trip}}} = 5.2 \text{ trucks}$$

Six trucks are required.

The answer is (C).

7. Calculate the average volume discarded of each waste component.

component	average of total discarded (%)	average mass discarded, m_{ave} (kg)	average curbside density, ρ_{ave} (kg/m^3)	average volume discarded, V_{ave} (m^3)
newspaper	16	524	200	2.62
aluminum	1.3	42.6	240	0.18
steel	1.9	62.2	405	0.15
plastic	5.7	187	265	0.71
glass	4.1	134	180	<u>0.74</u>
				4.4

Solid & Hazard.
Waste

The average mass discarded is

$$
m_{ave} = \left(\frac{\text{discarded}\%}{100\%} \right) \left(1.3 \ \frac{\text{kg}}{\text{person·d}} \right) (7 \ \text{d})
$$
$$
\times \left(3.6 \ \frac{\text{persons}}{\text{stop}} \right) (100 \ \text{stops})
$$

The equation for average volume discarded is

$$
V_{ave} = \frac{m_{ave}}{\rho_{ave}}
$$

Summing the average volumes of all the components, the total volume discarded is 4.4 m^3/wk.

The truck volume needed is 5 m^3.

The answer is (D).

Solid Waste Treatment and Disposal

Content in blue refers to the *NCEES Handbook*.

PRACTICE PROBLEMS

1. The chemical characterization of a municipal solid waste is summarized in the following table. The chemical characterization is based on typical published values for the waste components listed.

waste component	dry mass (kg/100 kg)	dry elemental chemical composition (%)					
		C	H	O	N	S	ash
food	4.9	48.0	6.4	37.6	2.6	0.4	5.0
glass/metal	3.2	–	–	–	–	–	100
paper	12.6	43.5	6.0	44.0	0.3	0.2	6.0
plastic	8.7	60.0	7.2	22.8	–	–	10.0
wood debris	2.1	49.5	6.0	42.7	0.2	0.1	1.5
yard clippings	29.5	47.8	6.0	38.0	3.4	0.3	4.5

What is the chemical formula of the waste if sulfur is included?

(A) $C_{50}H_{132}O_{61}N_5S$

(B) $C_{84}H_{265}O_{93}N_{12}S$

(C) $C_{259}H_{848}O_{376}N_{16}S$

(D) $C_{523}H_{1795}O_{795}N_{20}S$

2. The chemical characterization of a municipal solid waste is summarized in the following table. The chemical characterization is based on typical published values for the waste components listed.

waste component	dry mass (kg/100kg)	dry elemental chemical composition (%)					
		C	H	O	N	S	ash
food	4.9	48.0	6.4	37.6	2.6	0.4	5.0
glass/metal	3.2	–	–	–	–	–	100
paper	12.6	43.5	6.0	44.0	0.3	0.2	6.0
plastic	8.7	60.0	7.2	22.8	–	–	10.0
wood debris	2.1	49.5	6.0	42.7	0.2	0.1	1.5
yard clippings	29.5	47.8	6.0	38.0	3.4	0.3	4.5

What is the chemical formula of the waste if sulfur is excluded?

(A) $C_5H_{32}O_6N$

(B) $C_{11}H_{45}O_{18}N$

(C) $C_{27}H_{92}O_{41}N$

(D) $C_{50}H_{132}O_{61}N$

3. Municipal solid waste is being evaluated for composting with thickened waste-activated sludge. The city generates 300 000 kg of solid waste daily, 32% of which is compostable. The discarded moisture content of the solid waste is 38%, and the waste C:N ratio is 12:1. The waste-activated sludge is thickened to 18% solids, and the C:N ratio of the solids is 90:1. Most nearly, how much sludge must be mixed with the solid waste to produce a C:N ratio of 30:1?

(A) 22 000 kg/d

(B) 29 000 kg/d

(C) 220 000 kg/d

(D) 290 000 kg/d

4. Municipal solid waste is being evaluated for composting with thickened waste-activated sludge. The city generates 300 000 kg of solid waste daily, 32% of which is compostable. The discarded moisture content of the solid waste is 38%, and the waste-activated sludge is thickened to 18% solids. The solid waste is mixed with 220 000 kg/d of sludge to produce a C:N ratio of 30:1. The resulting moisture content of the mixture is most nearly

(A) 24%

(B) 30%

(C) 56%

(D) 69%

5. Municipal solid waste is being evaluated for composting with thickened waste-activated sludge. The city generates 300 000 kg of solid waste daily, 32% of which is compostable. The solid waste is mixed with 220 000 kg/d of sludge to produce a C:N ratio of 30:1 and a moisture content of 69%. Most nearly, how much water must be added or removed to bring the moisture content of the sludge-solid waste mixture, blended to meet the C:N ratio of 30:1, to 60%?

(A) remove 71 000 kg/d

(B) remove 13 000 kg/d

(C) add 99 000 kg/d

(D) add 120 000 kg/d

6. The chemical characterization of a municipal solid waste is summarized in the following table. The chemical characterization is based on typical published values for the waste components listed.

Waste Component	Dry Mass (kg/100 kg)	Dry Elemental Chemical Composition (%)					
		C	H	O	N	S	Ash
Food	4.9	48.0	6.4	37.6	2.6	0.4	5.0
Glass/Metal	3.2	–	–	–	–	–	100
Paper	12.6	43.5	6.0	44.0	0.3	0.2	6.0
Plastic	8.7	60.0	7.2	22.8	–	–	10
Wood Debris	2.1	49.5	6.0	42.7	0.2	0.1	1.5
Yard Clippings	29.5	47.8	6.0	38.0	3.4	0.3	4.5

If the energy content of the solid waste is represented by $333C + 1428(H - O/8) + 95S$, the ash-free energy content of the waste is most nearly

(A) 2100 kJ/kg

(B) 12 000 kJ/kg

(C) 110 000 kJ/kg

(D) 240 000 kJ/kg

7. A municipality landfills 5.3×10^7 kg/yr of solid waste. The landfilled waste is compacted to a maximum in-place density of 850 kg/m³. The soil cover-to-compacted waste ratio is 1:5 by volume. The annual in-place volume of the waste requiring disposal in the landfill is most nearly

(A) 2.4 ha·m/yr

(B) 6.2 ha·m/yr

(C) 12 ha·m/yr

(D) 31 ha·m/yr

8. A municipality landfills its solid waste. The annual in-place volume of compacted waste is 6.2 ha·m. The soil cover to compacted waste ratio is 1:5 by volume. The annual volume of soil cover required for the landfill is most nearly

(A) 0.48 ha·m/yr

(B) 0.70 ha·m/yr

(C) 1.2 ha·m/yr

(D) 6.2 ha·m/yr

9. A municipality landfills its solid waste. The annual in-place volume of compacted waste is 6.2 ha·m and the annual volume of cover is 1.2 ha·m. The operating life of the landfill is 30 yr. The total volume required for the landfill over its operating life is most nearly

(A) 86 ha·m

(B) 220 ha·m

(C) 430 ha·m

(D) 1100 ha·m

10. What is the likely composition of the gas produced by a typical landfill?

(A) 15% CO_2, 25% CH_4, and 60% others

(B) 20% CO_2, 35% CH_4, and 45% others

(C) 30% CO_2, 40% CH_4, and 30% others

(D) 40% CO_2, 60% CH_4, and <1% others

11. A landfill located in an arid region of the United States has an in-place waste capacity of 3.0×10^8 kg. The current methane gas production rate is 5.87×10^{-9} m³/min·kg. The current average rate at which gas is produced by the landfill is most nearly

(A) 6.5×10^2 m³/yr

(B) 2.5×10^3 m³/yr

(C) 9.3×10^5 m³/yr

(D) 1.8×10^9 m³/yr

12. A landfill currently produces 9.3×10^5 m³ of methane gas annually. For a decay rate constant of 0.05 yr^{-1}, the average rate at which gas is produced by the landfill 15 yr in the future is most nearly

(A) 3.1×10^2 m³/yr

(B) 1.2×10^3 m³/yr

(C) 4.4×10^5 m³/yr

(D) 8.5×10^8 m³/yr

13. The per capita solid waste generation rate for a city of 100,000 residents is 5 lbf/day at typical moisture content. The unit energy content of the waste is 6000 Btu/lbf. Most nearly, what is the stoichiometric combustion air requirement for incineration of the waste?

(A) 2.0×10^6 lbf/day

(B) 2.4×10^6 lbf/day

(C) 3.1×10^6 lbf/day

(D) 3.4×10^6 lbf/day

14. The residents of a city produce 1.7×10^7 kg/yr of solid waste. The discarded moisture content of the waste is 32%, and the chemical formula of the waste is $C_{40}H_{86}O_{37}N$. Methane production from solid waste can be estimated using the following equation.

$$C_aH_bO_cN_d + (0.25)(4a - b - 2c + 3d)H_2O \rightarrow$$
$$(0.125)(4a + b - 2c - 3d)CH_4$$
$$+ (0.125)(4a - b + 2c + 3d)CO_2 + dNH_3$$

a, b, c, d = moles of carbon, hydrogen, oxygen, and nitrogen, respectively.

The annual total volume of methane gas, at 1 atmosphere pressure and 25°C, potentially produced from decomposition of the waste is most nearly

(A) 5.4×10^4 m³/yr

(B) 7.3×10^6 m³/yr

(C) 2.1×10^7 m³/yr

(D) 7.7×10^9 m³/yr

SOLUTIONS

1. Use a 100 kg sample as the basis for calculations.

(a)

waste component	dry mass (kg/100kg)	dry elemental chemical composition (%)				
		C	H	O	N	S
food	4.9	2.4	0.3	1.8	0.1274	0.0196
glass/metal	3.2					
paper	12.6	5.5	0.8	5.5	0.0378	0.0252
plastic	8.7	5.2	0.6	2.0		
wood debris	2.1	1.0	0.1	0.9	0.0042	0.0021
yard clippings	31.5	14.1	1.82	10.27	0.17	0.05

The moisture content is

$$100 \text{ kg} - 61 \text{ kg} = 39 \text{ kg}$$

The chemical content of the moisture is

$$\text{hydrogen} = \left(\frac{2}{18}\right)(39 \text{ kg}) = 4.3 \text{ kg}$$

$$\text{oxygen} = \left(\frac{16}{18}\right)(39 \text{ kg}) = 34.7 \text{ kg}$$

The total hydrogen is

$$3.6 \text{ kg} + 4.3 \text{ kg} = 7.9 \text{ kg}$$

The total oxygen is

$$21 \text{ kg} + 34.7 \text{ kg} = 56 \text{ kg}$$

From table (a), the following table can be derived.

(b)

element	mass (kg)	mole weight (kg/kmol)	kmol	% mass
carbon	28	12	2.3	30
hydrogen	7.9	1	7.9	8.5
oxygen	56	16	3.5	60
nitrogen	1.2	14	0.086	1.3
sulfur	0.14	32	0.0044	0.15
	93			99.95

From kmol in table (b),

element	mole ratio S = 1
carbon	523
hydrogen	1795
oxygen	795
nitrogen	20
sulfur	1

The chemical formula is $C_{523}H_{1795}O_{795}N_{20}S$.

The answer is (D).

2. Find the mass of each component using a 100 kg sample as the basis for calculations.

waste component	dry mass (kg/100kg)	dry elemental chemical composition (%)			
		C	H	O	N
food	4.9	2.4	0.31	1.8	0.13
glass/metal	3.2	–	–	–	–
paper	12.6	5.5	0.76	5.5	0.038
plastic	8.7	5.2	0.63	2.0	–
wood debris	2.1	1.0	0.13	0.90	0.0042
yard clippings	29.5	14.1	1.77	11.21	1.003
	61	28	3.6	21	1.2

The moisture content is

$$100 \text{ kg} - 61 \text{ kg} = 39 \text{ kg}$$

The chemical content of moisture is

$$\text{hydrogen} = \left(\frac{2}{18}\right)(39 \text{ kg}) = 4.3 \text{ kg}$$

$$\text{oxygen} = \left(\frac{16}{18}\right)(39 \text{ kg}) = 34.7 \text{ kg}$$

The total hydrogen is

$$3.6 \text{ kg} + 4.3 \text{ kg} = 7.9 \text{ kg}$$

The total oxygen is

$$21 \text{ kg} + 34.7 \text{ kg} = 56 \text{ kg}$$

From the values in the table of mass of each component in the 100 kg sample, the following table can be derived.

element	mass (kg)	mole weight (kg/kmol)	kmol	% mass
carbon	28	12	2.3	30
hydrogen	7.9	1	7.9	8.5
oxygen	56	16	3.5	60
nitrogen	1.2	14	0.086	1.3
sulfur	0.14	32	0.0044	0.15
	93			99.95

From kmol in the table,

element	mole ratio N = 1
carbon	27
hydrogen	92
oxygen	41
nitrogen	1

The chemical formula is $C_{27}H_{92}O_{41}N$.

The answer is (C).

3. The equation for the C:N ratio is

$$\frac{\left((0.32)\left(300\,000 \ \frac{\text{kg}}{\text{d}}\right)\left(\frac{1}{12}\right) + (\dot{m}_{\text{sludge}})\left(\frac{1}{90}\right) \right)}{(0.32)\left(300\,000 \ \frac{\text{kg}}{\text{d}}\right) + \dot{m}_{\text{sludge}}} = 1/30$$

Solving for the mass flow rate of the sludge gives 216 000 kg/d (220 000 kg/d).

The answer is (C).

4. Assume the moisture content of the compostable solid waste is 38%.

$$\frac{\left((0.32)(0.38)\left(300\,000 \ \frac{\text{kg}}{\text{d}}\right) + (1 - 0.18)\left(220\,000 \ \frac{\text{kg}}{\text{d}}\right) \right)}{(0.32)\left(300\,000 \ \frac{\text{kg}}{\text{d}}\right) + 220\,000 \ \frac{\text{kg}}{\text{d}}} = 0.686$$

The moisture content of the mixture is 69%.

The answer is (D).

5. The total mass of the compostable mixture of waste and sludge at 69% moisture is

$$(0.32)\left(300\,000\ \frac{\text{kg}}{\text{d}}\right) + 220\,000\ \frac{\text{kg}}{\text{d}} = 316\,000\ \text{kg/d}$$

The dry mass of the mixture is

$$(1 - 0.690)\left(316\,000\ \frac{\text{kg}}{\text{d}}\right) = 97\,960\ \text{kg/d}$$

The total mass of the mixture at 69% moisture is

$$\frac{97\,960\ \dfrac{\text{kg}}{\text{d}}}{1 - 0.6} = 244\,900\ \text{kg/d}$$

The mass of moisture removed is

$$316\,000\ \frac{\text{kg}}{\text{d}} - 244\,900\ \frac{\text{kg}}{\text{d}}$$
$$= 71\,100\ \text{kg/d} \quad (71\,000\ \text{kg/d})$$

The answer is (A).

6. The energy content in kJ/kg is

$$333\text{C} + (1428)\left(\text{H} - \frac{\text{O}}{8}\right) + 95\text{S}$$

$$\text{C, H, O, S} = \text{elements, } \% \text{ of total}$$

Find the mass of each component using a 100 kg sample as the basis for calculations.

Waste Component	Dry Mass (kg/100kg)	Dry Elemental Chemical Composition (%)			
		C	H	O	N
Food	4.9	2.4	0.31	1.8	0.13
Glass/Metal	3.2	–	–	–	–
Paper	12.6	5.5	0.76	5.5	0.038
Plastic	8.7	5.2	0.63	2.0	–
Wood Debris	2.1	1.0	0.13	0.90	0.0042
Yard Clippings	–	–	–	–	–
	61	28	3.6	21	1.2

The moisture content is

$$100\ \text{kg} - 61\ \text{kg} = 39\ \text{kg}$$

The chemical content of moisture is

$$\text{hydrogen} = \left(\frac{2}{18}\right)(39\ \text{kg}) = 4.3\ \text{kg}$$

$$\text{oxygen} = \left(\frac{16}{18}\right)(39\ \text{kg}) = 34.7\ \text{kg}$$

The total hydrogen is

$$3.6\ \text{kg} + 4.3\ \text{kg} = 7.9\ \text{kg}$$

The total oxygen is

$$21\ \text{kg} + 34.7\ \text{kg} = 56\ \text{kg}$$

From the values in the table of mass of each component in the 100 kg sample, the following table can be derived.

element	mass (kg)	mole weight (kg/kmol)	kmol	% mass
carbon	28	12	2.3	30
hydrogen	7.9	1	7.9	8.5
oxygen	56	16	3.5	60
nitrogen	1.2	14	0.086	1.3
sulfur	0.14	32	0.0044	0.15
	93			99.95

From the percent mass in the table, the energy content is

$$(333)(30) + (1428)\left(8.5 - \frac{60}{8}\right) + (95)(0.15)$$
$$= 11432\ \text{kJ/kg} \quad (12\,000\ \text{kJ/kg})$$

The answer is (B).

7. The annual in-place volume of waste landfilled is

$$\frac{5.3 \times 10^7\ \dfrac{\text{kg}}{\text{yr}}}{\left(850\ \dfrac{\text{kg}}{\text{m}^3}\right)\left(10\,000\ \dfrac{\text{m}^2}{\text{ha}}\right)} = 6.2\ \text{ha·m/yr}$$

The answer is (B).

8. The annual cover volume is

$$\frac{6.2\ \dfrac{\text{ha·m}}{\text{yr}}}{5} = 1.24\ \text{ha·m/yr} \quad (1.2\ \text{ha·m/yr})$$

The answer is (C).

9. The total landfill volume is

$$\left(6.2\ \frac{\text{ha·m}}{\text{yr}} + 1.2\ \frac{\text{ha·m}}{\text{yr}}\right)(30\ \text{yr})$$

$$= 222\ \text{ha·m} \quad (220\ \text{ha·m})$$

The answer is (B).

10. The likely composition of the gas produced by a landfill, assuming it to be typical of other landfills, is 40% CO_2, 60% CH_4, and less than 1% others.

The answer is (D).

11. The current methane production rate is

$$\left(5.87 \times 10^{-9}\ \frac{\text{m}^3}{\text{min·kg}}\right)(3.0 \times 10^8\ \text{kg})$$

$$\times \left(1440\ \frac{\text{min}}{\text{d}}\right)\left(365\ \frac{\text{d}}{\text{yr}}\right) = 9.3 \times 10^5\ \text{m}^3/\text{yr}$$

The answer is (C).

12. The decay rate constant units indicate a first-order decay rate.

$$t_1 = 0\ \text{yr}$$
$$t_2 = 15\ \text{yr}$$
$$C_{A0} = \text{methane generation rate at time } t_1$$
$$\quad\ = 9.3 \times 10^5\ \text{m}^3/\text{yr}$$
$$C_A = \text{methane generation rate at time } t_2, \text{m}^3/\text{yr}$$
$$k = \text{decay constant} = 0.05\ \text{yr}^{-1}$$

First-Order Irreversible Reaction Kinetics

$$\ln(C_A/C_{A0}) = -kt$$

$$\frac{C_A}{C_{A0}} = \exp(-kt)$$

$$\frac{C_A}{C_{A0}} = \exp\left(-K(t_2 - t_1)\right)$$

$$C_A = \left(9.3 \times 10^5\ \frac{\text{m}^3}{\text{yr}}\right)\exp\left(\begin{array}{c}-(0.05\ \text{yr}^{-1}) \\ \times (15\ \text{yr} - 0\ \text{yr})\end{array}\right)$$

$$= 4.4 \times 10^5\ \text{m}^3/\text{yr}$$

The answer is (C).

13. The total energy content of the waste is

$$(100{,}000\ \text{people})\left(5\ \frac{\text{lbf}}{\text{person-day}}\right)$$

$$\times \left(6000\ \frac{\text{Btu}}{\text{lbf}}\right) = 3.0 \times 10^9\ \text{Btu/day}$$

The stoichiometric combustion air needed per million Btu is 684.6 lbf air/10^6 Btu. [Useful Relationships]

The total stoichiometric combustion air requirement for incineration of the waste is

$$\left(3.0 \times 10^9\ \frac{\text{Btu}}{\text{day}}\right)\left(\frac{684.6\ \text{lbf air}}{10^6\ \text{Btu}}\right)$$

$$= 2.0 \times 10^6\ \text{lbf air/day}$$

The answer is (A).

14. The reaction equation for the decomposition of the waste becomes

$$C_{40}H_{86}O_{37}N$$
$$+(0.25)\big((4)(40) - (86)(1) - (2)(37) + (3)(1)\big)H_2O \rightarrow$$
$$(0.125)\big((4)(40) + (86)(1) - (2)(37) - (3)(1)\big)CH_4$$
$$+(0.125)\big((4)(40) - (86)(1) + (2)(37) + (3)(1)\big)CO_2$$
$$+1NH_3$$
$$C_{40}H_{86}O_{37}N + 0.75H_2O \rightarrow 21CH_4 + 19CO_2 + 1NH_3$$

1 mole of solid waste as $C_{40}H_{86}O_{37}N$ yields 21 moles of methane gas as CH_4.

Find the molecular weight of the waste. [Periodic Table of Elements]

$$(40)\left(12\ \frac{\text{g}}{\text{mol}}\right) + (86)\left(1\ \frac{\text{g}}{\text{mol}}\right)$$

$$+(37)\left(16\ \frac{\text{g}}{\text{mol}}\right) + (1)\left(14\ \frac{\text{g}}{\text{mol}}\right)$$

$$= 1172\ \text{g/mol}$$

The amount of methane produced in a year is

$$\frac{\left(1.7 \times 10^7\ \frac{\text{kg waste}}{\text{yr}}\right)\left(21\ \frac{\text{kmol methane}}{\text{kmol waste}}\right)}{1172\ \frac{\text{kg waste}}{\text{kmol waste}}}$$

$$= 3.0 \times 10^5\ \text{kmol methane/yr}$$

Assume that methane behaves as an ideal gas so that the ideal gas equation applies.

$$P = \text{pressure} = 1\ \text{atm}$$
$$V = \text{volume of methane, m}^3$$
$$n = \text{moles of methane} = 3.0 \times 10^5\ \text{kmol/yr}$$
$$R = \text{universal gas law constant}$$
$$\quad\ = 8.2 \times 10^{-2}\ \text{L·atm/mol·K}$$
$$T = \text{temperature} = 298\text{K}$$

Ideal Gas Constants

$$PV = nRT$$

$$V = \frac{nRT}{P}$$

$$= \frac{\left(\begin{array}{c} \left(3.0 \times 10^5 \; \dfrac{\text{kmol}}{\text{yr}} \right) \\[6pt] \times \left(8.2 \times 10^{-2} \; \dfrac{\text{L·atm}}{\text{mol·K}} \right) (298\text{K}) \end{array} \right)}{(1 \text{ atm}) \left(\dfrac{1 \text{ kmol}}{1000 \text{ mol}} \right) (1000 \text{ L/m}^3)}$$

$$= 7.3 \times 10^6 \text{ m}^3/\text{yr}$$

The answer is (B).

# Hazardous Waste Storage, Collection, Transportation

PRACTICE PROBLEMS

1. What National Fire Protection Association (NFPA) standard addresses above-ground storage tank placement, specification, spill control, venting, and testing?

(A) NFPA 10

(B) NFPA 20

(C) NFPA 30

(D) NFPA 40

2. Which federal statute established standards applicable to hazardous waste generators and transporters and to owners and operators of treatment, storage, and disposal facilities (TSDFs)?

(A) Comprehensive Environmental Response, Compensation, and Liability Act (CERCLA), through the National Contingency Plan

(B) Resource Conservation and Recovery Act (RCRA), through Subtitle C

(C) RCRA, through Subtitle I

(D) Superfund Amendment and Reauthorization Act (SARA), through Title III

3. What waste management issue does Subtitle I of the Resource Conservation and Recovery Act (RCRA) address?

(A) citizen suits

(B) community right-to-know

(C) solid waste management units (SWMUs)

(D) underground storage tanks (USTs)

SOLUTIONS

1. National Fire Protection Association Standard 30 (NFPA 30), the *Flammable and Combustible Liquids Code*, defines standards to reduce hazards associated with the storage, handling, and use of flammable and combustible liquids. This includes above-ground storage tanks. NFPA 10 is the *Standard for Portable Fire Extinguishers*, NFPA 20 is the *Standard for the Installation of Stationary Pumps for Fire Protection*, and NFPA 40 is the *Standard for the Storage and Handling of Cellulose Nitrate Film*.

The answer is (C).

2. The Resource Conservation and Recovery Act Subtitle C defines the national hazardous waste management program, which includes standards applicable to hazardous waste generators and transporters and to owners and operators of treatment, storage, and disposal facilities (TSDFs).

The answer is (B).

3. Subtitle I was added to the 1984 edition of the Resource Conservation and Recovery Act (RCRA 1984) to address leaking underground storage tanks (USTs). It required the EPA to set standards for USTs and to apply the standards retroactively. The standards define UST design, operation, clean-up, administration, and closure requirements.

The answer is (D).

Solid & Hazard. Waste

42 Hazardous Waste Treatment and Disposal

Content in blue refers to the *NCEES Handbook*.

PRACTICE PROBLEMS

1. The principal organic hazardous constituents (POHC) mass feed rate to an incinerator is 4.12 kg/h. The POHC mass rate from the incinerator to air pollution control equipment is 0.43 kg/h, and the POHC mass rate from the stack is 0.00074 kg/h. The destruction and removal efficiency (DRE) for the POHC is most nearly

(A) 10.26%

(B) 89.56%

(C) 99.83%

(D) 99.98%

2. A city uses a rotary kiln incinerator for municipal solid waste disposal. The incinerator is characterized by the following.

> mean residence time = 30 min
> internal length-to-diameter ratio = 8
> kiln rake slope = 0.3 in/ft

Most nearly, at what rotational speed should the incinerator operate?

(A) 0.50 rpm

(B) 1.0 rpm

(C) 2.0 rpm

(D) 2.4 rpm

3. A schematic of a RCRA Subtitle C landfill liner system is shown in the figure.

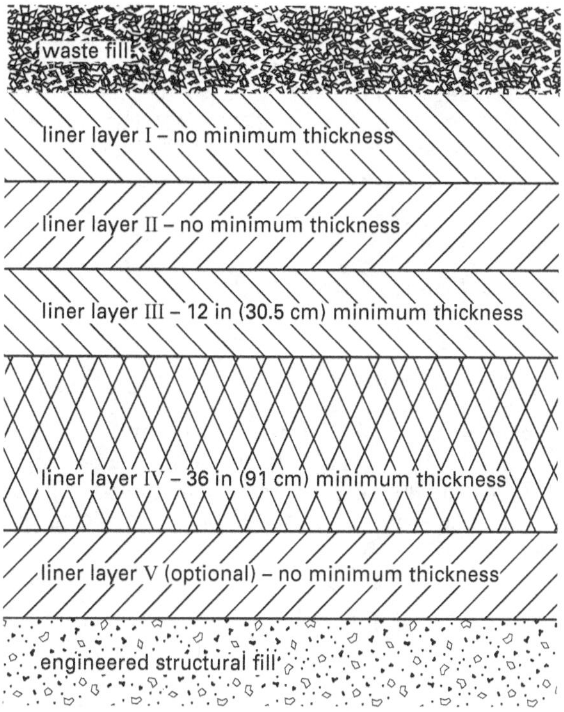

Which liner layer represents the leachate collection, detection, and removal system (LCDRS)?

(A) liner layer I

(B) liner layer II

(C) liner layer III

(D) liner layer IV

4. A city currently landfills its solid waste and flares 7.3×10^6 m³/yr of methane collected from landfill gases to the atmosphere. The city buys natural gas at \$0.21/m³. Most nearly, what are the potential annual savings, ignoring all other costs, if natural gas purchases are offset by methane gas recovered from the landfill?

(A) \$710,000/yr

(B) \$1,500,000/yr

(C) \$4,400,000/yr

(D) \$1,600,000,000/yr

5. Is methane gas usable as a replacement fuel for natural gas directly upon recovery from the landfill?

(A) yes

(B) No; the gas needs to be blended with propane or butane to increase its energy value.

(C) No; the gas needs to be dried to remove water vapor and scrubbed to remove gases that are noncombustible.

(D) No; the gas needs to be odorized and stabilized by blending with nitrogen gas.

6. An organic solvent waste is to be incinerated at a waste feed rate of 1200 L/h for 24 h/d. The waste contains 20% water and 80% organic solvent by weight. The specific gravity of the waste is 0.95. The mass waste feed rate is most nearly

(A) 60 kg/h

(B) 1140 kg/h

(C) 1260 kg/h

(D) 24 000 kg/h

7. An organic solvent waste is to be incinerated at a mass flow rate of 1140 kg/h. The waste contains 20% water and 80% organic solvent by weight. The composition of the organic solvent fraction is shown in the table.

organic solvent component	weight (%)
carbon	78
hydrogen	10
oxygen	10
chloride	2

Assuming complete combustion, the mass flows of CO_2 and HCl produced as combustion products are most nearly

(A) $CO_2 = 710$ kg/h, HCl $= 23$ kg/h

(B) $CO_2 = 920$ kg/h, HCl $= 980$ kg/h

(C) $CO_2 = 2600$ kg/h, HCl $= 19$ kg/h

(D) $CO_2 = 3300$ kg/h, HCl $= 25$ kg/h

8. An organic solvent waste is to be incinerated at a mass flow rate of 1140 kg/h. The waste contains 20% water and 80% organic solvent by weight. The composition of the organic solvent fraction is shown in the table.

organic solvent component	weight (%)
carbon	78
hydrogen	10
oxygen	10
chloride	2

The stoichiometric mass flow requirement of O_2 for complete combustion of the carbon in the waste is most nearly

(A) 710 kg/h

(B) 940 kg/h

(C) 1200 kg/h

(D) 1900 kg/h

9. A schematic of a RCRA Subtitle C landfill liner system is shown in the figure.

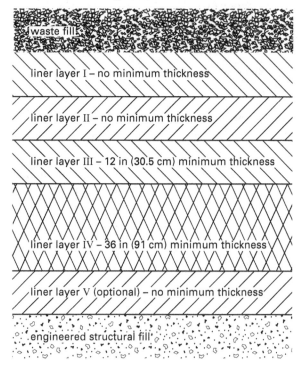

Which liner layer represents the geomembrane?

(A) liner layer I

(B) liner layer II

(C) liner layer III

(D) liner layer IV

10. A schematic of a RCRA Subtitle C landfill liner system is shown in the figure.

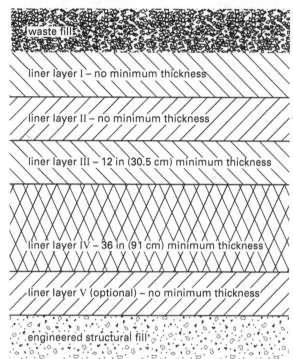

Which liner layer represents the soil liner?

   (A)   liner layer I

   (B)   liner layer II

   (C)   liner layer III

   (D)   liner layer IV

11. A schematic of a RCRA Subtitle C landfill liner system is shown in the figure.

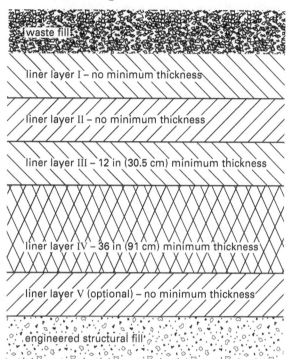

Which liner layer represents the leachate collection and removal system (LCRS)?

   (A)   liner layer I

   (B)   liner layer II

   (C)   liner layer III

   (D)   liner layer IV

SOLUTIONS

1. Find the destruction and removal efficiency.

W_{in} = feed mass flow rate, kg/h

W_{out} = principal organic hazardous constituents
mass flow rate from the stack, kg/h

Incineration

$$DRE = \frac{W_{in} - W_{out}}{W_{in}} \times 100\%$$

$$= \left(\frac{4.12 \frac{kg}{h} - 0.00074 \frac{kg}{h}}{4.12 \frac{kg}{h}} \right) \times 100\%$$

$$= 99.98\%$$

The answer is (D).

2. Use the equation for kiln retention time, and solve for the rotational speed.

N = rotational speed, rpm

t = mean residence time = 30 min

L/D = internal length-to-diameter ratio = 8

S = kiln rake slope = 0.3 in/ft

Kiln Retention Time

$$t = \frac{2.28 L/D}{SN}$$

$$N = \frac{2.28 \left(\frac{L}{D} \right)}{St} = \frac{(2.28)(8)}{\left(0.3 \frac{in}{ft} \right)(30 \text{ min})}$$

$$= 2.0 \text{ rpm}$$

The answer is (C).

3.

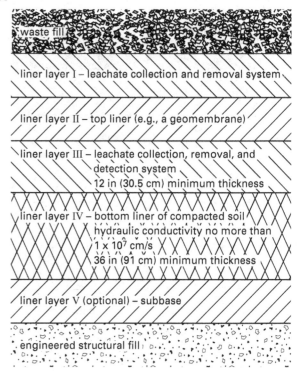

From the figure, liner layer III represents the leachate collection, detection, and removal system (LCDRS).

The answer is (C).

4. The potential annual savings is

$$\left(7.3 \times 10^6 \frac{m^3}{yr} \right) \left(\frac{\$0.21}{m^3} \right)$$

$$= \$1,533,000/yr \quad (\$1,500,000/yr)$$

The answer is (B).

5. Landfill gas contains methane at about 60% by volume. To be used as a replacement fuel for natural gas, the landfill gas must be dried to remove water vapor and scrubbed to remove gases that are noncombustible.

The answer is (C).

6. The mass waste feed rate is

$$\left(1200 \frac{L}{h} \right) (0.95) \left(1 \frac{kg}{L} \right) = 1140 \text{ kg/h}$$

The answer is (B).

7. The mass flow of CO_2 produced is

$$(0.80) \left(1140 \frac{kg}{h} \right) \left(0.78 \frac{g \text{ C}}{g \text{ solids}} \right) \left(\frac{44 \text{ g } CO_2}{12 \text{ g C}} \right)$$

$$= 2608 \text{ kg/h} \quad (2600 \text{ kg/h})$$

Solid & Hazard. Waste

The mass flow of HCl produced is

$$(0.80)\left(1140\ \frac{\text{kg}}{\text{h}}\right)\left(0.02\ \frac{\text{g Cl}}{\text{g solids}}\right)\left(\frac{36.5\ \text{g HCl}}{35.5\ \text{g Cl}}\right)$$
$$= 18.8\ \text{kg/h}\quad(19\ \text{kg/h})$$

The answer is (C).

8. The stoichiometric mass flow of O_2 required for combustion of carbon is

$$(0.80)\left(1140\ \frac{\text{kg}}{\text{h}}\right)\left(0.78\ \frac{\text{g C}}{\text{g solids}}\right)\left(\frac{32\ \text{g }O_2}{12\ \text{g C}}\right)$$
$$= 1897\ \text{kg/h}\quad(1900\ \text{kg/h})$$

The answer is (D).

9.

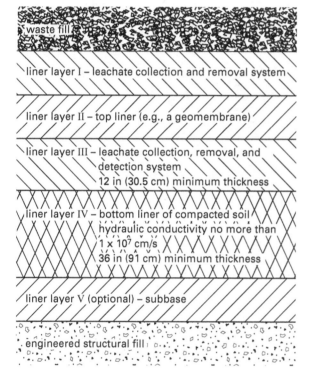

From the figure, liner layer II represents the geomembrane.

The answer is (B).

10.

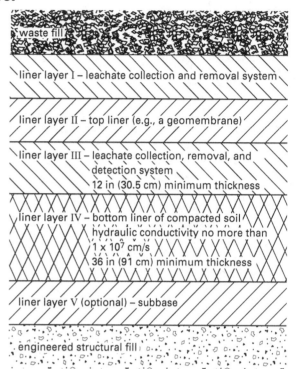

From the figure, liner layer IV represents the soil liner.

The answer is (D).

11.

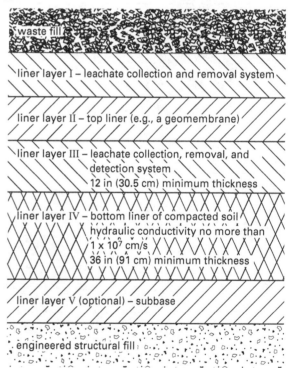

From the figure, liner layer I represents the leachate collection and removal system (LCRS).

The answer is (A).

Solid & Hazard. Waste

Topic IV Site Assessment and Remediation

43 Codes, Standards, Regulations, Guidelines

Content in blue refers to the *NCEES Handbook*.

PRACTICE PROBLEMS

1. The Superfund process occurs in nine steps. Which of the following cleanup actions occur in step six of the Superfund process?

(A) Remedial Investigation/Feasibility Study

(B) Remedy Selection

(C) Remedial Design

(D) Remedial Action

2. The acronym SWMU is used in federal regulations addressing site contamination. What does the acronym SWMU stand for?

(A) Soil and Water Mass Units

(B) Solid Waste Management Unit

(C) Superfund Waste Material Utilization

(D) Superfund Watch Memorandum of Understanding

3. Which Hazard Ranking System (HRS) score is sufficient to place a contaminated site on the National Priorities List (NPL)?

(A) ≥ 22

(B) > 28.5

(C) > 49

(D) ≥ 100

4. Which federal law and associated regulations govern the Corrective Action program?

(A) Comprehensive Environmental Response, Compensation, and Liability Act (CERCLA) under 40 CFR 307

(B) Resource Conservation and Recovery Act (RCRA) under 40 CFR 264 Subpart F

(C) Superfund Amendments and Reauthorization Act (SARA) under 40 CFR 300 Subpart C

(D) Toxic Substances Control Act (TSCA) under 40 CFR 761

SOLUTIONS

1. The nine steps defining the Superfund cleanup process are shown.

step 1: Discovery

step 2: Preliminary Assessment/Site Inspection

step 3: Hazard Ranking System/National Priorities List

step 4: Remedial Investigation/Feasibility Study

step 5: Remedy Selection

step 6: Remedial Design

step 7: Remedial Action

step 8: Operation and Maintenance

step 9: Removal from the NPL

The answer is (C).

2. SWMU is defined in the Resource Conservation and Recovery Act (RCRA) as a Solid Waste Management Unit.

The answer is (B).

3. An HRS score greater than 28.5 will place a contaminated site on the NPL.

The answer is (B).

4. The Corrective Action program is included in the Resource Conservation and Recovery Act (RCRA) under 40 CFR 264 Subpart F.

The answer is (B).

Chemistry and Biology

Content in blue refers to the *NCEES Handbook.*

PRACTICE PROBLEMS

1. What organic chemical is represented by the condensed structural formula shown?

$Cl_2C=CHCl$

(A) c-1,2-dichloroethene

(B) trichloromethane

(C) 1,1,2-trichloroethane

(D) trichloroethene

2. An oxidation sequence for families of organic chemicals is shown.

hydrocarbons → alcohols → aldehydes

What chemical family would follow aldehydes in the oxidation sequence?

(A) amines

(B) esters

(C) ketones

(D) organic acids

SOLUTIONS

1. The double bond between the two carbon atoms indicates the compound is an ethene. The three chlorine atoms are trichloro-, making the chemical trichloroethene.

The answer is (D).

2. The oxidation sequence is hydrocarbons → alcohols → aldehydes and ketones → organic acids.

The answer is (D).

Content in blue refers to the *NCEES Handbook*.

PRACTICE PROBLEMS

1. A well constructed in a confined aquifer and screened through the entire aquifer thickness of 18 m was pumped at 0.75 m³/min for 48 h. Time-drawdown observations at a well located 61 m and 100 m away were recorded and the data plotted. The aquifer transmissivity is 65 m²/d. The hydraulic conductivity of the aquifer is most nearly

(A) 0.18 m/d

(B) 0.36 m/d

(C) 1.8 m/d

(D) 3.6 m/d

2. An aquifer is characterized by a hydraulic conductivity of 0.42 m/d, a gradient of 0.010 m/m, and an effective porosity of 0.34. The actual groundwater flow velocity in the aquifer is most nearly

(A) 0.0072 m/d

(B) 0.0093 m/d

(C) 0.012 m/d

(D) 0.019 m/d

3. The intrinsic permeability of a soil is 1.4×10^{-6} in². The groundwater gradient is 0.00035, and the soil effective porosity is 0.42. If the groundwater temperature is 45°F, the rate of advection is most nearly

(A) 0.62 ft/day

(B) 1.5 ft/day

(C) 15 ft/day

(D) 210 ft/day

4. If the distribution coefficient for hexavalent chromium is 0.59 mL/g, will infiltration through contaminated overlying soils be a continuing source of hexavalent chromium to the groundwater?

(A) Yes, hexavalent chromium will leach slowly from the soil to the groundwater.

(B) Yes, hexavalent chromium will leach rapidly from the soil to the groundwater.

(C) No, hexavalent chromium will not leach from the soil.

(D) No, hexavalent chromium will be rapidly reduced to trivalent chromium.

5. Groundwater monitoring wells have been constructed on a site as shown in the figure. The site's hydrogeologic setting is characterized by an unconfined aquifer with silty sand to a depth of 10 m. The average hydraulic conductivity for the aquifer is 0.42 m/d and the effective porosity is 0.34. Groundwater elevation data for the monitoring wells are provided in the accompanying table.

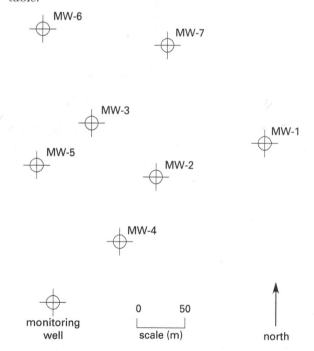

well	casing top elevation (m above mean sea level)	groundwater depth below casing top (m)
MW-1	49.77	4.74
MW-2	49.74	5.66
MW-3	49.59	5.59
MW-4	49.60	5.95
MW-5	49.09	5.57
MW-6	49.31	5.25
MW-7	49.63	4.62

The direction of the groundwater gradient is most nearly

(A) N 45° E

(B) N 45° W

(C) S 45° E

(D) S 45° W

6. A pumping test conducted in an unconfined aquifer is characterized by the following.

level ground surface elevation = 439 ft

aquifer thickness = 21 ft

steady-state discharge = 17 gpm

static water level below ground surface
    in discharge well at steady state = 8 ft

well diameter = 8 in

observation well distance from midpoint
    of discharge well = 14 ft

static water level below ground surface
    in observation well at steady state = 1.6 ft

Most nearly, what is the hydraulic conductivity of the aquifer?

(A) 6.3 ft/day

(B) 11 ft/day

(C) 19 ft/day

(D) 60 ft/day

7. An aquifer has a hydraulic conductivity of 110 ft/day. The change in hydraulic head in the aquifer is 0.74 ft over a distance of 100 ft. Most nearly, what is the flow rate in the aquifer per unit area of 1 ft^2?

(A) 0.81 ft^3/day

(B) 1.5 ft^3/day

(C) 67 ft^3/day

(D) 81 ft^3/day

8. An accident released 4500 gal of a heavy fuel oil to the ground surface. The ground is mostly bare with some areas of light vegetation. Site investigations reveal that the fuel oil penetrated to a maximum depth of 2.6 ft over a total area of 720 ft^2. Most nearly, what is the retention capacity of the soil for the fuel oil?

(A) 0.46

(B) 3.1

(C) 16

(D) 120

9. Groundwater monitoring wells have been constructed on a site as shown in the figure. The site's hydrogeologic setting is characterized by an unconfined aquifer with silty sand to a depth of 10 m. The average hydraulic conductivity for the aquifer is 0.42 m/d and the effective porosity is 0.34. Groundwater elevation data for the monitoring wells are provided in the accompanying table.

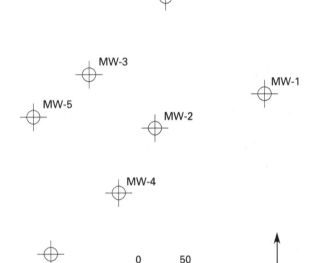

well	casing top elevation (m above mean sea level)	groundwater depth below casing top (m)
MW-1	49.77	4.74
MW-2	49.74	5.66
MW-3	49.59	5.59
MW-4	49.60	5.95
MW-5	49.09	5.57
MW-6	49.31	5.25
MW-7	49.63	4.62

Site Assessment & Remediation

The slope of the groundwater gradient is most nearly

(A) 0.0058

(B) 0.0075

(C) 0.010

(D) 0.015

10. The extraction rate from an unconfined aquifer serving a large metropolitan area exceeds the recharge rate, as demonstrated by the static water table's decline of 4.4 m over a 48-month monitoring period. The aquifer characteristics are as shown.

aquifer horizontal surface area	512 km^2
aquifer thickness	38 m
average porosity	0.43
hydraulic conductivity	0.38 cm/s
storativity	0.21

Most nearly, how much water is lost from aquifer storage during the monitoring period?

(A) 2.1×10^6 m^3

(B) 4.7×10^8 m^3

(C) 9.7×10^8 m^3

(D) 3.6×10^9 m^3

SOLUTIONS

1. Use the equation for transmissivity to find the hydraulic conductivity.

$$K = \text{hydraulic conductivity}, \text{m/d}$$
$$b = \text{aquifer thickness} = 18 \text{ m}$$
$$T = \text{transmissivity} = 65 \text{ m}^2/\text{d}$$

Transmissivity

$$T = Kb$$

$$K = \frac{T}{b} = \frac{65 \dfrac{\text{m}^2}{\text{d}}}{18 \text{ m}}$$
$$= 3.6 \text{ m/d}$$

The answer is (D).

2. Find the specific discharge.

$$v = \text{actual groundwater velocity}, \text{m/d}$$
$$K = \text{hydraulic conductivity} = 0.42 \text{ m/d}$$
$$n = \text{effective porosity} = 0.34$$
$$q = \text{specific discharge}$$
$$i = \text{gradient} = 0.010 \text{ m/m}$$

Specific Discharge

$$q = -K(dh/dx)$$

Take K as positive and let $dh/dx = i$.

$$v = \frac{q}{n} = \frac{Ki}{n} = \frac{\left(0.42 \dfrac{\text{m}}{\text{d}}\right)(0.010)}{0.34}$$
$$= 0.012 \text{ m/d}$$

The answer is (C).

3. Convert the groundwater temperature to degrees Celsius.

Temperature Conversions

$$\frac{(45°\text{F} - 32)}{1.8} = 7.2°\text{C} \quad (7°\,\text{C})$$

Find the hydraulic conductivity. Use a table of

Site Assessment & Remediation

thermophysical properties of air and water to extrapolate approximate values for density and dynamic viscosity. [Properties of Water (SI Metric Units)]

K = hydraulic conductivity, ft/sec

g = gravitational constant = 32.2 ft/sec^2

k = intrinsic permeability = 1.4×10^{-6} in^2

ρ = water density at 7°C = 1000 kg/m^3

  = 62.4 lbm/ft^3

μ = dynamic viscosity of water at 7°C

  = 1.41×10^{-3} kg/m·s = 0.00094 lbm/ft-sec

Hydraulic Conductivity

$$K = \rho g k / \mu$$

$$= \frac{\left(62.4 \ \dfrac{\text{lbm}}{\text{ft}^3}\right)\left(32.2 \ \dfrac{\text{ft}}{\text{sec}^2}\right)(1.4 \times 10^{-6} \ \text{in}^2)}{\left(0.00094 \ \dfrac{\text{lbm}}{\text{ft-sec}}\right)\left(144 \ \dfrac{\text{in}^2}{\text{ft}^2}\right)}$$

$$= 0.020 \ \text{ft/sec}$$

Use the equation for specific discharge to find the average seepage velocity.

q = specific discharge, $K(dh/dx)$

v = average seepage velocity

n = effective porosity = 0.42

dh/dx = groundwater gradient = 0.00035

Specific Discharge

$$v = q/n$$

$$= \frac{K\left(\dfrac{dh}{dx}\right)}{n}$$

$$= \frac{\left(0.020 \ \dfrac{\text{ft}}{\text{sec}}\right)(0.00035)\left(86{,}400 \ \dfrac{\text{sec}}{\text{day}}\right)}{0.42}$$

$$= 1.44 \ \text{ft/day} \quad (1.5 \ \text{ft/day})$$

The answer is (B).

4. Infiltration through contaminated overlying soils will be a continuing source of hexavalent chromium to the groundwater because the relatively large distribution coefficient will allow the hexavalent chromium to leach slowly from the soil.

The answer is (A).

5. Record and plot the data for the groundwater elevation above mean sea level as shown.

well	casing top elevation (m above mean sea level)	groundwater depth below casing top (m)	groundwater elevation (m above mean sea level)
MW-1	49.77	4.74	45.03
MW-2	49.74	5.66	44.08
MW-3	49.59	5.59	44.00
MW-4	49.60	5.95	43.65
MW-5	49.09	5.57	43.52
MW-6	49.31	5.25	44.06
MW-7	49.63	4.62	45.01

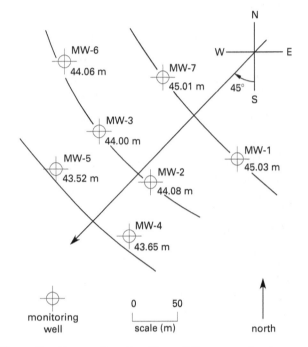

From the figure, the direction of the groundwater gradient is S 45° W.

The answer is (D).

6. Use Dupuit's formula, and solve for the hydraulic conductivity.

k = hydraulic conductivity, ft/d

Q = steady-state discharge = 17 gpm

h_1 = height of water surface above bottom of aquifer at discharge well

  = 21 ft − 8 ft = 13 ft

h_2 = height of water surface above bottom of aquifer at observation well

  = 21 ft − 1.6 ft = 19.4 ft

r_1 = discharge well radius = 4 in

r_2 = observation well distance from midpoint of discharge well = 14 ft

Dupuit's Formula

$$Q = \frac{\pi k (h_2^2 - h_1^2)}{\ln\left(\frac{r_2}{r_1}\right)}$$

$$k = \frac{Q \ln \frac{r_2}{r_1}}{\pi (h_2^2 - h_1^2)}$$

$$= \frac{\left(\dfrac{\left(17 \dfrac{\text{gal}}{\text{min}}\right)\left(1440 \dfrac{\text{min}}{\text{day}}\right)}{7.481 \dfrac{\text{gal}}{\text{ft}^3}}\right)}{\pi\left((19.4 \text{ ft})^2 - (13 \text{ ft})^2\right)}$$

$$\times \ln \frac{(14 \text{ ft})\left(12 \dfrac{\text{in}}{\text{ft}}\right)}{4 \text{ in}}$$

$$= 18.8 \text{ ft/day} \quad (19 \text{ ft/day})$$

The answer is (C).

7. Use Darcy's law, taking hydraulic conductivity, K, as positive.

$K =$ hydraulic conductivity $= 110$ ft/day

$dh/dx = 0.74 \text{ ft}/100 \text{ ft} = 0.0074$

$A =$ cross-sectional area of flow $= 1 \text{ ft}^2$

Darcy's Law

$$Q = KA(dh/dx)$$

$$= \left(110 \frac{\text{ft}}{\text{day}}\right)(1 \text{ ft}^2)(0.0074)$$

$$= 0.81 \text{ ft}^3/\text{day}$$

The answer is (A).

8. Use the equation for depth of penetration, and solve for the retention capacity.

$Rv =$ retention capacity

$D =$ maximum depth of penetration $= 2.6$ ft

$V =$ volume of infiltrating fuel oil $= 4500$ gal

$A =$ ground area covered by spill $= 720 \text{ ft}^2$

Vadose Zone Penetration

$$D = \frac{RvV}{A}$$

$$Rv = \frac{DA}{V} = \frac{(2.6 \text{ ft})(720 \text{ ft}^2)}{\left(\dfrac{4500 \text{ gal}}{7.481 \dfrac{\text{gal}}{\text{ft}^3}}\right)}$$

$$= 3.1$$

The answer is (B).

9. Data for the groundwater elevation above mean sea level is recorded in the table and plotted in the accompanying figure.

well	casing top elevation (m above mean sea level)	groundwater depth below casing top (m)	groundwater elevation (m above mean sea level)
MW-1	49.77	4.74	45.03
MW-2	49.74	5.66	44.08
MW-3	49.59	5.59	44.00
MW-4	49.60	5.95	43.65
MW-5	49.09	5.57	43.52
MW-6	49.31	5.25	44.06
MW-7	49.63	4.62	45.01

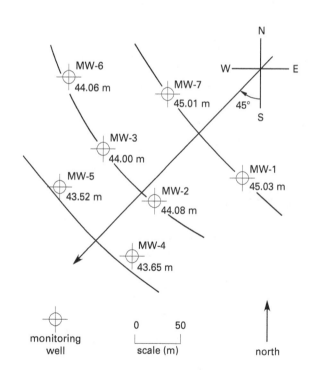

Site Assessment & Remediation

From the figure, the gradient is

$$\frac{\Delta\text{elevation}}{\Delta\text{distance}} = \frac{45.0 \text{ m} - 43.5 \text{ m}}{150 \text{ m}} = 0.010$$

The answer is (C).

10. The volume of water lost to storage is the product of the drained aquifer volume and the aquifer storativity.

V_d = water volume lost from storage, m^3
A_a = horizontal surface area of the aquifer = 512 km^2
S = storativity = 0.21
Δh = change in water table elevation = 4.4 m

$$V_d = \Delta h S A_a$$
$$= (4.4 \text{ m})(0.21)(512 \text{ km}^2)\left(\frac{1000 \text{ m}}{1 \text{ km}}\right)^2$$
$$= 473\,088\,000 \text{ m}^3 \quad (4.7 \times 10^8 \text{ m}^3)$$

The answer is (B).

46 Sampling and Measurement Methods

Content in blue refers to the *NCEES Handbook*.

PRACTICE PROBLEMS

1. The characteristics of an organic chemical in solution with water are shown.

   influent concentration = 1.57 mg/L

   required effluent concentration = 0.05 mg/L

   vapor pressure = 0.11 atm at 20°C

   solubility in water = 1250 mg/L at 20°C

   molecular weight = 87 g/mol

   temperature = 20°C

The value of Henry's constant in unitless form is most nearly

- (A) 8.8×10^{-5}
- (B) 7.7×10^{-3}
- (C) 0.32
- (D) 93

2. Properties of different organic chemicals are given in the following table.

	solubility in water (mg/L)	specific gravity
chemical 1	1170	0.92
chemical 2	infinite	0.97
chemical 3	1740	1.13
chemical 4	59	0.78

Which chemical will exist as light nonaqueous phase liquid (LNAPL) if present in the saturated zone at a concentration of 1080 mg/L?

- (A) chemical 1
- (B) chemical 2
- (C) chemical 3
- (D) chemical 4

3. Groundwater resources in the United States are classified as Class I, II, or III waters based on water quality indicators. What classifies a water source as Class III?

- (A) current or potential source of drinking water and having other beneficial uses
- (B) highest potential beneficial use is drinking water and located in a potentially vulnerable setting
- (C) not a potential drinking water source and having limited other beneficial uses
- (D) not a potential drinking water source but having other potentially beneficial uses

4. A chemical is at equilibrium in a soil-groundwater system. The soil is an organic loam and the chemical is present in the soil at a concentration of 783 mg/kg. The concentration of the chemical in the groundwater is 1231 mg/L. The value of the organic carbon partition coefficient is most nearly

- (A) 0.32
- (B) 0.64
- (C) 0.96
- (D) 1.6

5. The results of a soil adsorption isotherm test using groundwater contaminated with an organic chemical are shown in the illustration.

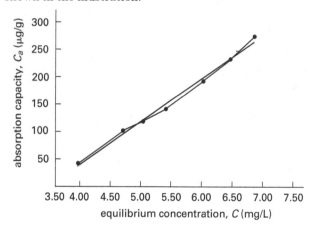

Site Assessment & Remediation

The soil effective porosity is 0.43, the soil bulk density is 1.68 g/cm^3, and the soil total organic carbon (TOC) is 271 mg/kg. The soil-water partition coefficient is most nearly

(A) 30 cm^3/g

(B) 37 cm^3/g

(C) 49 cm^3/g

(D) 77 cm^3/g

6. Tetrachloroethene (PERC), 1,1,1-trichloroethane (TCA), and trichloroethene (TCE) have been discovered in the groundwater of an unconfined aquifer. The effective porosity and bulk density of the aquifer soil are 0.34 and 1.83 g/cm^3, respectively. The soil total organic carbon (TOC) concentration is 148 mg/kg. What are the maximum possible solute concentrations at or near the release point for PERC, TCA, and TCE?

(A) 5 μg/L, 20 μg/L, 5 μg/L

(B) 150 mg/L, 1500 mg/L, 1100 mg/L

(C) no maximum concentration limits exist

(D) unable to determine from the information provided

7. Tetrachloroethene (PERC), 1,1,1-trichloroethane (TCA), and trichloroethene (TCE) have been discovered in the groundwater of an unconfined aquifer. Which of the three chemicals will move away from the source at the highest velocity?

(A) PERC

(B) TCA

(C) TCE

(D) all will move at the same velocity

8. Tetrachloroethene (PERC) has been discovered in the groundwater of an unconfined aquifer. The effective porosity and bulk density of the aquifer soil are 0.34 and 1.83 g/cm^3, respectively. The soil organic carbon fraction is 148 mg/kg. The soil-water partition coefficient is most nearly

(A) 0.025 mL/g

(B) 0.054 mL/g

(C) 0.54 mL/g

(D) 2.5 mL/g

9. Soil-water partition coefficients, Henry's constants, and solubility for selected chlorinated organic solvents are summarized in the table.

compound	K_{oc} (mL/g)	H (atm·m^3/mol)	solubility (mg/L)
chloroethane	42	0.0085	5740
chloroform	34	0.0038	9300
1,1-dichloroethene	217	0.021	2730
cis-1,2-dichloroethene	34	0.0037	3500
trans-1,2-dichloroethene	39	0.38	6300
methylene chloride	25	0.0032	16 700
perchloroethene	303	0.018	150
trichloroethene	152	0.010	1080
vinyl chloride	8400	2.8	1100

Which compound in the table is most likely to exist as a nonaqueous phase liquid?

(A) 1,1-dichloroethene

(B) methylene chloride

(C) perchloroethene

(D) vinyl chloride

10. If the distribution coefficient for hexavalent chromium is 0.59 mL/g, will a "pump and treat" remedial alternative be successful in quickly removing the hexavalent chromium from the soil/groundwater system?

(A) Yes, because the distribution coefficient is relatively small.

(B) Yes, because the distribution coefficient is relatively large.

(C) No, because the distribution coefficient is relatively small.

(D) No, because the distribution coefficient is relatively large.

SOLUTIONS

1. Find Henry's constant.

H = Henry's constant, unitless

P_A^* = vapor pressure = 0.11 atm at 20°C

MW = molecular weight = 87 g/mol

C_{AL} = solubility in water

= 1250 mg/L at 20°C

R = universal gas constant

= 8.2×10^{-5} atm·m^3/mol·K

T = temperature = 293K

Convection: Overall Coefficients

$$p_A^* = HC_{AL}$$

$$H = \frac{p_A^*}{C_{AL}}$$

For unitless H,

$$H = \frac{P_A^* \text{MW}}{C_{AL} RT}$$

$$= \frac{(0.11 \text{ atm})\left(87 \dfrac{\text{g}}{\text{mol}}\right)\left(1000 \dfrac{\text{mg}}{\text{g}}\right)}{\left(1250 \dfrac{\text{mg}}{\text{L}}\right)\left(8.2 \times 10^{-5} \dfrac{\text{atm·m}^3}{\text{mol·K}}\right)}$$

$$\times (293\text{K})\left(1000 \dfrac{\text{L}}{\text{m}^3}\right)$$

$$= 0.32$$

The answer is (C).

2. For a chemical to exist as LNAPL, it must have a specific gravity less than 1.0 and be present at a concentration greater than its solubility in water. Chemical 4 is the only chemical that satisfies both criteria.

The answer is (D).

3. The EPA applies the following classification to groundwater resources:

Class I: Special Groundwaters—groundwater resources with a high beneficial use (drinking water) and located in a potentially vulnerable setting

Class II: Current and Potential Sources of Drinking Water and Waters Having Other Beneficial Uses—all groundwaters that are neither Class I nor Class III

Class III: Groundwaters Not Considered Potential Sources of Drinking Water and of Limited Beneficial Use—usually limited to waters with TDS greater than 10 000 mg/L

The answer is (C).

4. Find the value of the organic carbon partition coefficient. Assume all the soil is organic since the organic fraction is not given and the soil is an organic loam. Assume the density of water is 1 kg/L.

Organic Carbon Partition Coefficient K_{oc}

$$K_{oc} = C_{\text{soil}}/C_{\text{water}}$$

$$K_{oc} = \frac{\left(783 \dfrac{\text{mg}}{\text{kg}}\right)\left(1 \dfrac{\text{kg}}{\text{L}}\right)}{1231 \dfrac{\text{mg}}{\text{L}}} = 0.64$$

The answer is (B).

5. Find the distribution coefficient.

K_d = distribution coefficient

(slope of the isotherm plot coefficient), cm^3/g

X = chemical concentration in soil, μg/g

C = concentration of chemical in water, mg/L

Soil-Water Partition Coefficient $K_d = K_p$

$$K_d = X/C = \frac{\Delta X}{\Delta C}$$

Using corresponding (C, X) pairs from the isotherm plot,

$$K_d = \frac{\left(250 \dfrac{\mu\text{g}}{\text{g}} - 50 \dfrac{\mu\text{g}}{\text{g}}\right)\left(\dfrac{1 \text{ mg}}{1000 \, \mu\text{g}}\right)}{6.7 \dfrac{\text{mg}}{\text{L}} - 4.1 \dfrac{\text{mg}}{\text{L}}}$$

$$\times \left(\dfrac{1000 \text{ mL}}{\text{L}}\right)\left(1 \dfrac{\text{cm}^3}{\text{mL}}\right)$$

$$= 76.9 \text{ cm}^3/\text{g} \quad (77 \text{ cm}^3/\text{g})$$

The answer is (D).

6. The maximum possible solute concentration will occur at the approximate water solubilities of tetrachloroethene (PERC), 1,1,1-trichloroethane (TCA), and trichloroethene (TCE). From a table of water solubility values for selected chemicals, water solubility for PERC is 150 mg/L, for TCA is 1500 mg/L, and for TCE is 1100 mg/L. [Water Solubility, Vapor Pressure, Henry's Law Constant, K_{oc}, and K_{ow} Data for Selected Chemicals]

The answer is (B).

Site Assessment & Remediation

7. From a table of coefficients for organic carbon in selected chemicals, K_{oc} for PERC is 364 mL/g, for TCA is 152 mL/g, and for TCE is 126 mL/g. [Water Solubility, Vapor Pressure, Henry's Law Constant, K_{oc}, and K_{ow} Data for Selected Chemicals]

As the organic carbon coefficient K_{oc} increases, solute velocity decreases. Therefore, chemicals with smaller K_{oc} values will move more quickly.

With the smallest K_{oc}, TCE will move away from the source at the highest velocity.

The answer is (C).

8. From a table of coefficients for organic carbon in selected chemicals, K_{oc} for PERC is 364 mL/g. [Water Solubility, Vapor Pressure, Henry's Law Constant, K_{oc}, and K_{ow} Data for Selected Chemicals]

Find the distribution coefficient.

$$K_d = \text{distribution coefficient}$$
$$K_{oc} = \text{organic carbon partition coefficient}$$
$$= 364 \text{ mL/g}$$
$$f_{oc} = \text{organic carbon fraction} = 148 \text{ mg/kg}$$
$$= 1.48 \times 10^{-4} \text{ g/g}$$

Soil-Water Partition Coefficient $K_d = K_p$

$$K_d = K_{oc}f_{oc}$$
$$= \left(364 \ \frac{\text{mL}}{\text{g}}\right)\left(1.48 \times 10^{-4} \ \frac{\text{g}}{\text{g}}\right)$$
$$= 0.0539 \text{ mL/g} \quad (0.054 \text{ mL/g})$$

The answer is (B).

9. As solubility decreases, the potential for a compound to exist as a nonaqueous phase liquid increases. The compound in the table with the lowest solubility is perchloroethene.

The answer is (C).

10. A "pump and treat" remedial alternative will probably not be successful in quickly removing the hexavalent chromium from the soil/groundwater system because the distribution coefficient is relatively large.

The answer is (D).

Site Assessment & Remediation

Site Assessment and Characterization

Content in blue refers to the *NCEES Handbook*.

PRACTICE PROBLEMS

1. What phenomena or properties are most responsible for the dilution of a solute as it is transported away from its source by groundwater flow?

(A) conductivity and transmissivity

(B) dispersion and diffusion

(C) retardation and adsorption

(D) solubility and density

2. Characteristics of interbedded soil layers in an aquifer are given in the table shown.

layer	thickness (cm)	hydraulic conductivity (cm/s)
1	91	0.23
2	180	0.086
3	207	0.062

The average horizontal hydraulic conductivity for the layer aquifer strata is most nearly

(A) 0.10 cm/s

(B) 0.23 cm/s

(C) 0.46 cm/s

(D) 0.60 cm/s

3. A cluster of three wells are each screened in a successively deeper aquifer. Each aquifer is separated by a competent aquitard. What is the effect on the vertical distribution of a dense solute in the upper aquifer when water is pumped from the deepest of the three wells?

(A) The solute will be rapidly drawn through the aquitards into the deeper aquifers.

(B) The solute will migrate slowly into the second and then the third aquifer.

(C) The solute will move at a higher velocity in the upper aquifers due to a steeper gradient.

(D) The solute will remain confined to the upper aquifer.

4. An above-ground storage tank failure results in the release of 17 m³ of gasoline onto bare soil. The soil is a medium to fine sand, and the spill covers an area of 83 m². Most nearly, how deep into the ground does the spilled gasoline likely penetrate?

(A) 5 m

(B) 16 m

(C) 18 m

(D) 27 m

5. The following figure presents a schematic of a landfill and cap, and the accompanying table summarizes the cap and drainage layer material, fill thicknesses, and hydraulic conductivities. The maximum desired leachate head is 30 cm and the leachate collection laterals are 150 m long and are placed on the bottom of the drainage layer at 25 m intervals. Assume a self-cleaning velocity on the laterals of 0.6 m/s.

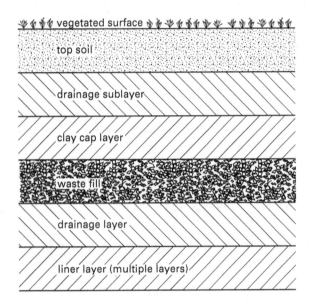

material	thickness (cm)	hydraulic conductivity (cm/s)
top soil	30	10^{-3}
drainage sub-layer	15	10^{-2}
clay cap layer	60	10^{-6}
waste fill	3200	10^{-3}
drainage layer	30	10^{-1}

The overall hydraulic conductivity for all layers, including the waste fill, is most nearly

(A) 1.7×10^{-6} cm/s

(B) 5.3×10^{-5} cm/s

(C) 580 cm/s

(D) 980 cm/s

SOLUTIONS

1. In a soil-groundwater system, solute tends to spread out as it moves away from the source. The spreading may occur in all three dimensions, but is most pronounced longitudinally. This spreading phenomenon is called hydrodynamic dispersion. It is a combination of mechanical dispersion, commonly referred to simply as dispersion, and diffusion. Dispersion and diffusion cause dilution of the solute.

The answer is (B).

2. The average horizontal conductivity is

$$K_h\text{avg} = \text{average horizontal hydraulic conductivity}$$
$$K_{hm} = \text{individual hydraulic conductivity of the } m\text{th layer}$$
$$b_m = \text{thickness of the } m\text{th layer}$$
$$b = \text{total thickness of all layers}$$

Average Horizontal Conductivity (Parallel to Layering)

$$
\begin{aligned}
K_h\text{avg} &= \sum_{m=1}^{n} \frac{K_{hm}b_m}{b} \\
&= \frac{\left(0.23\ \dfrac{\text{cm}}{\text{s}}\right)(91\ \text{cm}) + \left(0.086\ \dfrac{\text{cm}}{\text{s}}\right)(180\ \text{cm}) + \left(0.062\ \dfrac{\text{cm}}{\text{s}}\right)(207\ \text{cm})}{91\ \text{cm} + 180\ \text{cm} + 207\ \text{cm}} \\
&= 0.103\ \text{cm/s} \quad (0.10\ \text{cm/s})
\end{aligned}
$$

The answer is (A).

3. Because the aquitards are competent, there is no hydraulic communication among the three aquifers. Pumping from the lower aquifers will not influence solute transport in the upper aquifer.

The answer is (D).

4. To find the vadose zone penetration, the constant Rv must be found. Rv is a constant reflecting retention capacity of the soil and the viscosity of the product released. For gasoline on a medium to fine sand, $Rv = 80$. [Typical Values of Rv]

The vadose zone penetration is

$$D = \text{maximum depth of penetration, m}$$
$$V = \text{volume of gasoline released} = 17\ \text{m}^3$$
$$A = \text{area of spill} = 83\ \text{m}^2$$

Vadose Zone Penetration

$$D = \frac{RvV}{A} = \frac{(80)(17 \text{ m}^3)}{83 \text{ m}^2} = 16 \text{ m}$$

The answer is (B).

5. Find the average vertical hydraulic conductivity.

$K_v \text{ avg} = $ average vertical hydraulic conductivity for
       all layers, cm/s
$b = $ total aquifer thickness
       $(d_1 + d_2 + ...), \text{cm}$
$b_m = $ thickness of each layer, cm
$K_{vm} = $ vertical hydraulic conductivity of each
       layer, cm/s

Overall Vertical Hydraulic Conductivity (Perpendicular to
Layering)

$$
\begin{aligned}
K_v \text{ avg} &= \frac{b}{\displaystyle\sum_{m=1}^{n} \frac{b_m}{K_{vm}}} \\[2mm]
&= \frac{b}{\dfrac{b_1}{K_1} + \dfrac{b_2}{K_2} + \dfrac{b_3}{K_3} + \dfrac{b_4}{K_4} + \dfrac{b_5}{K_5}} \\[2mm]
&= \frac{30 \text{ cm} + 15 \text{ cm} + 60 \text{ cm} + 3200 \text{ cm} + 30 \text{ cm}}{\left(\begin{array}{c} \dfrac{30 \text{ cm}}{10^{-3} \text{ cm·s}^{-1}} + \dfrac{15 \text{ cm}}{10^{-2} \text{ cm·s}^{-1}} \\[2mm] + \dfrac{60 \text{ cm}}{10^{-6} \text{ cm·s}^{-1}} + \dfrac{3200 \text{ cm}}{10^{-3} \text{ cm·s}^{-1}} \\[2mm] + \dfrac{30 \text{ cm}}{10^{-1} \text{ cm·s}^{-1}} \end{array} \right)} \\[2mm]
&= 5.3 \times 10^{-5} \text{ cm/s}
\end{aligned}
$$

The answer is (B).

Risk Assessment

PRACTICE PROBLEMS

1. Which of the following is NOT among the four steps that traditionally define the risk assessment process?

(A) dose-response assessment

(B) exposure assessment

(C) health effects assessment

(D) risk characterization

2. In risk assessment, to what does the term "hazard" refer?

(A) identification of a risk

(B) quantification of a risk

(C) existence of a toxin

(D) occurrence of exposure

3. What are the four basic elements of risk assessment?

(A) hazard identification, population characterization, chemical assessment, risk characterization

(B) hazard identification, dose-response assessment, exposure assessment, risk characterization

(C) dose-response assessment, chemical assessment, population characterization, risk characterization

(D) dose-response assessment, chemical assessment, exposure assessment, risk characterization

4. Subchronic animal studies are performed to evaluate the noncarcinogenic toxic effects of a chemical compound. The lowest observed adverse effect level (LOAEL) of the chemical compound is 287 mg/kg·d. The uncertainty factors are shown in the table.

description	factor
uncertainty factor for population effects	10
uncertainty factor for extrapolating animal data to humans	10
uncertainty factor for using NOAEL from subchronic instead of chronic studies	10
uncertainty factor for using LOAEL instead of NOAEL	10
safety factor for other issues based on professional judgment	5

The administered oral reference dose is most nearly

(A) 0.0057 mg/kg·d

(B) 0.029 mg/kg·d

(C) 0.057 mg/kg·d

(D) 0.11 mg/kg·d

5. Subchronic animal studies are performed to evaluate the noncarcinogenic toxic effects of a chemical compound. The chemical compound is administered to the animals by ingestion of food with a 20% absorption efficiency. The administered oral reference dose is 0.0087 mg/kg·d. The absorbed oral reference dose for the chemical compound by ingestion of food is most nearly

(A) 0.0017 mg/kg·d

(B) 0.0088 mg/kg·d

(C) 0.034 mg/kg·d

(D) 0.44 mg/kg·d

6. Subchronic animal studies are performed to evaluate the noncarcinogenic toxic effects of a chemical compound. The chemical compound is administered to the animals by ingestion of food with a 20% absorption efficiency. If administered by ingestion of drinking water, the absorption efficiency is estimated to increase to 90%. The absorbed oral reference dose is 0.0013 mg/kg·d. The equivalent administered oral reference dose for the chemical compound by ingestion of drinking water is most nearly

(A) 0.0014 mg/kg·d

(B) 0.012 mg/kg·d

(C) 0.060 mg/kg·d

(D) 0.073 mg/kg·d

7. Subchronic animal studies are performed to evaluate the noncarcinogenic toxic effects of three chemical compounds. The chemicals are administered to the animals by ingestion of drinking water.

chemical	NOAEL (mg/kg·d)	intake (mg/kg·d)	*RfD* (mg/kg·d)
1	435	0.020	0.0017
2	287	0.0092	0.0011
3	329	0.015	0.0013

Based on the hazard index, which chemicals should be targeted for remediation if the exposure route is ingestion of drinking water?

(A) chemicals 1 and 2 only

(B) chemicals 1, 2, and 3

(C) chemicals 2 and 3 only

(D) chemical 3 only

8. Rural drinking water supplies in some western United States communities draw from groundwater that contains naturally occurring arsenic and fluoride at average concentrations of 109 μg/L and 1.7 mg/L, respectively. The groundwater is also used for watering vegetable gardens.

Residents of these communities tend to live out their lives within a few miles of the homes of their grandparents, parents, and siblings. Commodities from home food production, including canned home-grown fruits and vegetables, make up a large part of their diet. Groundwater conditions and the toxicity characteristics of arsenic and fluoride are shown in the table.

For groundwater,

dissolved oxygen	0.8 mg/L
pH	8.0
temperature	16°C
specific conductivity	860 μS/cm
TDS	650 mg/L

For arsenic,

relative specie toxicity	As(V) less toxic than As(III)
MCL	10 μg/L
slope factor	5×10^{-5} $(\mu\text{g/L})^{-1}$
oral RfD	1×10^{-3} mg/kg·d
bioconcentration factor	44 mL/g

For fluoride,

MCL	4 mg/L
oral RfD	6×10^{-2} mg/kg·d

Are the residents likely to be exposed to a significant cancer risk from ingesting the arsenic in their drinking water?

(A) No, the cancer risk is less than 1 in 10^6.

(B) No, the cancer risk is greater than 1 in 10^6.

(C) Yes, the cancer risk is less than 1 in 10^6.

(D) Yes, the cancer risk is greater than 1 in 10^6.

9. Rural drinking water supplies in some western United States communities draw from groundwater that contains naturally occurring arsenic and fluoride at average concentrations of 109 μg/L and 1.7 mg/L, respectively. The groundwater is also used for watering vegetable gardens.

Residents of these communities tend to live out their lives within a few miles of the homes of their grandparents, parents, and siblings. Commodities from home food production, including canned home-grown fruits and vegetables, make up a large part of their diet. Groundwater conditions and the toxicity characteristics of arsenic and fluoride are shown in the table.

For groundwater,

dissolved oxygen	0.8 mg/L
pH	8.0
temperature	16°C
specific conductivity	860 μS/cm
TDS	650 mg/L

For arsenic,

relative specie toxicity	As(V) less toxic than As(III)
MCL	10 μg/L
slope factor	5×10^{-5} $(\mu\text{g/L})^{-1}$
oral RfD	1×10^{-3} mg/kg·d
bioconcentration factor	44 mL/g

For fluoride,

MCL	4 mg/L
oral RfD	6×10^{-2} mg/kg·d

Are the residents likely to be exposed to significant non-carcinogenic health risks from ingesting the arsenic in their drinking water?

(A) No, the hazard index, *HI*, is less than 1.0.

(B) No, the hazard index, *HI*, is greater than 1.0.

(C) Yes, the hazard index, *HI*, is less than 1.0.

(D) Yes, the hazard index, *HI*, is greater than 1.0.

10. Rural drinking water supplies in some western United States communities draw from groundwater that contains naturally occurring arsenic and fluoride at average concentrations of 109 μg/L and 1.7 mg/L, respectively. The groundwater is also used for watering vegetable gardens.

Residents of these communities tend to live out their lives within a few miles of the homes of their grandparents, parents, and siblings. Commodities from home food production, including canned home-grown fruits and vegetables, make up a large part of their diet. Groundwater conditions and the toxicity characteristics of arsenic and fluoride are shown in the table.

For groundwater,

dissolved oxygen	0.8 mg/L
pH	8.0
temperature	16°C
specific conductivity	860 μS/cm
TDS	650 mg/L

For arsenic,

relative specie toxicity	As(V) less toxic than As(III)
MCL	10 μg/L
slope factor	5×10^{-5} $(\mu\text{g/L})^{-1}$
oral RfD	1×10^{-3} mg/kg·d
bioconcentration factor	44 mL/g

For fluoride,

MCL	4 mg/L
oral RfD	6×10^{-2} mg/kg·d

Are significant negative health consequences to the residents likely to result from ingesting the fluoride with the drinking water?

(A) No, the hazard index, *HI*, is less than 1.0 and the concentration is below the MCL.

(B) No, the hazard index, *HI*, is greater than 1.0 and the concentration is below the MCL.

(C) Yes, the hazard index, *HI*, is less than 1.0 and the concentration exceeds common fluoridation levels.

(D) Yes, the hazard index, *HI*, is greater than 1.0 and the concentration exceeds common fluoridation levels.

11. A population of 25,000 is exposed over a period of 23 yr to the following volatile organic compounds (VOCs) in their drinking water supply.

chemical	concentration (μg)/L	slope factor (mg/kg · d)$^{-1}$
trichloroethene	120 μg/L	1.1×10^{-2}
1,1-dichloroethene	80 μg/L	6.0×10^{-1}
methylene chloride	50 μg/L	7.5×10^{-3}

The incremental lifetime cancer risk to the average female adult from using the water supply is most nearly

(A) 5.4×10^{-8}

(B) 5.4×10^{-4}

(C) 1.4×10^{-3}

(D) 3.5×10^{-2}

Site Assessment & Remediation

12. If the incremental lifetime cancer risk from drinking contaminated water is 4.7×10^{-4}, is the risk considered acceptable by EPA criteria?

(A) Yes, because the risk is less than 1 in 10,000.

(B) Yes, because the risk is greater than 1 in 1 million.

(C) No, because the risk is less than 1 in 1 million.

(D) No, because the risk is greater than 1 in 10,000.

13. In any given year, most nearly how many additional cancers would be expected to result in a population of 25,000 people exposed to an incremental lifetime cancer risk of 4.7×10^{-4}?

(A) 1.9×10^{-5} cancers/yr

(B) 0.16 cancers/yr

(C) 35 cancers/yr

(D) 1600 cancers/yr

14. Low levels of dioxins are emitted from a hazardous waste incinerator that results in annual average ambient air concentrations, measured as 2,3,7,8-TCDD, of 0.2 pg/m³ in the adjacent community of 80,000 residents. The slope factor for 2,3,7,8-TCDD is 1.5×10^5 (mg/kg·d)$^{-1}$. The incremental lifetime risk from the dioxin exposure to the average adult male residing in the community for 35 yr is most nearly

(A) 1.2×10^{-7}

(B) 2.5×10^{-7}

(C) 2.7×10^{-6}

(D) 8.7×10^{-6}

15. Low levels of dioxins are emitted from a hazardous waste incinerator that results in annual average ambient air concentrations, measured as 2,3,7,8-TCDD, of 0.2 pg/m³ in the adjacent community of 80,000 residents. The slope factor for 2,3,7,8-TCDD is 1.5×10^5 (mg/kg·d)$^{-1}$. The incremental lifetime risk from the dioxin exposure to a child living in the community from birth to age 6 yr is most nearly

(A) 3.6×10^{-9}

(B) 2.2×10^{-8}

(C) 8.6×10^{-7}

(D) 1.2×10^{-6}

16. Low levels of dioxins are emitted from a hazardous waste incinerator that results in annual average ambient air concentrations, measured as 2,3,7,8-TCDD, of 0.2 pg/m³ in the adjacent community of 80,000 residents. The slope

factor for 2,3,7,8-TCDD is 1.5×10^5 (mg/kg·d)$^{-1}$. If a 1 in 1 million or less risk is considered acceptable, the maximum permissible annual average ambient air concentration of dioxin as 2,3,7,8-TCDD for an adult male exposed over a 35 yr period is most nearly

(A) 0.0034 pg/m³

(B) 0.018 pg/m³

(C) 0.073 pg/m³

(D) 0.16 pg/m³

17. Toxicity values and concentrations for four chemicals found in a groundwater sample are shown in the table.

chemical	concentration (μg/L)	slope factor (mg/L)$^{-1}$	reference dose (mg/L)
1,1-DCE	173	1.7×10^{-2}	0.315
MeCl	207	2.1×10^{-4}	2.1
PCE	879	1.5×10^{-3}	0.35
1,1,2-TCA	764	1.6×10^{-3}	0.14

The total carcinogenic risk factor from exposure to the chemicals is most nearly

(A) 0.0054

(B) 0.10

(C) 1.3

(D) 8.6

18. Toxicity values and concentrations for four chemicals found in a groundwater sample are shown in the table.

chemical	concentration (μg/L)	slope factor (mg/L)$^{-1}$	reference dose, (mg/L)
1,1-DCE	173	1.7×10^{-2}	0.315
MeCl	207	2.1×10^{-4}	2.1
PCE	879	1.5×10^{-3}	0.35
1,1,2-TCA	764	1.6×10^{-3}	0.14

Which chemicals could be eliminated from further evaluation of carcinogenic risk?

(A) none of the chemicals

(B) MeCl

(C) 1,1-DCE and MeCl

(D) 1,1-DCE, MeCl, and PCE

19. Toxicity values and concentrations for four chemicals found in a groundwater sample are shown in the table.

chemical	concentration $(\mu g/L)$	slope factor, SF $(mg/L)^{-1}$	reference dose (mg/L)
1,1-DCE	173	1.7×10^{-2}	0.315
MeCl	207	2.1×10^{-4}	2.1
PCE	879	1.5×10^{-3}	0.35
1,1,2-TCA	764	1.6×10^{-3}	0.14

Which chemicals could be eliminated from further evaluation of noncarcinogenic risk?

(A) none of the chemicals

(B) MeCl

(C) 1,1-DCE and MeCl

(D) 1,1-DCE, MeCl, and PCE

20. What is the weight-of-evidence category for "probable human carcinogen"?

(A) Group A

(B) Group B

(C) Group C

(D) Group D

21. For noncarcinogens, which among the following represents the highest dose?

(A) threshold dose

(B) reference dose (RfD)

(C) dose equal to the no observed adverse effects level (NOAEL)

(D) dose equal to the lowest observed adverse effects level (LOAEL)

SOLUTIONS

1. The four steps that define the risk assessment process are:

Step 1: Hazard Identification. Does a particular chemical present the potential for adverse health effects? The process of evaluating the potential available routes of exposure and the potential consequences of the exposure.

Step 2: Dose-Response Assessment. The process of finding a mathematical relationship between the amount of chemical to which a human is exposed and the risk that there will be an unhealthy response to that dose.

Step 3: Exposure Assessment. The process of determining actual pathways that allow transport of the chemical from the source to the point of contact and estimating the amount of contact likely to occur through ingestion, inhalation, and dermal contact.

Step 4: Risk Characterization. The process of bringing all the previous steps together to define an overall risk to a specific population.

Health effects assessment may be included within one or more steps, but it does not represent a stand-alone step in the traditional definition of risk assessment.

The answer is (C).

2. In risk assessment, the term "hazard" refers to the existence of a toxin, independent of exposure. Risk cannot be identified or quantified unless exposure to a toxin occurs. [Safety and Prevention]

The answer is (C).

3. The four basic elements of risk assessment are hazard identification, dose-response assessment, exposure assessment, and risk characterization.

The answer is (B).

4. Find the oral reference dose.

$$RfD = \text{reference dose, mg/kg·d administered}$$
$$NOAEL = \text{no observed adverse effect level, mg/kg·d}$$

Reference Dose

$$RfD = \frac{NOAEL}{UF}$$
$$= \frac{NOAEL}{(UF_1)(UF_2)(UF_3)(UF_4)(\text{SF})}$$
$$= \frac{287 \ \frac{mg}{kg \cdot d}}{(10)(10)(10)(10)(5)}$$
$$= 0.0057 \ mg/kg \cdot d$$

The answer is (A).

5. The oral reference dose absorbed through ingestion of food is the product of the administered oral reference dose and the ingestion absorption efficiency.

$$RfD_{\text{absorbed}} = (RfD)\left(\frac{20\% \text{ absorbed}}{100\% \text{ administered}}\right)$$

$$= 0.0087 \left(\frac{\text{mg}}{\text{kg·d}}\right)\left(\frac{20\% \text{ absorbed}}{100\% \text{ administered}}\right)$$

$$= 0.0017 \text{ mg/kg·d}$$

The answer is (A).

6. Assume the dose absorbed via food and water is the same. The equivalent administered oral reference dose equals the absorbed reference dose divided by the ingestion absorption efficiency.

$$RfD_{\text{via water}} = (RfD)\left(\frac{100\% \text{ administered}}{90\% \text{ absorbed}}\right)$$

$$= 0.0013 \left(\frac{\text{mg}}{\text{kg·d}}\right)\left(\frac{100\% \text{ administered}}{90\% \text{ absorbed}}\right)$$

$$= 0.0014 \text{ mg/kg·d}$$

The answer is (A).

7. Find the hazard index for each chemical.

HI_i = hazard index for each chemical i

CDI_i = chronic daily intake for each chemical i, mg/kg·d

RfD_i = absorbed oral reference dose for each
        chemical i by ingestion of water, mg/kg·d

Noncarcinogens

$$HI_i = \frac{CDI_i}{RfD_i}$$

$$HI_1 = \frac{0.020 \dfrac{\text{mg}}{\text{kg·d}}}{0.0017 \dfrac{\text{mg}}{\text{kg·d}}} = 11.8$$

$$HI_2 = \frac{0.0092 \dfrac{\text{mg}}{\text{kg·d}}}{0.0011 \dfrac{\text{mg}}{\text{kg·d}}} = 8.4$$

$$HI_3 = \frac{0.015 \dfrac{\text{mg}}{\text{kg·d}}}{0.0013 \dfrac{\text{mg}}{\text{kg·d}}} = 10.8$$

The hazard index for chemicals 1, 2, and 3 is greater than 1.0; therefore, all three chemicals should be targeted for remediation.

The answer is (B).

8. From an EPA table of recommended values for estimating intake, the average lifetime is 75 yr. [Intake Rates]

Find the chronic daily intake using the variables taken from an EPA table of residential exposure equations for various pathways.

CDI = chronic daily intake, μg/L

CW = concentration = 109 μg/L

ED = exposed duration
    = 75 yr (for lifetime resident)

EF = exposure frequency (assume 100%) = 1

CSF = slope factor = 5×10^{-5} $(\mu$g/L$)^{-1}$

AT = averaging time (lifetime)
    = 75 yr (standard USEPA exposure value)

Exposure

$$CDI = \frac{(CW)(IR)(EF)(ED)}{(BW)(AT)}$$

For this application, because CSF units are in $(\mu$g/L$)^{-1}$, CDI units are in μg/L. The revised equation for chronic daily intake is

$$CDI = \frac{(CW)(EF)(ED)}{(AT)}$$

$$= \frac{\left(109 \dfrac{\mu g}{\text{L}}\right)(1)(75 \text{ yr})}{75 \text{ yr}}$$

$$= 109 \text{ } \mu g/\text{L}$$

Find the risk.

Carcinogens

$$\text{Risk} = CDI \times CSF$$

$$= \left(109 \dfrac{\mu g}{\text{L}}\right)\left(5 \times 10^{-5}\left(\dfrac{\mu g}{\text{L}}\right)^{-1}\right)$$

$$= 0.0055 \quad (55 \text{ in } 10^4)$$

Acceptable cancer risk is typically considered to be 1 in 10^6 or less, therefore, 55 in 10^4 would be considered unacceptable. The residents are likely to be exposed to a significant cancer risk from ingesting the arsenic in their drinking water.

The answer is (D).

9. Exposure occurs over a full lifetime so adult values are used for the average lifetime and average amount of water ingested. From an EPA table of recommended values for estimating intake, the average amount of water ingested for adults is 2.3 L/d and the average lifetime is 75 yr. Since the population is both male and female, the value used for body weight is the average of the values for a male adult (78 kg) and a female adult (65.4 kg). [Intake Rates]

Find the chronic daily intake using the variables taken from an EPA table of residential exposure equations for various pathways.

CDI = chronic daily intake, mg/kg·day

CW = concentration = 109 μg/L

IR = concentration = 2.3 L/d

ED = exposed duration

= 75 yr (for lifetime resident)

EF = exposure frequency $= \dfrac{350 \dfrac{\text{days}}{\text{yr}}}{365 \dfrac{\text{days}}{\text{yr}}} = 0.96$

BW = average body weight (male-female average)

= 72 kg

AT = averaging time (lifetime)

= 75 yr

Exposure

$$CDI = \frac{(CW)(IR)(EF)(ED)}{(BW)(AT)}$$

$$= \frac{\left(109 \dfrac{\mu g}{L}\right)\left(2.3 \dfrac{L}{d}\right)(0.96)(75 \text{ yr})}{(72 \text{ kg})(75 \text{ yr})\left(10^3 \dfrac{\mu g}{mg}\right)}$$

$$= 0.00334 \text{ mg/kg·d}$$

Find the hazard index.

HI = hazard index

RfD = reference dose = 1×10^{-3} mg/kg·d

Noncarcinogens

$$HI = CDI_{\text{noncarcinogen}}/RfD$$

$$= \frac{0.00334 \dfrac{mg}{kg \cdot d}}{1 \times 10^{-3} \dfrac{mg}{kg \cdot d}}$$

$$= 3.3$$

When an HI is greater than 1.0, potential noncarcinogenic health hazard concerns exist. Since the HI of 3.3 exceeds 1.0, the residents are likely to be exposed to a significant noncarcinogenic health risk from ingesting the arsenic in their drinking water.

The answer is (D).

10. Exposure occurs over a full lifetime so adult values are used for the average lifetime and average amount of water ingested. From an EPA table of recommended values for estimating intake, the average amount of water ingested for adults is 2.3 L/d and the average lifetime is 75 yr. Since the population is both male and female, the value used for body weight is the average of the values for a male adult (78 kg) and a female adult (65.4 kg). [Intake Rates]

Find the chronic daily intake using the variables taken from an EPA table of residential exposure equations for various pathways.

CDI = chronic daily intake, mg/kg·day

CW = concentration = 1.7 mg/L

IR = concentration = 2.3 L/d

ED = exposed duration

= 75 yr (for lifetime resident)

EF = exposure frequency $= \dfrac{350 \dfrac{\text{days}}{\text{yr}}}{365 \dfrac{\text{days}}{\text{yr}}} = 0.96$

BW = average body weight (male-female average)

= 72 kg

AT = averaging time (lifetime)

= 75 yr

Exposure

$$CDI = \frac{(CW)(IR)(EF)(ED)}{(BW)(AT)}$$

$$= \frac{\left(1.7 \dfrac{mg}{L}\right)\left(2.3 \dfrac{L}{d}\right)(0.96)(75 \text{ yr})}{(72 \text{ kg})(75 \text{ yr})}$$

$$= 0.0521 \text{ mg/kg·d}$$

Find the hazard index.

HI = hazard index

RfD = reference dose = 6×10^{-2} mg/kg·day

Noncarcinogens

$$HI = CDI_{\text{noncarcinogen}}/RfD$$

$$= \frac{0.0521 \dfrac{mg}{kg \cdot d}}{6 \times 10^{-2} \dfrac{mg}{kg \cdot d}}$$

$$= 0.868 \quad (0.9)$$

The HI of 0.9 is less than 1.0, indicating that no likely health hazard exists. Also, given that municipal water supplies are commonly fluoridated to provide a fluoride

Site Assessment & Remediation

concentration of 1.0 mg/L and that the maximum contaminant level (MCL) for fluoride is 4 mg/L, a concentration of 1.7 mg/L will not likely present any significant negative health consequences to the residents from ingesting the fluoride with the drinking water.

The answer is (A).

11. From an EPA table of recommended values for estimating intake, the amount of water ingested for female adults is 2.3 L/d. The average body weight of a female adult is 65.4 kg and the average lifetime is 75 yr. [Intake Rates]

Find the chronic daily intake of each VOC using the variables taken from an EPA table of residential exposure equations for various pathways.

CDI = chronic daily intake, mg/kg·d

CW = concentration, mg/L

IR = daily intake = 2.3 L/d

ED = exposed duration = 23 yr

EF = fraction of time exposed (assume 350 days/yr)

BW = body weight = 65.4 kg

AT = averaging time = ×365 days/yr

Exposure

$$CDI = \frac{(CW)(IR)(EF)(ED)}{(BW)(AT)}$$

$$TCE\ CDI = \frac{\left(120\ \frac{\mu g}{L}\right)\left(\frac{1\ mg}{1000\ \mu g}\right)}{\times\left(2.3\ \frac{L}{d}\right)(23\ yr)(1)}{(65.4\ kg)(75\ yr)}$$
$$= 0.00129\ mg/kg \cdot d$$

$$1,1\text{-}DCE\ CDI = \frac{\left(80\ \frac{\mu g}{L}\right)\left(\frac{1\ mg}{1000\ \mu g}\right)}{\times\left(2.3\ \frac{L}{d}\right)(23\ yr)(1)}{(65.4\ kg)(75\ yr)}$$
$$= 0.00086\ mg/kg \cdot d$$

$$MeCl\ CDI = \frac{\left(50\ \frac{\mu g}{L}\right)\left(\frac{1\ mg}{1000\ \mu g}\right)}{\times\left(2.3\ \frac{L}{d}\right)(23\ yr)(350\ days/yr)}{(65.4\ kg)(365\ days/yr)}$$
$$= 0.00052\ mg/kg \cdot d$$

Find the risk of cancer.

Carcinogens

$$Risk = CDI \times CSF$$

$$risk\ TCE = \left(0.00129\ \frac{mg}{kg \cdot d}\right)\left(1.1 \times 10^{-2}\left(\frac{mg}{kg \cdot d}\right)^{-1}\right)$$
$$= 1.4 \times 10^{-5}$$

$$risk\ 1,1\text{-}DCE = \left(0.00086\ \frac{mg}{kg \cdot d}\right)$$
$$\times \left(6.0 \times 10^{-1}\left(\frac{mg}{kg \cdot d}\right)^{-1}\right)$$
$$= 5.2 \times 10^{-4}$$

$$risk\ MeCl = \left(0.00052\ \frac{mg}{kg \cdot d}\right)\left(7.5 \times 10^{-3}\left(\frac{mg}{kg \cdot d}\right)^{-1}\right)$$
$$= 3.9 \times 10^{-6}$$

Calculating the total risk gives

$$total\ risk = risk\ TCE + risk\ 1,1\text{-}DCE + risk\ MeCl$$
$$= 1.4 \times 10^{-5} + 5.2 \times 10^{-4} + 3.9 \times 10^{-6}$$
$$= 5.4 \times 10^{-4}$$

The answer is (B).

12. The EPA considers an acceptable risk for carcinogens to be within the range of 10^{-4} (1 in 10,000) to 10^{-6} (1 in 1 million). The contaminated water produces a risk of 4.7×10^{-4} (1 in approximately 2200), which is greater than 1.0×10^{-4} and, therefore, considered unacceptable. [Carcinogens]

The answer is (D).

13. From an EPA table of recommended values for estimating intake, the average lifetime is 75 years. [Intake Rates]

The number of cancers contracted per year from drinking water is

$$\frac{(risk)(population)}{lifetime}$$
$$= \frac{\left(4.7 \times 10^{-4}\ \frac{cancers}{person}\right)(25{,}000\ people)}{75\ yr}$$
$$= 0.16\ cancers/yr$$

The answer is (B).

14. From an EPA table of recommended values for estimating intake, the amount of air breathed for male adults is 15.2 m³/d. The average body weight for a male adult is 78 kg, and the average lifetime is 75 yr. [Intake Rates]

Find the chronic daily intake using the variables from the EPA table of residential exposure equations for various pathways.

CDI = chronic daily intake, mg/kg·d

CA = concentration = 0.2 pg/m^3

IR = inhalation rate, 15.2 m^3/d

ET = exposure time (assume 100% or 1)

ED = exposed duration = 35 yr

EF = exposure frequency (assume 100% or 1)

BW = body weight = 78 kg

AT = averaging time = 75 yr

Exposure

$$CDI = \frac{(CA)(IR)(ET)(EF)(ED)}{(BW)(AT)}$$

$$= \frac{\left(0.2 \ \frac{pg}{m^3}\right)\left(15.2 \ \frac{m^3}{d}\right)(1)(35 \ yr)(1)}{(78 \ kg)(75 \ yr)}$$

$$= 0.018 \ pg/kg \cdot d$$

Find the risk.

Carcinogens

$$Risk = CDI \times CSF$$

$$= \left(0.018 \ \frac{pg}{kg \cdot d}\right)\left(1.5 \times 10^5 \left(\frac{mg}{kg \cdot d}\right)^{-1}\right)\left(10^{-9} \ \frac{mg}{pg}\right)$$

$$= 2.7 \times 10^{-6}$$

The answer is (C).

15. From an EPA table of recommended values for estimating intake, the amount of air breathed for a child 3–5 yr of age is 8.3 m^3/d. The average body weight of a child 1–5 yr of age is 16 kg, and the average lifetime is 75 yr. [Intake Rates]

Find the chronic daily intake using the variables from the EPA table of residential exposure equations for various pathways.

CDI = chronic daily intake, 1 mg/kg·d

CA = concentration = 0.2 pg/m^3

IR = inhalation rate, 8.3 m^3/d

ET = exposure time (assume 100% or 1)

ED = exposed duration = 6 yr

EF = exposure frequency (assume 100% or 1)

BW = body weight = 16 kg

AT = averaging time = 75 yr

Exposure

$$CDI = \frac{(CA)(IR)(ET)(EF)(ED)}{(BW)(AT)}$$

$$= \frac{\left(0.2 \ \frac{pg}{m^3}\right)\left(8.3 \ \frac{m^3}{d}\right)(1)(6 \ yr)(1)}{(16 \ kg)(75 \ yr)}$$

$$= 0.0083 \ pg/kg \cdot d$$

Find the risk.

Carcinogens

$$Risk = CDI \times CSF$$

$$= \left(0.0083 \ \frac{pg}{kg \cdot d}\right)\left(1.5 \times 10^5 \left(\frac{mg}{kg \cdot d}\right)^{-1}\right)\left(10^{-9} \ \frac{mg}{pg}\right)$$

$$= 1.2 \times 10^{-6}$$

The answer is (D).

16. From an EPA table of recommended values for estimating intake, the daily amount of air breathed for male adults is 15.2 m^3/d. The average body weight for a male adult is 78 kg, and the average lifetime is 75 yr. [Intake Rates]

Use the equation for risk to find the concentration, CA. The variables used to calculate the chronic daily intake, CDI, care taken from an EPA table of residential exposure equations for various pathways.

CDI = chronic daily intake, mg/kg·d

IR = inhalation rate, 15.2 m^3/d

ET = exposure time (assume 100% or 1)

ED = exposed duration = 35 yr

EF = exposure frequency (assume 100% or 1)

BW = body weight = 78 kg

AT = averaging time = 75 yr

Site Assessment & Remediation

Carcinogens

$$Risk = CDI \times CSF$$
$$= \frac{(CA)(IR)(ET)(EF)(ED)}{(BW)(AT)}(CSF)$$

$$CA = \frac{(risk)(BW)(AT)}{(IR)(ET)(EF)(ED)(CSF)}$$

$$= \frac{(10^{-6})(78 \text{ kg})(75 \text{ yr})\left(10^9 \, \dfrac{pg}{mg}\right)}{\left(\left(15.2 \, \dfrac{m^3}{d}\right)(1)(35 \text{ yr})(1) \times \left(1.5 \times 10^5 \left(\dfrac{mg}{kg \cdot d}\right)^{-1}\right)\right)}$$

$$= 0.073 \text{ pg/m}^3$$

The answer is (C).

17. The carcinogenic risk factor is the product of the concentration ($C \times 10^{-3}$) and the slope factor. The carcinogenic risk factors are summarized in the table.

chemical	concentration, C (μg/L)	slope factor, SF (mg/L)$^{-1}$	carcinogenic risk factor, R
1,1-DCE	173	1.7×10^{-2}	0.0029
MeCl	207	2.1×10^{-4}	0.000043
PCE	879	1.5×10^{-3}	0.0013
1,1,2-TCA	764	1.6×10^{-3}	0.0012

The total carcinogenic risk factor is

$$0.0029 + 0.000043 + 0.0013 + 0.0012 = 0.0054$$

The answer is (A).

18. Calculate the carcinogenic risk factor, R, and relative risk factors, R_r, for each chemical.

$$R = (C \times 10^{-3})(SF)$$
$$R_r = \frac{R}{\sum R}$$

The carcinogenic risk factors and relative risk factors are summarized in the table.

chemical	concentration, C (μg/L)	slope factor, SF (mg/L)$^{-1}$	carcinogenic risk factor, R	relative risk, R_r
1,1-DCE	173	1.7×10^{-2}	0.0029	0.53
MeCl	207	2.1×10^{-4}	0.000043	0.0080
PCE	879	1.5×10^{-3}	0.0013	0.24
1,1,2-TCA	764	1.6×10^{-3}	0.0012	0.22

Typically, chemicals that present a relative risk of 0.01 (1% of total risk) or less are eliminated from further evaluation of carcinogenic risk. Based on this criterion, MeCl is the only chemical that could be eliminated from further study.

The answer is (B).

19. Calculate the noncarcinogenic risk factor, R, and relative risk factors, R_r, for each chemical.

$$R = \frac{C \times 10^{-3}}{RfD}$$
$$R_r = \frac{R}{\sum R}$$

The noncarcinogenic risk factors and relative risk factors are summarized in the table.

chemical	concentration, C (μg/L)	reference dose, RfD (mg/L)	risk factor, R	relative risk, R_r
1,1-DCE	173	0.315	0.55	0.064
MeCl	207	2.1	0.099	0.012
PCE	879	0.35	2.5	0.29
1,1,2-TCA	764	0.14	5.5	0.64

Typically, chemicals that present a relative risk of 0.01 (1% of total risk) or less are eliminated from further evaluation of noncarcinogenic risk. Based on this criterion, none of the chemicals could be eliminated from further study.

The answer is (A).

20. The weight-of-evidence category for "probable human carcinogen" is Group B. Group B chemicals are divided into one of the two subgroups, B1 or B2, depending on the type of evidence suggesting carcinogenicity. The Group A category is "human carcinogen," Group C is "possible human carcinogen," and Group D is "not classified."

The answer is (B).

21. In increasing order, the four doses listed for noncarcinogens are: (1) reference dose (RfD), (2) dose equal to the no observed adverse effects level (NOAEL), (3) threshold dose, and (4) dose equal to the lowest observed adverse effects level (LOAEL). [Reference Dose]

The answer is (D).

Fate and Transport

PRACTICE PROBLEMS

1. Leaks have been discovered in piping from a fuel storage depot. The leaking pipes convey no. 2 fuel oil to tanker truck fill stations at the facility. The site is characterized by an unconfined aquifer with an average hydraulic conductivity of 0.42 m/d, a gradient of 0.022, and an effective porosity of 0.28. The average groundwater temperature is 8°C. The density and dynamic viscosity of no. 2 fuel are 900 kg/m³ and 6.5×10^{-3} kg/m·s, respectively. The hydraulic conductivity for no. 2 fuel oil is 0.081 m/d. The average actual velocity of the no. 2 fuel oil as a nonaqueous phase liquid is most nearly

(A) 3.3×10^{-2} m/d

(B) 6.4×10^{-3} m/d

(C) 6.6×10^{-5} m/d

(D) 7.4×10^{-8} m/d

2. Leaks have been discovered in piping from a fuel storage depot. The leaking pipes convey no. 2 fuel oil to tanker truck fill stations at the facility. The site is characterized by an unconfined aquifer with an average hydraulic conductivity of 0.42 m/d, a gradient of 0.022, and an effective porosity of 0.28. The average groundwater temperature is 8°C. The density and dynamic viscosity of no. 2 fuel are 900 kg/m³ and 6.5×10^{-3} kg/m·s, respectively. The intrinsic permeability of the soil is most nearly

(A) 1.4×10^{-13} m²

(B) 3.5×10^{-13} m²

(C) 5.6×10^{-13} m²

(D) 6.9×10^{-13} m²

3. Leaks have been discovered in piping from a fuel storage depot. The leaking pipes convey no. 2 fuel oil to tanker truck fill stations at the facility. The site is characterized by an unconfined aquifer with an average hydraulic conductivity of 0.42 m/d, a gradient of 0.022, and an effective porosity of 0.28. The intrinsic permeability of the soil is 6.9×10^{-13} m². The average groundwater temperature is 8°C. The density and dynamic

viscosity of no. 2 fuel are 900 kg/m³ and 6.5×10^{-3} kg/m·s, respectively. The hydraulic conductivity of the aquifer for no. 2 fuel oil as a nonaqueous phase liquid is most nearly

(A) 9.4×10^{-7} m/d

(B) 8.4×10^{-4} m/d

(C) 0.081 m/d

(D) 0.42 m/d

4. A manufacturer has historically used a chromic acid-based solution to impregnate timbers and other structural wood products. The chromic acid solution was allowed to drain from the impregnated timbers onto bare soil thereby contaminating the underlying shallow unconfined aquifer. Soil samples were collected at the site, and a soil leaching test was conducted. The saturated soil porosity and soil bulk density are 0.34 and 1.8 g/cm³, respectively. The results of the leaching test are shown in the table.

test sample	Cr(VI) concentration in leachate (mg/L)	Cr(VI) adsorbed onto soil (mg/kg)
1	31	17
2	20	10
3	13	7.5
4	10	7.0
5	7.1	4.5

The retardation factor of the hexavalent chromium in the soil/groundwater system is most nearly

(A) 0.5

(B) 1.5

(C) 3.0

(D) 4.0

5. Leaks have been discovered in piping from an underground tank used to store tetrachloroethylene (PCE) at a commercial dry cleaning business. The site's hydrogeologic setting is characterized by an unconfined aquifer with an average groundwater velocity of 0.37 m/d and a retardation factor for PCE of 1.08. The average velocity of the PCE plume center of mass is most nearly

(A) 0.30 m/d

(B) 0.34 m/d

(C) 0.37 m/d

(D) 0.40 m/d

6. Tetrachloroethylene (PCE) has leaked from underground piping into an unconfined aquifer. The aquifer conditions are

$$\text{soil bulk density} = 1.85 \text{ g/cm}^3$$
$$\text{soil porosity} = 0.26$$
$$\text{organic carbon content} = 210 \text{ mg/kg}$$
$$\text{average groundwater velocity} = 0.37 \text{ m/d}$$

PCE organic carbon partition coefficient = 200 cm^3/g

The value of the retardation factor for the PCE in the aquifer is most nearly

(A) 1.3

(B) 2.1

(C) 2.7

(D) 3.6

7. The degradation sequence for perchloroethene to dichloroethene isomers is as follows.

perchloroethene (CCl$_2$-CCl$_2$)

$\Downarrow$

trichloroethene (CHCl-CCl$_2$)

$\Downarrow$

dichloroethene isomers
(CH$_2$-CCl$_2$, *cis*- and *trans*-CHCl-CHCl)

What is NOT a significant pathway for degradation of perchloroethene in a soil-groundwater system?

(A) biological oxidation

(B) chemical reduction

(C) deamination

(D) hydrolysis

8. The degradation sequence for perchloroethene to dichloroethene isomers is as follows.

perchloroethene (CCl$_2$-CCl$_2$)

$\Downarrow$

trichloroethene (CHCl-CCl$_2$)

$\Downarrow$

dichloroethene isomers
(CH$_2$-CCl$_2$, *cis*- and *trans*-CHCl-CHCl)

What compound would follow the dichloroethene isomers in the degradation sequence for perchloroethene?

(A) chloroethane (CH$_3$-CH$_2$Cl)

(B) chloroform (CHCl$_3$)

(C) methylene chloride (CH$_2$Cl$_2$)

(D) vinyl chloride (CH$_2$-CHCl)

9. Soil-water partition coefficients, Henry's constants, and solubility for selected chlorinated organic solvents are summarized in the table.

compound	K_{oc} (mL/g)	H (atm·m^3/mol)	solubility (mg/L)
chloroethane	42	0.0085	5740
chloroform	34	0.0038	9300
1,1-dichloroethene	217	0.021	2730
cis-1,2-dichloroethene	34	0.0037	3500
trans-1,2-dichloroethene	39	0.38	6300
methylene chloride	25	0.0032	16 700
perchloroethene	303	0.018	150
trichloroethene	152	0.010	1080
vinyl chloride	8400	2.8	1100

Which compound in the table is likely to be most mobile in a soil-groundwater system?

(A) 1,1-dichloroethene

(B) methylene chloride

(C) perchloroethene

(D) vinyl chloride

10. Soil-water partition coefficients, Henry's constants, and solubility for selected chlorinated organic solvents are summarized in the table.

compound	K_{oc} (mL/g)	H (atm·m³/mol)	solubility (mg/L)
chloroethane	42	0.0085	5740
chloroform	34	0.0038	9300
1,1-dichloroethene	217	0.021	2730
cis-1,2-dichloroethene	34	0.0037	3500
trans-1,2-dichloroethene	39	0.38	6300
methylene chloride	25	0.0032	16 700
perchloroethene	303	0.018	150
trichloroethene	152	0.010	1080
vinyl chloride	8400	2.8	1100

Which compound in the table is most likely to partition to the vapor phase?

- (A) 1,1-dichloroethene
- (B) methylene chloride
- (C) perchloroethene
- (D) vinyl chloride

11. What is the primary mechanism responsible for weathering of heavier petroleum products, such as diesel fuel and lubricating oil, released to the environment?

- (A) biodegradation
- (B) dissolution
- (C) photolysis
- (D) volatilization

12. What is the primary mechanism responsible for weathering of lighter petroleum products, such as gasoline and aviation fuels, released to the environment?

- (A) biodegradation
- (B) dissolution
- (C) photolysis
- (D) volatilization

13. Which factors are most significant in reducing the degradation rate of petroleum products released to the marine environment?

- (A) dissolution
- (B) emulsification
- (C) storm activity
- (D) temperature increase

14. What distinguishes weathered petroleum products from those that are unweathered?

- (A) Weathering decreases the amount of lower molecular weight compounds.
- (B) Weathering decreases the bulk petroleum specific weight.
- (C) Weathering decreases the viscosity of residual petroleum.
- (D) Weathering increases the solubility of residual petroleum in water.

15. How does weathering affect toxicity of petroleum products released to the environment?

- (A) Weathering decreases toxicity through the loss of more soluble compounds.
- (B) Weathering decreases toxicity through increased mobility.
- (C) Weathering increases toxicity through enrichment of more complex compounds.
- (D) Weathering increases toxicity through the release of chloride and sulfur salts.

16. Bench test results for packing media evaluated to strip benzene from contaminated groundwater are summarized in the table. The tests are performed at 20°C.

elapsed time (min)	benzene concentration in sample (μg/L)
0	978
2	569
4	303
6	153
8	76
10	36

The mass transfer coefficient for the packing media and benzene at 8°C is most nearly

- (A) 0.19 min^{-1}
- (B) 0.23 min^{-1}
- (C) 0.30 min^{-1}
- (D) 0.40 min^{-1}

17. Toxicity values and concentrations for four chemicals found in a groundwater sample are shown in the table.

chemical	concentration (μg/L)	slope factor (mg/L)$^{-1}$	reference dose (mg/L)
1,1-DCE	173	1.7×10^{-2}	0.315
MeCl	207	2.1×10^{-4}	2.1
PCE	879	1.5×10^{-3}	0.35
1,1,2-TCA	764	1.6×10^{-3}	0.14

The total noncarcinogenic risk factor from exposure to the chemicals is most nearly

(A) 0.0054

(B) 0.10

(C) 1.3

(D) 8.6

18. Soil samples were collected below the water table from bore holes at a contaminated site. The soil effective porosity is 0.43, the soil bulk density is 1.68 g/cm^3, and the soil total organic carbon (TOC) is 271 mg/kg. The soil-water partition coefficient is 76.9 cm^3/g. The approximate velocity of the organic chemical relative to the groundwater velocity, v_{gw}, is most nearly

(A) $9.0 \times 10^{-7} v_{gw}$

(B) $3.3 \times 10^{-6} v_{gw}$

(C) $0.0033 v_{gw}$

(D) $0.95 v_{gw}$

19. Tetrachloroethene (PERC) has been discovered in the groundwater of an unconfined aquifer. The effective porosity and bulk density of the aquifer soil are 0.34 and 1.83 g/cm^3, respectively. The soil-water partition coefficient is 0.0539 mL/g. If the groundwater velocity is 0.23 m/d, the time it will take for the center of mass of the PERC plume to move 100 m from the source is most nearly

(A) 58 d

(B) 130 d

(C) 480 d

(D) 560 d

SOLUTIONS

1. Use the equations for specific discharge to find the average fuel velocity.

q = specific discharge

K = hydraulic conductivity for no. 2 fuel oil
$\quad = 0.081$ m/d

$\dfrac{dh}{dx}$ = gradient = 0.022

v = fuel velocity, m/d

n = effective porosity = 0.28

Specific Discharge

$$q = -K(dh/dx)$$

$$v = q/n = \dfrac{-K\left(\dfrac{dh}{dx}\right)}{n}$$

Take K as positive.

$$
\begin{aligned}
v &= \dfrac{K\left(\dfrac{dh}{dx}\right)}{n} \\
&= \dfrac{\left(0.081\ \dfrac{\text{m}}{\text{d}}\right)(0.022)}{0.28} \\
&= 6.4 \times 10^{-3}\ \text{m/d}
\end{aligned}
$$

The answer is (B).

2. Use the equation for finding the hydraulic conductivity to solve for the intrinsic permeability.

K = hydraulic conductivity with water as the fluid
$\quad = 0.42$ m/d

ρ = fluid density = 1000 kg/m^3 for water

g = gravitational acceleration = 9.81 m/s^2

k = intrinsic permeability, m^2

μ = dynamic viscosity
$\quad = 1.39 \times 10^{-3}$ kg/m·s for water at 8°C

Hydraulic Conductivity

$$K = \rho g k / \mu$$

$$k = \frac{K\mu}{\rho g}$$

$$= \frac{\left(0.42 \ \frac{m}{d}\right)\left(1.39 \times 10^{-3} \ \frac{kg}{m \cdot s}\right)\left(\frac{1 \ d}{86\,400 \ s}\right)}{\left(1000 \ \frac{kg}{m^3}\right)\left(9.81 \ \frac{m}{s^2}\right)}$$

$$= 6.9 \times 10^{-13} \ m^2$$

The answer is (D).

3. Find the hydraulic conductivity of the aquifer, K, for no. 2 fuel oil.

ρ = fluid density = 900 kg/m^3

g = gravitational acceleration = 9.81 m/s^2

k = intrinsic permeability = 6.9×10^{-13} m^2

μ = dynamic viscosity

$\quad = 6.5 \times 10^{-3}$ kg/m·s

Hydraulic Conductivity

$$K = \rho g k / \mu$$

$$= \frac{(6.9 \times 10^{-13} \ m^2)\left(900 \ \frac{kg}{m^3}\right)\left(9.81 \ \frac{m}{s^2}\right)}{\left(6.5 \times 10^{-3} \ \frac{kg}{m \cdot s}\right)\left(\frac{1 \ d}{86\,400 \ s}\right)}$$

$$= 0.081 \ m/d$$

The answer is (C).

4. The distribution coefficient for each test sample is the Cr(VI) adsorbed onto soil divided by the Cr(VI) concentration in leachate.

test sample	Cr(VI) concentration in leachate (mg/L)	Cr(VI) adsorbed onto soil (mg/kg)	distribution coefficient (mL/g)
1	31	17	0.55
2	20	10	0.50
3	13	7.5	0.58
4	10	7.0	0.70
5	7.1	4.5	0.63

The distribution coefficient for all test samples is

$$\frac{\left(\begin{array}{c} 0.55 \ \frac{mL}{g} + 0.50 \ \frac{mL}{g} + 0.58 \ \frac{mL}{g} \\ + 0.70 \ \frac{mL}{g} + 0.63 \ \frac{mL}{g} \end{array}\right)}{5} = 0.59 \ mL/g$$

Find the retardation factor.

R = retardation factor, unitless

ρ = soil density = 1.8 g/cm^3

η = saturated soil porosity = 0.34

K_d = distribution coefficient = 0.59 mL/g

Retardation Factor R

$$R = 1 + (\rho/\eta)K_d$$

$$= 1 + \frac{\left[1.8 \ \frac{g}{cm^3}\right]\left(0.59 \ \frac{mL}{g}\right)\left(\frac{1 \ cm^3}{1 \ mL}\right)}{0.34}$$

$$= 4.12 \quad (4.0)$$

The answer is (D).

5. Find the solute velocity.

v_c = solute velocity, m/d

v_x = water velocity = 0.37 m/d

R = retardation factor, unitless = 1.08

$$v_c = \frac{v_x}{R} = \frac{0.37 \ \frac{m}{d}}{1.08}$$

$$= 0.34 \ m/d$$

The answer is (B).

6. Find the distribution coefficient using the equation for the soil-water partition coefficient.

K_d = distribution coefficient, cm^3/g

K_{oc} = organic carbon partition coefficient = 200 cm^3/g

f_{oc} = fraction of organic carbon in the soil = 210 mg/kg

Soil-Water Partition Coefficient $K_d = K_p$

$$K_d = K_{oc}f_{oc} = \left(200 \ \frac{cm^3}{g}\right)\left(210 \ \frac{mg}{kg}\right)\left(10^{-6} \ \frac{kg}{mg}\right)$$

$$= 0.042 \ cm^3/g$$

Find the retardation factor.

R = retardation factor

ρ = soil bulk density = 1.85 g/cm^3

η = soil porosity = 0.26

Site Assessment
& Remediation

Retardation Factor R

$$R = 1 + (\rho/\eta)K_d = 1 + \left(\frac{1.85 \; \frac{g}{cm^3}}{0.26}\right)\left(0.042 \; \frac{cm^3}{g}\right)$$

$$= 1.3$$

The answer is (A).

7. Biological oxidation, chemical reduction, and hydrolysis are possible significant pathways for degradation of perchloroethene in a soil-groundwater system, but deamination is not.

The answer is (C).

8. The degradation sequence from the dichloroethene isomers would be to a monochloroethene. Therefore, each dichloroethene will degrade to vinyl chloride (CH_2-$CHCl$).

The answer is (D).

9. Mobility in a soil-groundwater system generally increases with decreasing values of the soil-water partition coefficient. The compound in the table with the lowest soil-water partition coefficient is methylene chloride.

The answer is (B).

10. The potential to partition to the vapor phase, or volatility, increases with increasing values of Henry's constant. The compound in the table with the greatest Henry's constant is vinyl chloride.

The answer is (D).

11. Biodegradation is the primary mechanism responsible for weathering of heavier petroleum products, such as diesel fuel and lubricating oil, released to the environment. Photolysis may also occur if heavy petroleum products are exposed to direct sunlight (but to a lesser degree), and dissolution and volatilization are minor mechanisms, if they occur at all.

The answer is (A).

12. The primary mechanism responsible for weathering of lighter petroleum products, such as gasoline and aviation fuels, released to the environment is volatilization to the atmosphere or, if the release is subsurface, to soil vapor. Dissolution may also occur, but would not be as significant as volatilization. Volatilization of lighter fractions would occur relatively rapidly and allow less time for significant biodegradation or, if exposed to direct sunlight, photolysis.

The answer is (D).

13. Degradation of petroleum products released to the marine environment is most significant through biological, photolytic, and vaporization pathways, all of which are enhanced by surface spreading of the hydrocarbon to increase surface area. Consequently, factors such as

emulsification that reduce surface area are most significant in reducing the degradation rate of petroleum products released to the marine environment.

The answer is (B).

14. Weathered petroleum products are most readily distinguished from those that are unweathered by the loss of lower molecular weight compounds that are more easily volatilized, biodegraded, and dissolved.

The answer is (A).

15. Weathering reduces the toxicity of petroleum products released to the environment by the loss of the more water soluble, lower molecular weight compounds.

The answer is (A).

16. Under conditions of gas transfer from the liquid to the vapor phase, the equation for first-order kinetics applies.

$$C_A = \text{chemical concentration corresponding}$$
$$\text{to any future time, } \mu g/L$$
$$C_{A0} = \text{chemical concentration } t = 0 \text{ min, } \mu g/L$$
$$K_L a = \text{mass transfer coefficient, min}^{-1}$$
$$t = \text{time, min}$$

First-Order Irreversible Reaction Kinetics

$$\ln(C_A / C_{A0}) = -kt$$

Substitute $K_L a$ for k.

$$-K_L a = \frac{\ln \dfrac{C_{A0}}{C_A}}{t}$$

elapsed time (min)	benzene concentration in sample ($\mu g/L$)	$\ln \dfrac{C_{A0}}{C_A}$	mass transfer coefficient (min^{-1})
0	978	–	–
2	569	0.542	−0.271
4	303	1.17	−0.293
6	153	1.86	−0.310
8	76	2.55	−0.319
10	36	3.30	−0.330
			−1.523

$n = $ number of $K_L a$ calculations

$$-K_L a = \frac{-\sum K_L a_i}{n} = \frac{1.523 \text{ min}^{-1}}{5}$$

$$= 0.30 \text{ min}^{-1}$$

Find the mass transfer coefficient at 8°C.

T = temperature, °C

θ = temperature correction coefficient = 1.024 (typical)

Kinetic Temperature Corrections

$$k_T = k_{20}(\theta)^{T-20}$$

$$-K_L a \text{ at } 8°C = (-K_L a \text{ at } 20°C)\theta^{T-20}$$
$$= (0.30 \text{ min}^{-1})(1.024^{8°C-20°C})$$
$$= 0.23 \text{ min}^{-1}$$

The answer is (B).

17. Calculate the noncarcinogenic risk factor, R, for each chemical.

$$R = \frac{C \times 10^{-3}}{RfD}$$

The noncarcinogenic risk factors are summarized in the table.

chemical	concentration, C (μg/L)	reference dose, RfD (mg/L)	risk factor, R
1,1-DCE	173	0.315	0.55
MeCl	207	2.1	0.099
PCE	879	0.35	2.5
1,1,2-TCA	764	0.14	5.5

The total noncarcinogenic risk factor is

$$0.55 + 0.099 + 2.5 + 5.5 = 8.6$$

The answer is (D).

18. Find the retardation factor.

R = retardation factor

ρ = bulk density = 1.68 g/cm^3

η = aquifer porosity = 0.34

Retardation Factor R

$$R = 1 + (\rho/\eta)K_d = 1 + \left(\frac{1.68 \dfrac{g}{cm^3}}{0.43}\right)\left(76.9 \dfrac{cm^3}{g}\right)$$

$$= 301$$

Find the velocity of the organic chemical relative to the groundwater velocity.

v_s = chemical velocity

v_{gw} = groundwater velocity

$$v_s = \frac{v_{gw}}{R} = \frac{v_{gw}}{301}$$
$$= 0.0033 v_{gw}$$

The answer is (C).

19. Find the retardation factor.

R = retardation factor

ρ = bulk density = 1.83 g/cm^3

η = 0.34

Retardation Factor R

$$R = 1 + (\rho/\eta)K_d$$
$$= 1 + \left(\frac{1.83 \dfrac{g}{cm^3}}{0.34}\right)\left(0.0539 \dfrac{mL}{g}\right)\left(1 \dfrac{cm^3}{mL}\right)$$

$$= 1.29$$

Find the contaminant velocity by dividing the groundwater velocity by the retardation factor.

v_c = contaminant velocity, m/d

v_{gw} = groundwater velocity, m/d

$$v_c = \frac{v_{gw}}{R}$$
$$= \frac{0.23 \dfrac{m}{d}}{1.29}$$
$$= 0.1783 \text{ m/d} \quad (0.18 \text{ m/d})$$

The time required to travel 100 m is

$$\frac{100 \text{ m}}{0.18 \dfrac{m}{d}} = 556 \text{ d} \quad (560 \text{ d})$$

The answer is (D).

Site Assessment & Remediation

50 Remediation Alternative Identification

PRACTICE PROBLEMS

1. Four alternative sites are being evaluated for construction of a RCRA-permitted hazardous waste landfill. The site characteristics are summarized as shown. WF is a weighting factor assigned to each criterion, scaled from 1 (most important) to 7 (least important). R is the site's rating for the given criterion, scaled from 1 (most satisfactory) to 7 (least satisfactory).

category	criteria	WF	site A R	site B R	site C R	site D R
soil	permeability	1	4	3	5	2
	heterogeneities	3	2	2	1	3
geology	seismic activity	3	5	5	5	4
groundwater	quality/use	7	6	4	3	5
	gradient/depth	5	7	4	4	3
hydrology	topography	7	3	5	2	4
	streams/lakes	4	4	3	6	4
community	population	3	6	7	3	3
	land uses	7	7	7	2	5

Which site best meets the desired criteria?

(A) site A

(B) site B

(C) site C

(D) site D

2. A lagooned bio-solid sludge contaminated with low concentrations of PCBs is currently being destroyed by incineration. Equipment is in place to dewater the sludge to 35% moisture and dry it to 5% moisture prior to incineration. Local community opposition to incineration has prompted the industry to investigate alternative destruction or disposal technologies. What resources would likely be most fruitful for identifying appropriate alternative technologies to incineration?

(A) alternative technology equipment vendor and trade publications

(B) publications of the USEPA Office of Research and Development

(C) publications of the USEPA Chemical Emergency Preparedness and Prevention Office

(D) publications of the USEPA Superfund Innovative Technology Evaluation (SITE) Program

3. Which of the following is NOT typically an important factor for *ex situ* bioremediation of petroleum contaminated soils?

(A) moisture content

(B) soil grain size distribution

(C) acclimated microbial population

(D) nutrient availability

4. Which of the following is NOT an appropriate *in situ* remedial alternative for heavier fuel oil-contaminated soils?

(A) thermal desorption and recovery

(B) ambient vapor extraction

(C) enhanced biodegradation

(D) containment

5. How is soil retention capacity for petroleum hydrocarbons affected by soil moisture?

(A) Retention capacity increases with increasing soil moisture.

(B) Retention capacity decreases with increasing soil moisture.

(C) Retention capacity increases as soil moisture approaches field capacity, then declines as soil moisture continues to increase.

(D) Retention capacity is not significantly influenced by soil moisture.

6. When might "no action" be acceptable as a remedial action alternative?

(A) when the contamination is widely dispersed, making remediation expensive

(B) when the contamination occurs in rural areas where property values are low

(C) when the contamination is confined and the contaminated soils are fine-grained

(D) when the contamination is volatile and the contaminated soils are coarse-grained

SOLUTIONS

1. Calculate the weighted ratings of the site characteristics.

$$\text{WR (weighted rating)} = (\text{WF})(R)$$

The site characteristics with weighted ratings are summarized as shown.

category	criteria	WF	site A R	WR	site B R	WR	site C R	WR	site D R	WR
soil	permeability	1	4	4	3	3	5	5	2	2
	heterogeneities	3	2	6	2	6	1	3	3	9
geology	seismic activity	3	5	15	5	15	5	15	4	12
groundwater	quality/use	7	6	42	4	28	3	21	5	35
	gradient/ depth	5	7	35	4	20	4	20	3	15
hydrology	topography	7	3	21	5	35	2	14	4	28
	streams/ lakes	4	4	16	3	12	6	24	4	16
community	population	3	6	18	7	21	3	9	3	9
	land uses	7	7	49	7	49	2	14	5	35
				206		189		125		161
	relative rating			1.65		1.51		1.00		1.29

Site C has the best rating.

The answer is (C).

2. Of the resources listed, publications of the USEPA Superfund Innovative Technology Evaluation (SITE) Program would be most fruitful for identifying appropriate alternative technologies to incineration. Publications of the USEPA Office of Research and Development and the USEPA Chemical Emergency Preparedness and Prevention Office would include limited, if any, information regarding alternative technologies. Alternative technology equipment vendor publications and trade publications typically do not provide objective information appropriate for evaluating alternative technologies.

The answer is (D).

3. Moisture content, an acclimated microbial population, and nutrient availability are all important factors for *ex situ* bioremediation of petroleum contaminated soils. Soil grain size distribution is much less important.

The answer is (B).

4. Heavier fuel oil-contaminated soils may be appropriately remediated by *in situ* alternatives such as thermal desorption and recovery, enhanced biodegradation, and containment. Ambient vapor extraction would not be effective for heavier fuel oils since they are low in volatile fractions.

The answer is (B).

5. Soil retention capacity for petroleum hydrocarbons decreases with increasing soil moisture.

The answer is (B).

6. "No action" may be acceptable as a remedial action alternative when the contamination is confined and the contaminated soils are fine-grained.

The answer is (C).

51 Remediation Technologies and Management

Content in blue refers to the *NCEES Handbook.*

PRACTICE PROBLEMS

1. An air stripping tower will treat 780 L/min of water at 15°C. The water contains benzene at a concentration of 1800 μg/L. The chemical must be removed to 1 μg/L. The tower uses a packing media with a mass transfer coefficient of 0.022 s^{-1}. The hydraulic loading rate is 1.2 m^3/m^2·min, and the stripping factor is 3.5. The required packing height is most nearly

(A) 5.6 m

(B) 6.8 m

(C) 9.1 m

(D) 10 m

2. Taste problems in a water supply have been traced to disinfection byproducts (DBP) at a cumulative concentration of 138 μg/L. For the Freundlich model, the intercept of the adsorption isotherm is 21 mg/g, and the slope is 0.54. The powdered activated carbon (PAC) dose required to reduce the DBP concentration to 5 μg/L is most nearly

(A) 110 mg/L

(B) 320 mg/L

(C) 1200 mg/L

(D) 2700 mg/L

3. A groundwater extraction system produces a flow of 2.0 m^3/min containing a mixture of five organic chemicals. The concentration of each chemical is shown in the table.

chemical	concentration (mg/L)
1,1,1-trichloroethane	2.90
trichloroethylene	1.40
1,1-dichloroethane	0.11
1,1-dichloroethylene	0.32
methylene chloride	0.61

Bench scale adsorption isotherm tests were conducted using the chemical mixture. The resulting isotherm equation is

$$\frac{X}{M} = 5.37 C_r^{0.58}$$

X is the chemical mass removed, and M is the GAC (granular activated carbon) mass used. The summed concentration of the five chemicals must be reduced to 0.005 mg/L in order to satisfy discharge requirements. The activated carbon use rate is most nearly

(A) 4.0 kg/d

(B) 7.0 kg/d

(C) 35 kg/d

(D) 62 kg/d

4. Granular activated carbon (GAC) is being considered for treatment of contaminated groundwater. The GAC use rate is 62 kg GAC/d. To ease operational requirements, the GAC change-out period should be between 60 and 90 days. If standard GAC adsorption vessel sizes are used, the GAC change-out period is most nearly

(A) 51 d

(B) 60 d

(C) 73 d

(D) 90 d

5. A granular activated carbon (GAC) vessel treating a contaminated groundwater flow of 2.0 m^3/min has a 4500 kg GAC capacity. For a GAC bulk density of 450 kg/m^3, the empty bed contact time is most nearly

(A) 2 min

(B) 5 min

(C) 10 min

(D) 12 min

SOLUTIONS

1. Calculate the number of transfer units.

$$NTU = \text{number of transfer units}$$
$$S = \text{stripping factor} = 3.5$$
$$C_o = \text{initial chemical concentration} = 1800 \ \mu g/L$$
$$C = \text{final chemical concentration} = 1 \ \mu g/L$$

$$NTU = \frac{S}{S-1} \ln \left(\frac{\left(\frac{C_o}{C} \right)(S-1)+1}{S} \right)$$

$$= \frac{3.5}{3.5-1} \ln \left(\frac{\left(\frac{1800 \ \frac{\mu g}{L}}{1 \ \frac{\mu g}{L}} \right)(3.5-1)+1}{3.5} \right)$$

$$= 10$$

Calculate the height of transfer units.

$$HLR_m = \text{mass hydraulic loading rate, kg/m}^2\text{·s}$$
$$HLR = \text{hydraulic loading rate} = 1.2 \ \text{m}^3/\text{m}^2\text{·min}$$
$$HTU = \text{height of transfer units, m}$$
$$K_L a = \text{mass transfer coefficient} = 0.022 \ \text{s}^{-1}$$
$$\rho_w = \text{water density} = 1000 \ \text{kg/m}^3$$

$$HLR_m = HLR \rho_w$$
$$HTU = \frac{HLR_m}{(K_L a)\rho_w} = \frac{HLR \rho_w}{(K_L a)\rho_w} = \frac{HLR}{K_L a}$$

$$= \frac{1.2 \ \frac{\text{m}^3}{\text{m}^2\text{·min}}}{(0.022 \ \text{s}^{-1}) \left(60 \ \frac{\text{s}}{\text{min}} \right)}$$

$$= 0.91 \ \text{m}$$

Calculate the packing height.

$$\text{packing height} = (NTU)(HTU) = (10)(0.91 \ \text{m})$$
$$= 9.1 \ \text{m}$$

The answer is (C).

2. The information given, such as the isotherm intercept and slope, suggests using the Freundlich isotherm equation. The alternative would be the Langmuir isotherm,

but it includes empirical constants and not the isotherm intercept and slope terms.

Use the Freundlich equation to find the carbon dose.

$$C_e = \text{DBP concentration remaining in solution, mg/L}$$
$$C_o = \text{DBP concentration before adsorption, mg/L}$$
$$q_e = \text{equilibrium loading on the PAC, mg chemical/g PAC}$$
$$K = \text{isotherm intercept, mg/g}$$
$$M = \text{powdered activated carbon (PAC) dose, g/L}$$
$$1/n = \text{isotherm slope}$$

Resistivity of a Medium

$$q_e = KC_e^{1/n} = \frac{C_o - C_e}{M}$$

$$M = \frac{C_o - C_e}{KC_e^{1/n}}$$

$$= \frac{\left(138 \ \frac{\mu g}{L} \right)\left(\frac{1 \ \text{mg}}{(10)^3 \ \mu g} \right) - \left(5 \ \frac{\mu g}{L} \right)\left(\frac{1 \ \text{mg}}{(10)^3 \ \mu g} \right)}{\left(21 \ \frac{\text{mg}}{\text{g}} \right)\left(\left(5 \ \frac{\mu g}{L} \right)\left(\frac{1 \ \text{mg}}{(10)^3 \ \mu g} \right) \right)^{0.54}}$$

$$= 0.11 \ \text{g/L water treated} \quad (110 \ \text{mg/L})$$

The answer is (A).

3. Find the adsorption capacity.

$$X = \text{chemical mass removed, kg}$$
$$M = \text{GAC mass used, kg}$$
$$C_r = 0.005 \ \text{mg/L}$$

$$\frac{X}{M} = 5.37 C_r^{0.58}$$

$$= (5.37)\left(0.005 \ \frac{\text{mg}}{L} \right)^{0.58}$$

$$= 0.25 \ \text{kg chemical/kg GAC}$$

Find the chemical mass removed.

$$\sum_{\text{influent concentration}} = 5.34 \ \text{mg/L}$$
$$X = \text{chemical mass removed}$$
$$= \sum_{\text{influent concentration}} - 0.005 \ \text{mg/L}$$
$$Q = \text{flow rate} = 2.0 \ \text{m}^3/\text{min}$$

$$X = \left(5.34 \ \frac{\text{mg}}{L} - 0.005 \ \frac{\text{mg}}{L} \right)\left(2.0 \ \frac{\text{m}^3}{\text{min}} \right)$$
$$\times \left(1000 \ \frac{L}{\text{m}^3} \right)\left(1440 \ \frac{\text{min}}{\text{d}} \right)\left(10^{-6} \ \frac{\text{kg}}{\text{mg}} \right)$$

$$= 15.4 \ \text{kg/d}$$

Find the GAC mass used.

$$M = \dfrac{15.4 \ \dfrac{\text{kg chemical}}{\text{d}}}{0.25 \ \dfrac{\text{kg chemical}}{\text{kg GAC}}}$$

$$= 61.6 \ \text{kg GAC/d} \quad (62 \ \text{kg/d})$$

The answer is (D).

4. The GAC change-out period is defined as from 60 d to 90 d. For 60 d, the required GAC mass is

$$\left(62 \ \dfrac{\text{kg GAC}}{\text{d}}\right)(60 \ \text{d}) = 3720 \ \text{kg GAC}$$

For 90 d, the required GAC mass is

$$\left(62 \ \dfrac{\text{kg GAC}}{\text{d}}\right)(90 \ \text{d}) = 5580 \ \text{kg GAC}$$

Select a standard GAC vessel with a capacity between 3720 kg GAC and 5580 kg GAC. The Calgon Model 7.5 has a GAC capacity of 4500 kg GAC. Therefore, select the Calgon Model 7.5 or equivalent.

$$\text{GAC capacity} = 4500 \ \text{kg GAC}$$

The GAC change-out period is

$$\dfrac{4500 \ \text{kg}}{62 \ \dfrac{\text{kg}}{\text{d}}} = 72.6 \ \text{d} \quad (73 \ \text{d})$$

The answer is (C).

5. The empty bed contact time (EBCT) is

$$\dfrac{4500 \ \text{kg}}{\left(450 \ \dfrac{\text{kg}}{\text{m}^3}\right)\left(2.0 \ \dfrac{\text{m}^3}{\text{min}}\right)} = 5 \ \text{min}$$

The answer is (B).

Topic V Environmental Health and Safety

52 Health and Safety

PRACTICE PROBLEMS

1. A 3.0 L air sample was collected under field conditions of 0.96 atm at 14°C. The sample volume under standard conditions is most nearly

(A) 2.74 L

(B) 2.97 L

(C) 3.03 L

(D) 3.29 L

2. The concentration of a gas in the atmosphere at 20°C and 1 atm is measured as 1.7×10^5 ppmv. The molecular weight of the gas is 84 g/mol. The concentration in units of mass/volume is most nearly

(A) 0.25 kg/m^3

(B) 0.59 kg/m^3

(C) 0.72 kg/m^3

(D) 0.86 kg/m^3

3. A work crew is exposed to benzene vapor at the concentrations and exposure durations shown.

duration (h)	concentration (mg/m^3)
2	110
3	260
3	320
1	180
1	80

The time-weighted average of the exposure is most nearly

(A) 190 mg/m^3

(B) 220 mg/m^3

(C) 230 mg/m^3

(D) 280 mg/m^3

4. Emergency planning and notification are required for facilities at which an amount of an extremely hazardous substance or CERCLA hazardous substance is present in excess of the regulated quantity. Emergency planning and notification requirements are specified in 40 CFR 355. When is emergency response notification NOT required?

(A) when the release occurs as abandonment of intact closed containers such as barrels

(B) when the release results in exposure to persons solely within the boundaries of the facility

(C) when the release does not involve a CERCLA hazardous substance

(D) when the release is less than the threshold planning quantity

5. Emergency planning and notification are required for facilities at which an amount of an extremely hazardous substance or CERCLA hazardous substance is present in excess of the regulated quantity. Emergency planning and notification requirements are specified in 40 CFR 355. What information is NOT required to be provided as part of emergency release notification?

(A) the identity or name of the released material

(B) an estimate of the quantity of the released material

(C) the name and title of the person reporting the release

(D) the time and duration of the release

6. Emergency planning and notification are required for facilities at which an amount of an extremely hazardous substance or CERCLA hazardous substance is present in excess of the regulated quantity. Emergency planning and notification requirements are specified in 40 CFR 355. What penalties may result from failure to report a release subject to the regulations?

(A) administrative penalties only

(B) administrative and civil penalties only

(C) monetary, civil, and administrative penalties only

(D) all allowable penalties, including criminal penalties

SOLUTIONS

1. Calculate the volume under standard conditions.

V_S = volume at standard conditions, L
V_F = volume at field conditions = 3.0 L
T_S = standard temperature = 0°C
T_F = field temperature = 14°C
p_s = standard pressure = 1.0 atm
p_F = field pressure = 0.96 atm

Correcting Gas Streams for Standard Conditions

$$Q_S = Q_A\left(\frac{P_A}{P_S}\right)\left(\frac{T_S}{T_A}\right)\left(\frac{1}{1-y_{H_2O}(g)}\right)$$

$$\times\left(\frac{12\%\ CO_2\ \text{by volume in dry gas stream}}{X\%\ CO_2\ \text{by volume in dry gas stream}}\right)$$

The standard unit for flue gas flow is CFM with other parameters expressed in U.S. units. Because the given information is limited to pressure and temperature, other parameters are ignored and with flow rate, Q, replaced by volume, V, the equation reduces to

$$V_S = V_F\left(\frac{p_F}{p_s}\right)\left(\frac{T_S}{T_F}\right)$$

$$= (3.0\ \text{L})\left(\frac{0.96\ \text{atm}}{1.0\ \text{atm}}\right)\left(\frac{0°C+273°}{14°C+273°}\right) = 2.74\ \text{L}$$

The answer is (A).

2. Calculate the mass/volume concentration. The ideal gas constant, R, is 0.08205 L·atm/mol·K. [Ideal Gas Constants]

C = concentration
M = molecular weight = 84 g/mol
p = pressure = 1 atm
T = temperature = 20°C + 273° = 293K

$$C_{kg/m^3} = \left(\frac{C_{ppmv}}{10^{-6}}\right)\left(\frac{MP}{RT}\right)$$

$$= \left(\frac{1.7\times10^5\ \text{ppmv}}{10^{-6}}\right)\left(\frac{\left(84\ \frac{g}{mol}\right)(1\ \text{atm})\times\left(1000\ \frac{L}{m^3}\right)}{\left(0.08205\ \frac{L\cdot atm}{mol\cdot K}\right)\times(293K)\left(1000\ \frac{g}{kg}\right)}\right)$$

$$= 0.59\ \text{kg/m}^3$$

The answer is (B).

3. Calculate the time-weighted average using each duration, t_i, and concentration, c_i.

Time-Weighted Average (TWA)

$$\text{TWA} = \frac{\sum_{t=1}^{n} c_i t_i}{\sum_{i=1}^{n} t_i}$$

$$= \frac{(2\ h)\left(110\ \frac{mg}{m^3}\right) + (3\ h)\left(260\ \frac{mg}{m^3}\right)}{2\ h + 3\ h + 3\ h + 1\ h + 1\ h}$$
$$+ (3\ h)\left(320\ \frac{mg}{m^3}\right)$$
$$+ (1\ h)\left(180\ \frac{mg}{m^3}\right) + (1\ h)\left(80\ \frac{mg}{m^3}\right)$$

$$= 222\ \text{mg/m}^3\quad(220\ \text{mg/m}^3)$$

The answer is (B).

4. As indicated in 40 CFR 355.31, emergency response notification is not required when the release results in exposure to persons solely within the boundaries of the facility.

The answer is (B).

5. The name and title of the person reporting the release are not included as part of the emergency release notification under 40 CFR 355.40.

The answer is (C).

6. Failure to report a release subject to the regulations may result in all of the sanctions allowed by CERCLA, including criminal penalties as specified in 40 CFR 302.7.

The answer is (D).

53 Security, Emergency Plans, Incident Response

gas	partial pressure (atm)
propane	0.49
methane	0.39
ethylene	0.12

PRACTICE PROBLEMS

1. A liquid has a flash point of 87°C. What is the United States Department of Transportation (USDOT) classification of the liquid?

(A) combustible

(B) flammable

(C) hazardous ignitability

(D) ignitable

2. What OSHA standard applies to worker safety during hazardous clean-up, emergency response, and corrective actions?

(A) 29 CFR 1910.119

(B) 29 CFR 1910.132(d)

(C) 29 CFR 1910.146

(D) 29 CFR 1910.1096

3. A gas mixture contains methane at 43% by volume, propane at 50% by volume, and ethylene at 7% by volume. The lower flammability limit for the mixture is most nearly

(A) 2.1

(B) 2.9

(C) 3.1

(D) 3.3

4. A mixture of gases is present in a confined space at 1 atm total pressure. The gases in the mixture are shown.

Most nearly, what is the lower flammability limit of the mixture?

(A) 2.5

(B) 2.8

(C) 3.3

(D) 3.5

Environmental Health & Safety

SOLUTIONS

1. Under Title 49 of the United States Code of Federal Regulations (49 CFR) 173.120(b), the United States Department of Transportation (USDOT) defines a combustible liquid as "any liquid that does not meet the definition of any other hazard class... and (1) has a flash point above 60.5°C and below 93°C, or (2) is a flammable liquid with a flash point at or above 37.8°C."

A USDOT flammable liquid is one having a flash point below 60.5°C, or at or above 37.8°C that is intentionally heated and offered for transportation or is transported at its flash point. There is no such thing as a "hazardous ignitability" or "ignitable" liquid.

The liquid is classified as combustible.

The answer is (A).

2. The OSHA regulation that provides standards for worker safety during hazardous clean-up, emergency response, and corrective actions is found in Title 29 of the United States Code of Federal Regulations (29 CFR) 1910.119.

29 CFR 1910.132(d) addresses personal protective equipment, 29 CFR 1910.146 addresses confined spaces, and 29 CFR 1910.1096 addresses radon in air.

The answer is (A).

3. The lower flammability limit (LFL) for each gas in the mixture is as shown. [Flammability]

gas	LFL
propane	2.1
methane	5
ethylene	2.7

Calculate the LFL for the mixture.

Predicting Lower Flammable Limits of Mixtures of Flammable Gases (Le Chatelier's Rule)

$$\text{LFL}_m = \frac{100}{\displaystyle\sum_{i=1}^{n}(C_{fi}/\text{LFL}_i)}$$

$$= \frac{100}{\left(\dfrac{43\%}{5}\right) + \left(\dfrac{50\%}{2.1}\right) + \left(\dfrac{7\%}{2.7}\right)}$$

$$= 2.9$$

The answer is (B).

4. From Dalton's law of partial pressure, the percent of the total volume occupied by each gas in the mixture is shown.

gas	partial pressure (atm)	volume (%)
propane	0.49	49
methane	0.39	39
ethylene	0.12	12

The lower flammability limit of each gas is shown. [Flammability]

gas	LFL
propane	2.1
methane	5
ethylene	2.7

LFL_m = lower flammability limit of the gas mixture
LFL_i = lower flammability limit of each gas
C_{fi} = volume percent of each gas in the mixture

Predicting Lower Flammable Limits of Mixtures of Flammable Gases (Le Chatelier's Rule)

$$\text{LFL}_m = \frac{100}{\displaystyle\sum_{i=1}^{n}(C_{fi}/\text{LFL}_i)}$$

$$= \frac{100}{\dfrac{49}{2.1} + \dfrac{39}{5} + \dfrac{12}{2.7}}$$

$$= 2.8$$

The answer is (B).

54 Codes, Standards, Regulations, Guidelines

Content in blue refers to the *NCEES Handbook*.

PRACTICE PROBLEMS

1. Which agency or organization provides research and recommendations for workplace exposure limits to protect workers?

(A) American Conference of Governmental Industrial Hygienists (ACGIH)

(B) U.S. Environmental Protection Agency (USEPA)

(C) National Institute for Occupational Safety and Health (NIOSH)

(D) Occupational Safety and Health Administration (OSHA)

2. The annual limit on the occupational radiation dose to an adult is the lower of (a) the total effective dose equivalent that is equal to 5 rems per year, or (b) the sum of the deep-dose equivalent and the committed dose equivalent to any organ or tissue that is equal to 50 rems per year. What is the occupational dose to minors?

(A) no acceptable occupational exposure

(B) 10% of the adult exposure

(C) 50% of the adult exposure

(D) same as the adult exposure

3. What hazards are represented in the fire/hazard diamond shown?

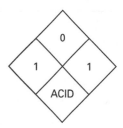

(A) normal material, will ignite if preheated, unstable if heated, acid

(B) normal material, will not burn, stable and nonreactive with water, acid

(C) slightly hazardous, will not burn, unstable if heated, acid

(D) stable and nonreactive with water, slightly hazardous, will not burn, acid

SOLUTIONS

1. The National Institute for Occupational Safety and Health (NIOSH) provides research and recommendations for workplace exposure limits, known as recommended exposure limits.

The American Conference of Governmental Industrial Hygienists (ACGIH) provides recommendations to industrial hygienists about workplace exposure. The U.S. Environmental Protection Agency (USEPA) provides exposure limits and risk factors for environmental cleanup projects. The Occupational Safety and Health Administration (OSHA) sets legally enforceable workplace exposure limits.

The answer is (C).

2. The annual occupational radiation dose for minors is limited to 10% of the adult annual occupational dose.

The answer is (B).

3. The fire/hazard diamond gives the following hazard data. [Hazard Assessment]

      position A = 1, slightly hazardous
      position B = 0, will not burn
      position C = 1, unstable if heated
      position D = acid

The answer is (C).

55 Industrial Hygiene

Content in blue refers to the *NCEES Handbook*.

PRACTICE PROBLEMS

1. What are the consequences if waste elemental alkali or alkaline earth metals are mixed with water or mixtures containing water?

(A) explosion

(B) flammable gas and heat generation

(C) innocuous and nonflammable gas generation

(D) toxic gas generation

2. What specific information is required in safety data sheet (SDS) Section 8 for exposure controls and personal protection?

(A) appropriate engineering controls, PELs, PPE, TLVs

(B) LELs/UELs, MCLs, PPE, TLVs

(C) LELs/UELs, PPE, TLVs, WBGT

(D) MCLs, PELs, PPE, WBGT

3. Which of the following is a characteristic of hydrogen sulfide that presents a hazard when encountered in poorly ventilated spaces?

(A) At concentrations above 10 ppm, it develops strong rotten egg odor causing a severe gag reflex.

(B) At concentrations above 100 ppm, it creates a smoky haze that limits visibility and distorts vision.

(C) Because it is heavier than air, it accumulates in lower areas where potential for contact increases.

(D) It forms sulfurous acid when mixed with water vapor present in humid environments.

4. Who is the target audience for OSHA regulations in Title 29 CFR 1910.120?

(A) employers of contaminated site workers

(B) local emergency planning committees

(C) individual citizens

(D) hazardous waste generators

5. Which of the following is NOT included in the OSHA description of a confined space?

(A) contains a material that has the potential to create a slip hazard or otherwise interferes with an entrant's ability to retain balance while standing

(B) contains a material that has the potential to engulf an entrant

(C) contains or has the potential to contain a hazardous atmosphere

(D) has walls that converge inward or floors that slope downward and taper into a smaller area that could trap or asphyxiate an entrant

6. A food processing facility operates seasonally where workers work two 10 hr shifts, 6 days per week, for 20 wk per year. During the last season, 17 work-related injuries and illnesses were reported among the facility's 439 workers. What is the total injury/illness incidence rate for the season?

(A) 1.6

(B) 2.6

(C) 3.2

(D) 3.9

7. Workers on 8 hr shifts are required to lift 600 items per hour. The vertical distance of the hands from the floor varies from 22 in to 28 in. What frequency multiplier must be used when calculating the recommended weight limit?

(A) 0.00

(B) 0.13

(C) 0.26

(D) 0.45

8. Workers wearing cotton coveralls are subjected to a work/rest regimen of 75% work, 25% rest each hour. The work load is moderate. What is the corrected permissible threshold limit value (TLV) for these workers?

(A) 79°F

(B) 80°F

(C) 82°F

(D) 86°F

SOLUTIONS

1. Mixing elemental alkali or alkaline earth metals with water or mixtures containing water may result in flammable gas and heat generation. [Hazardous Waste Compatibility Chart]

The answer is (B).

2. Section 8 requires appropriate engineering controls, permissible exposure limits (PELs), personal protective equipment (PPE), and threshold limit values (TLVs). [Safety Data Sheet (SDS)]

The answer is (A).

3. Hydrogen sulfide has the following characteristics. [Confined Space Safety]

- irritating at 10 ppm and deadly at 500 ppm

- accumulates at lower levels and in corners where circulation is minimal

- rotten egg odor

Of the four answer options, only option C accurately describes one of these characteristics.

The answer is (C).

4. The target audience for OSHA regulations in Title 29 CFR 1910.120 are employers of contaminated site workers.

The answer is (A).

5. OSHA describes a confined space as one that has any of the following characteristics. [Confined Space Safety]

- contains or has the potential to contain a hazardous atmosphere

- contains a material that has the potential to engulf an entrant

- has walls that converge inward or floors that slope downward and taper into a smaller area that could trap or asphyxiate an entrant

- contains any other recognized safety hazard such as unguarded machinery, exposed live wires, or heat stress

OSHA does not explicitly include a slip hazard in its description of a confined space.

The answer is (A).

6. Calculate T, the total hours worked by all employees during the season.

$$T = (2 \text{ shifts})\left(10 \ \frac{\text{hr}}{\text{worker-shift-day}}\right)$$
$$\times \left(6 \ \frac{\text{day}}{\text{wk}}\right)\left(20 \ \frac{\text{wk}}{\text{yr}}\right)(439 \text{ workers})$$
$$= 1{,}053{,}600 \text{ hr/yr}$$

$IR = $ total injury/illness incidence rate
$N = $ number of injuries and illnesses $= 17$

Incidence Rates
$$IR = N \times 200{,}000 \div T$$

In the calculation, 200,000 represents the total number of hours 100 employees typically work in a year. It is the standard base value.

$$IR = (17)\left(\frac{200{,}000}{1{,}053{,}600}\right)$$
$$= 3.2$$

The answer is (C).

7. Use the frequency multiplier table for the NIOSH formula for calculating the recommended weight limit.

$F = $ frequency $= 600$ items/hr $= 10$ items/min
  $= 10 \text{ min}^{-1}$
$V = $ vertical distance of the hands from the floor
  $= 22 \text{ in to } 28 \text{ in} \le 30 \text{ in}$

For an 8 hr shift, the frequency multiplier, FM, is 0.00. [Frequency Multiplier Table]

The answer is (A).

8. For a work/rest regimen of 75% work, 25% rest each hour, the TLV $= 28°$C for a moderate work load. When wearing clothing that provides insulation, such as cotton coveralls, the TLV is reduced by a clothing correction. [OSHA-Permissible Heat Exposure Threshold Limit Value]

Cotton coveralls require a correction of $-2°$C. [OSHA WBGT Correction Factors in °C]

The corrected permissible TLV $= 28°$C $- 2°$C $= 26°$C.

Temperature Conversions
$$(26°\text{C})(1.8) + 32° = 78.8°\text{F} \quad (79°\text{F})$$

The answer is (A).

Environmental Health & Safety

56 Exposure Assessments: Chemical and Biological

Content in blue refers to the *NCEES Handbook*.

PRACTICE PROBLEMS

1. A radiation exposure through ingestion results in an alpha particle absorbed dose of 0.012 rad and a beta particle absorbed dose of 0.32 rad. The radiation weighting factors for alpha and beta particles are 20 and 1, respectively. The dose equivalent from exposure to the alpha radiation is most nearly

(A)   0.012 rad

(B)   0.24 rem

(C)   0.32 rad

(D)   0.32 rem

2. A radiation exposure through ingestion results in an alpha particle absorbed dose of 0.012 rad and a beta particle absorbed dose of 0.32 rad. The radiation weighting factors for alpha and beta particles are 20 and 1, respectively. The dose equivalent from exposure to the beta radiation is most nearly

(A)   0.012 rad

(B)   0.24 rem

(C)   0.32 rad

(D)   0.32 rem

3. How are "absorbed dose" and "dose equivalent" related?

(A)   Absorbed dose and dose equivalent are synonymous terms used to describe radiation energy absorbed per unit time.

(B)   Dose equivalent is the product of absorbed dose and a unitless conversion factor specific to the type of radioactivity.

(C)   Dose equivalent is the sum of absorbed doses for all types of radioactivity to which exposure occurs.

(D)   They are unrelated; dose equivalent defines relative particle radioactivity and absorbed dose represents exposure.

4. The half-life for uranium-238 is 4.51×10^9 yr. The length of time required for 99% of the original uranium-238 atoms in a radioactive waste to degrade to another isotope is most nearly

(A)   20 min

(B)   25 d

(C)   10 800 yr

(D) 3.0×10^{10} yr

5. Rural drinking water supplies in some western United States communities draw from groundwater that contains naturally occurring arsenic and fluoride at average concentrations of 109 μg/L and 1.7 mg/L, respectively. The groundwater is also used for watering vegetable gardens.

Residents of these communities tend to live out their lives within a few miles of the homes of their grandparents, parents, and siblings. Commodities from home food production, including canned home-grown fruits and vegetables, make up a large part of their diet. Groundwater conditions and the toxicity characteristics of arsenic and fluoride are shown in the table.

For groundwater,

dissolved oxygen	0.8 mg/L
pH	8.0
temperature	16°C
specific conductivity	860 μS/cm
TDS	650 mg/L

For arsenic,

relative specie toxicity	As(V) less toxic than As(III)
MCL	10 μg/L
slope factor	5×10^{-5} $(\mu g/L)^{-1}$
oral RfD	1×10^{-3} mg/kg·d
bioconcentration factor	44 mL/g

For fluoride,

MCL	4 mg/L
oral RfD	6×10^{-2} mg/kg·d

Environmental Health & Safety

Is the arsenic likely to be present in the drinking water as less toxic arsenate (As(V)) or as more toxic arsenite (As(III))?

(A) arsenate, because the groundwater dissolved oxygen concentration suggests oxidizing conditions

(B) arsenate, because the groundwater dissolved oxygen concentration suggests reducing conditions

(C) arsenite, because the groundwater dissolved oxygen concentration suggests oxidizing conditions

(D) arsenite, because the groundwater dissolved oxygen concentration suggests reducing conditions

6. Rural drinking water supplies in some western United States communities draw from groundwater that contains naturally occurring arsenic and fluoride at average concentrations of 109 μg/L and 1.7 mg/L, respectively. The groundwater is also used for watering vegetable gardens.

Residents of these communities tend to live out their lives within a few miles of the homes of their grandparents, parents, and siblings. Commodities from home food production, including canned home-grown fruits and vegetables, make up a large part of their diet. Groundwater conditions and the toxicity characteristics of arsenic and fluoride are shown in the table.

For groundwater,

dissolved oxygen	0.8 mg/L
pH	8.0
temperature	16°C
specific conductivity	860 μS/cm
TDS	650 mg/L

For arsenic,

relative specie toxicity	As(V) less toxic than As(III)
MCL	10 μg/L
slope factor	5×10^{-5} $(\mu\text{g/L})^{-1}$
oral RfD	1×10^{-3} mg/kg·d
bioconcentration factor	44 mL/g

For fluoride,

MCL	4 mg/L
oral RfD	6×10^{-2} mg/kg·d

Is bioaccumulation of the arsenic in the residents' bodies over a lifetime of exposure likely to occur?

(A) Unknown; the bioconcentration factor applies to fish only and no conclusions can be made from the information provided regarding humans.

(B) No; although developed for exposure through ingestion of fish, the bioconcentration factor is low, suggesting unlikely bioaccumulation in humans through ingestion of water and vegetables.

(C) Yes, the bioconcentration factor suggests the potential for bioaccumulation and significant exposure exists through ingestion of water and vegetables.

(D) Yes, bioaccumulation always results from exposure to inorganic contaminants in drinking water and the food supply.

7. A stream is contaminated with a compound that will bioaccumulate in fish tissue according to the single compartment model. The contaminant is present in the stream water at a concentration of 184 μg/L. The uptake rate constant is 1.32 mL/g·h, and the depuration rate constant is 1.43×10^{-3} h^{-1}. The fish are exposed over a period of 100 d. The contaminant half-life in the fish tissue is most nearly

(A) 0.24 d

(B) 4.0 d

(C) 20 d

(D) 100 d

8. 96 h toxicity tests using rainbow trout were conducted over a range of six concentrations for two chemical herbicides. The results are shown in the table.

herbicide 1		herbicide 2	
concentration (μg/L)	survival (%)	concentration (μg/L)	survival (%)
1000	100	200	94
1200	88	220	82
1400	73	240	69
1600	58	260	53
1800	46	280	44
2000	34	300	29

Tissue concentrations for rainbow trout exposed to the herbicides are as shown.

	exposure concentration (μg/L)	tissue concentration (μg/kg)
herbicide 1	0.9	38
herbicide 2	0.1	64

The lethal concentration for 50% of the test species after a 96 h exposure (96 h LC_{50}) for each herbicide is most nearly

(A) 1685 μg/L, 254 μg/L

(B) 1706 μg/L, 273 μg/L

(C) 1733 μg/L, 267 μg/L

(D) 1837 μg/L, 271 μg/L

9. 96 h toxicity tests to assess the toxicity of two herbicides to rainbow trout produced LC_{50} results of 1200 μg/L for herbicide 1 and 500 μg/L for herbicide 2. Based on 96 h LC_{50}, which herbicide is more toxic?

(A) herbicide 1, because it has a lower LC_{50} than herbicide 2

(B) herbicide 1, because it has a higher LC_{50} than herbicide 2

(C) herbicide 2, because it has a lower LC_{50} than herbicide 1

(D) herbicide 2, because it has a higher LC_{50} than herbicide 1

10. 96 h toxicity tests using rainbow trout were conducted over a range of six concentrations for two chemical herbicides. The results are shown in the table.

herbicide 1		herbicide 2	
concentration (μg/L)	survival (%)	concentration (μg/L)	survival (%)
1000	100	200	94
1200	88	220	82
1400	73	240	69
1600	58	260	53
1800	46	280	44
2000	34	300	29

Tissue concentrations for rainbow trout exposed to the herbicides are as shown.

	exposure concentration (μg/L)	tissue concentration (μg/kg)
herbicide 1	0.9	38
herbicide 2	0.1	64

The bioconcentration factor for each herbicide is most nearly

(A) 0.024, 0.0016

(B) 34, 6.4

(C) 42, 640

(D) 240, 16

11. A stream is contaminated with a compound that will bioaccumulate in fish tissue according to the single compartment model. The contaminant is present in the stream water at a concentration of 184 μg/L. The uptake rate constant is 1.32 mL/g·h, and the depuration rate constant is 1.43×10^{-3} h^{-1}. The fish are exposed over a period of 100 d. The concentration of the contaminant in the fish tissue is most nearly

(A) 23 μg/g

(B) 150 μg/g

(C) 160 μg/g

(D) 200 μg/g

12. A stream is contaminated with a compound that will bioaccumulate in fish tissue according to the single compartment model. The contaminant is present in the stream water at a concentration of 184 μg/L and is present in the fish at 160 μg/g. The bioconcentration factor is most nearly

(A) 0.87

(B) 29

(C) 870

(D) 1200

13. The properties of polychlorinated biphenyls (PCBs) are shown.

$LD_{50} = 15\,000$ mg/kg

water solubility $= 0.031$ mg/L

organic carbon partition coefficient $= 5.3 \times 10^5$ mL/g

What do these characteristics reveal about PCBs?

(A) high human toxicity, high environmental mobility

(B) high human toxicity, low environmental mobility

(C) low human toxicity, high environmental mobility

(D) low human toxicity, low environmental mobility

Environmental Health & Safety

14. A workplace exposure survey has produced the following results.

	acetone	toluene
1 h exposure at concentration (ppm)	1210	NA
2 h exposure at concentration (ppm)	719	NA
3 h exposure at concentration (ppm)	NA	380
5 h exposure at concentration (ppm)	370	140
cumulative exposure (ppm)	560	230
8 h TWA PEL* (ppm)	1000	200

*TWA PEL is the time-weighted average peak exposure limit.

The cumulative exposure to the acetone and toluene mixture is most nearly

(A) 0.48

(B) 1.7

(C) 2.4

(D) 4.9

15. A workplace exposure survey has produced the following results.

	acetone	toluene
1 h exposure at concentration (ppm)	1210	NA
2 h exposure at concentration (ppm)	719	NA
3 h exposure at concentration (ppm)	NA	380
5 h exposure at concentration (ppm)	370	140

A peak toluene exposure of 480 ppm occurs for 10 min during the first 3 h. The cumulative exposure to toluene is most nearly

(A) 230 ppm

(B) 330 ppm

(C) 1000 ppm

(D) 3000 ppm

16. What are the human health effects of exposure to ammonia gas?

(A) no detectable effects to 10 ppm, irritating at 25 ppm, fatal above 500 ppm

(B) no detectable effects to 25 ppm, irritating at 50 ppm, fatal above 1000 ppm

(C) no detectable effects to 50 ppm, irritating at 50 ppm, fatal above 1000 ppm

(D) no detectable effects to 50 ppm, irritating at 100 ppm, fatal above 2000 ppm

SOLUTIONS

1. Find the dose equivalent from exposure to the alpha radiation.

h = dose equivalent, rem
D = absorbed dose = 0.012 rad
W_R = radiation weighting factor, unitless = 20

$$h = DW_R = (0.012 \text{ rad})(20)$$
$$= 0.24 \text{ rem}$$

The answer is (B).

2. Find the dose equivalent from exposure to the beta radiation.

h = dose equivalent, rem
D = absorbed dose = 0.32 rad
W_R = radiation weighting factor, unitless = 1

$$h = (0.32 \text{ rad})(1) = 0.32 \text{ rem}$$

The answer is (D).

3. Dose equivalent is the product of the absorbed dose and a unitless conversion factor, called the radiation weighting factor (previously called the "quality factor"), specific to the type of radioactivity.

The answer is (B).

4. Use the equation for half-life to find the rate constant.

τ = radioactive half-life = 4.51×10^9 yr
λ = rate constant, yr^{-1}

Half-Life (Radioactive Decay)

$$\tau = \frac{0.639}{\lambda}$$

$$\lambda = \frac{0.639}{\tau}$$

$$= \frac{0.639}{4.51 \times 10^9 \text{ yr}^{-1}}$$

$$= -1.54 \times 10^{-10} \text{ yr}$$

Find the time required for the uranium-238 atoms to degrade.

t = time, yr
k = rate constant = 1.54×10^{-10} yr
C_{A0} = initial concentration
C_A = concentration at 99% degradation
$$= C_{A0}(1 - 0.99) = 0.01 C_{A0}$$

First-Order Irreversible Reaction Kinetics

$$\ln(C_A/C_{A0}) = -kt$$

$$t = \frac{\ln\left(\dfrac{C_A}{C_{A0}}\right)}{k}$$

$$= \frac{\ln\left(\dfrac{0.01\,C_{A0}}{C_{A0}}\right)}{-1.54 \times 10^{-10}\ \text{yr}}$$

$$= 3.0 \times 10^{10}\ \text{yr}$$

The answer is (D).

5. The arsenic is likely to be present in the drinking water as the more toxic arsenite (As(III)) because the groundwater dissolved oxygen concentration of 0.8 mg/L suggests reducing conditions.

The answer is (D).

6. The bioconcentration factor is developed for exposures to fish, not direct ingestion of contaminated drinking water by humans. However, it does suggest that arsenic is absorbed by biological tissue. Considering the exposed dose through ingestion of drinking water and other exposure through ingestion of vegetables irrigated with the contaminated water, it is likely that some bioaccumulation of arsenic will occur in the residents' bodies over a lifetime of exposure.

The answer is (C).

7. Find the contaminant half-life.

τ = contaminant half-life in the fish tissue
λ = rate coefficient for contaminant depuration
$\quad = 1.43 \times 10^{-3}\ \text{h}^{-1}$

Half-Life (Radioactive Decay)

$$\tau = \frac{0.693}{\lambda}$$

$$= \frac{(0.693)\left(\dfrac{1\ \text{d}}{24\ \text{h}}\right)}{1.43 \times 10^{-3}\ \text{h}^{-1}}$$

$$= 20.2\ \text{d} \quad (20\ \text{d})$$

The answer is (C).

8. The 96 h LC_{50} is the median lethal concentration in water that is expected to kill 50% of a group of test animals when administered as a single exposure. [Dose-Response Curves]

For herbicide 1, at a concentration of 1600 μg/L, survival is 58%, and at a concentration of 1800 μg/L, survival is 46%. For herbicide 2, at a concentration of 260 μg/L, survival is 53%, and at a concentration of 280 μg/L, survival is 44%. For each herbicide, use interpolation to find the approximate concentration at which survival is 50%.

For herbicide 1, the 96 h LC_{50} is

$$1800\ \frac{\mu\text{g}}{\text{L}} - \frac{\left(1800\ \dfrac{\mu\text{g}}{\text{L}} - 1600\ \dfrac{\mu\text{g}}{\text{L}}\right)(50\% - 46\%)}{58\% - 46\%}$$

$$= 1733\ \mu\text{g/L}$$

For herbicide 2, the 96 h LC_{50} is

$$280\ \frac{\mu\text{g}}{\text{L}} - \frac{\left(280\ \dfrac{\mu\text{g}}{\text{L}} - 260\ \dfrac{\mu\text{g}}{\text{L}}\right)(50\% - 44\%)}{53\% - 44\%}$$

$$= 267\ \mu\text{g/L}$$

The answer is (C).

9. Herbicide 2 is more toxic than herbicide 1 because herbicide 2 has a lower 96 h LC_{50} than does herbicide 1.

The answer is (C).

10. Find the bioconcentration factor.

BCF = bioconentration factor
C_{org} = equilibirum concentration in organism
$\quad C$ = concentration in water

Bioconcentration Factor BCF

$$BCF = C_{org}/C$$

For herbicide 1, the bioconcentration factor is

$$BCF_1 = \frac{C_{org}}{C}$$

$$= \frac{38\ \dfrac{\mu\text{g}}{\text{kg}}}{\left(0.9\ \dfrac{\mu\text{g}}{\text{L}}\right)\left(1\ \dfrac{\text{L}}{\text{kg}}\right)}$$

$$= 42$$

For herbicide 2, the bioconcentration factor is

$$BCF_2 = \frac{C_{org}}{C}$$

$$= \frac{64\ \dfrac{\mu\text{g}}{\text{kg}}}{\left(0.1\ \dfrac{\mu\text{g}}{\text{L}}\right)\left(1\ \dfrac{\text{L}}{\text{kg}}\right)}$$

$$= 640$$

The answer is (C).

11. Find the bioconcentration. Since the BOD exertion equation defines a relationship between two concentrations (y_t or concentration C_{t+1} at time $t+1$, and L or

Environmental Health & Safety

concentration C_t and time t), the BOD exertion equation can be adapted to find the bioconcentration.

C_b = contaminant concentration in biota (fish), μ/g

C_m = contaminant concentration in medium (water)
= 184 μg/L

K_u = rate coefficient for contaminant uptake
= 1.32 mL/g·h

K_d = rate coefficient for contaminant depuration
= 1.43×10^{-3} h^{-1}

$t = (100 \text{ d})\left(24\ \dfrac{\text{h}}{\text{d}}\right) = 2400 \text{ h}$

BOD Exertion

$$y_t = L(1 - e^{-kt})$$

$$C_{t+1} = C_t(1 - e^{-kt})$$

$$C_b = C_m\left(\frac{K_u}{K_d}\right)(1 - e^{-K_d t})$$

$$= \frac{\left(184\ \frac{\mu\text{g}}{\text{L}}\right)\left(\dfrac{1.32\ \frac{\text{mL}}{\text{g·h}}}{1.43 \times 10^{-3}\ \text{h}^{-1}}\right)}{1000\ \dfrac{\text{mL}}{\text{L}}}$$

$$\times \left(1 - e^{-(1.43 \times 10^{-3}\ \text{h}^{-1})(2400\ \text{h})}\right)$$

$$= 164\ \mu\text{g/g} \quad (160\ \mu\text{g/g})$$

The answer is (C).

12. Find the bioconcentration factor. 1000g/L is the density of water.

BCF = bioconcentration factor, unitless

C_{org} = equilibrium concentration in organism = 160 μg/g

C = concentration in water = 184 μg/L

Bioconcentration Factor BCF

$$BCF = C_{\text{org}}/C$$

$$= \frac{\left(160\ \frac{\mu\text{g}}{\text{g}}\right)\left(1000\ \frac{\text{g}}{\text{L}}\right)}{184\ \frac{\mu\text{g}}{\text{L}}}$$

$$= 870$$

The answer is (C).

13. The LD_{50} for PCBs indicates a relatively low toxicity. PCBs are less toxic than ethanol ($LD_{50} = 10\,000$ mg/kg) and table salt ($LD_{50} = 4000$ mg/kg). For comparison, more toxic substances include strychnine ($LD_{50} = 2$ mg/kg) and botulinum toxin ($LD_{50} = 0.00001$ mg/kg). [Comparative Acutely Lethal Doses]

The low water solubility indicates movement with groundwater to be minimal, especially when coupled with the high organic carbon partition coefficient. The organic carbon partition coefficient indicates a PCB affinity for solid surfaces—a high adsorption potential.

In general, PCBs exhibit low toxicity to humans and low environmental mobility.

The answer is (D).

14. Calculate the cumulative exposure to the acetone and toluene mixture.

E_c = cumulative exposure for acetone and toluene, ppm

E_m = cumulative exposure to mixture, unitless

L = TWA PEL for acetone and for toluene, ppm

$$E_m = \frac{E_{c,\text{acetone}}}{L_{\text{acetone}}} + \frac{E_{c,\text{toluene}}}{L_{\text{toluene}}}$$

$$= \frac{560 \text{ ppm}}{1000 \text{ ppm}} + \frac{230 \text{ ppm}}{200 \text{ ppm}}$$

$$= 1.7$$

The answer is (B).

15. Find the time-weighted average.

TWA = time-weighted average (cumulative exposure), ppm

t_i = exposure duration of ith interval, h

c_i = concentration during ith interval, ppm

Time-Weighted Average (TWA)

$$TWA = \frac{\sum_{t=1}^{n} c_i t_i}{\sum_{i=1}^{n} t_i}$$

$$= \frac{c_1 t_1 + c_2 t_2 + c_3 t_3}{t_1 + t_2 + t_3}$$

$$= \frac{(480 \text{ ppm})\left(\dfrac{10 \text{ min}}{60 \dfrac{\text{min}}{\text{h}}}\right) + (380 \text{ ppm})\left(3 \text{ h} - \dfrac{10 \text{ min}}{60 \dfrac{\text{min}}{\text{h}}}\right) + (140 \text{ ppm})(5 \text{ h})}{8 \text{ h}}$$

$$= 232 \text{ ppm} \quad (230 \text{ ppm})$$

The answer is (A).

16. Ammonia is irritating at 50 ppm and lethal above 1000 ppm. [Confined Space Safety]

Ammonia has a threshold limit value (TLV) of 25 ppm. The TLV is the highest dose that the human body is able to detoxify without detectable effects. [Threshold Value]

The answer is (B).

Environmental Health & Safety

57 Exposure Assessments: Sound and Radiation

Content in blue refers to the *NCEES Handbook*.

PRACTICE PROBLEMS

1. What radioactive particle is characterized by high energy and very small mass?

(A) alpha particles

(B) beta particles

(C) gamma particles

(D) delta particles

2. What units are used to express radioactivity?

(A) rem

(B) rad

(C) curie

(D) roentgen

3. In what order would travel distances of radiation through air increase?

(A) alpha < beta < gamma

(B) beta < alpha < gamma

(C) gamma < beta < alpha

(D) gamma < alpha < beta

4. The characteristics of a workplace exposure to Co-60 through ingestion of drinking water are shown.

concentration	1.6×10^{-11} Ci/L
ingestion rate	1.4 L/d
exposure frequency	200 d/yr
exposure duration	30 yr

Assume acceptable risk is 1 in 1 million and acceptable occupational exposure is 5 rem.

The intake of Co-60 through ingestion of the drinking water is most nearly

(A) 1.3×10^{-7} Ci

(B) 1.6×10^{-7} Ci

(C) 3.1×10^{-7} Ci

(D) 5.6×10^{-7} Ci

5. The intake of Co-60 through ingestion of drinking water is 1.3×10^{-7} Ci and the whole-body committed effective dose-equivalent per unit intake is 9.5×10^6 mrem/Ci. The whole-body committed effective dose equivalent from the exposure is most nearly

(A) 1.2 mrem

(B) 1.5 mrem

(C) 2.9 mrem

(D) 5.3 mrem

6. The intake of Co-60 through ingestion of drinking water is 1.3×10^{-7} Ci and the slope factor is 1.5×10^{-11} pCi^{-1}. The total lifetime cancer risk from the exposure is most nearly

(A) 0.93×10^{-6}

(B) 2.0×10^{-6}

(C) 4.4×10^{-6}

(D) 7.9×10^{-6}

7. If exposure to Co-60 at a dose of 1.2 mrem results in a lifetime cancer risk of 1.8×10^{-6}, is the exposure level acceptable?

(A) Yes, the risk is less than one in one million, and the whole-body dose is less than 5 rem.

(B) No, the risk is less than one in one million, but the whole-body dose is greater than 5 rem.

(C) No, the risk is greater than one in one million even though the whole-body dose is less then 5 rem.

(D) No, the risk is greater than one in one million and the whole-body dose is greater than 5 rem.

8. Noise at a construction site located approximately 100 m from a residential area is characterized by a sound power of 50 W and a frequency of 1000 Hz. The reference sound power is 10^{-12} W. The sound power level is most nearly

(A) 100 dB

(B) 120 dB

(C) 130 dB

(D) 140 dB

9. Noise at a construction site located approximately 100 m from a residential area is characterized by a sound power of 50 W and a frequency of 1000 Hz. Assuming isotropic radiation of sound, the sound intensity in the residential area is most nearly

(A) 4.0×10^{-4} W/m^2

(B) 1.6×10^{-3} W/m^2

(C) 5.0×10^{-3} W/m^2

(D) 6.4×10^{-3} W/m^2

10. Noise at a construction site located approximately 100 m from a residential area is characterized by a sound power of 50 W and a frequency of 1000 Hz. The reference sound intensity is 10^{-12} W/m^2. The sound intensity is 4.0×10^{-4} W/m^2. The sound intensity level in the residential area is most nearly

(A) 86 dB

(B) 92 dB

(C) 96 dB

(D) 98 dB

11. Workers in a manufacturing facility are subjected to a noise level of 88 dB at a frequency of 1000 Hz. Over what time period of exposure will the average worker experience a 5 dB net hearing loss?

(A) never (there is no measurable hearing loss at the exposure level)

(B) 3 yr

(C) 8 yr

(D) 9 yr

12. Strontium-90 has a radioactive half-life of 28.8 yr and a biological half-life of 49.3 yr. Most nearly, what is the effective half-life of strontium-90?

(A) 18.2 yr

(B) 28.8 yr

(C) 39.1 yr

(D) 49.3 yr

SOLUTIONS

1. Alpha particles are relatively low energy particles that can be stopped by thick sheets of paper. Beta particles have higher energy than alpha particles but can be stopped by the equivalent of thin aluminum sheeting. Gamma particles, however, are characterized by high energy and very small mass and can only be stopped by lead sheeting or thick concrete.

The answer is (C).

2. The unit used to express radioactivity is the curie, Ci. Roentgen, R, is used for exposure, rad is used for absorbed dose, and rem is used for dose-equivalent.

The answer is (C).

3. In air, alpha particles can travel a few inches, beta particles can travel up to about 30 m, and gamma radiation can travel several thousand meters.

The answer is (A).

4. Find the intake of Co-60.

$$I = \text{intake, Ci}$$
$$C = \text{concentration} = 1.6 \times 10^{-11} \text{ Ci/L}$$
$$R_I = \text{ingestion rate} = 1.4 \text{ L/d}$$
$$f_E = \text{exposure frequency} = 200 \text{ d/yr}$$
$$D_t = \text{exposure duration} = 30 \text{ yr}$$

$$I = CR_I f_E D_t$$
$$= \left(1.6 \times 10^{-11} \frac{\text{Ci}}{\text{L}}\right)\left(1.4 \frac{\text{L}}{\text{d}}\right)\left(200 \frac{\text{d}}{\text{yr}}\right)(30 \text{ yr})$$
$$= 1.34 \times 10^{-7} \text{ Ci} \quad (1.3 \times 10^{-7} \text{ Ci})$$

The answer is (A).

5. Find the whole-body committed effective dose-equivalent from the exposure.

$$H_{E,50} = \text{whole-body committed effective}$$
$$\text{dose equivalent, mrem}$$
$$h_{E,50} = \text{whole-body committed effective}$$
$$\text{dose equivalent per unit intake} =$$
$$9.5 \times 10^6 \text{ mrem/Ci}$$
$$I = 1.3 \times 10^{-7} \text{ Ci}$$

$$H_{E,50} = Ih_{E,50}$$
$$= (1.3 \times 10^{-7} \text{ Ci})\left(9.5 \times 10^6 \frac{\text{mrem}}{\text{Ci}}\right)$$
$$= 1.24 \text{ mrem} \quad (1.2 \text{ mrem})$$

The answer is (A).

6. Find the risk from exposure.

$$\text{SF} = \text{slope factor} = 1.5 \times 10^{-11} \text{ pCi}^{-1}$$
$$I = \text{intake} = 1.3 \times 10^{-7} \text{ Ci}$$

$$\text{risk} = (\text{SF})I$$
$$= \left(1.5 \times 10^{-11} \text{ pCi}^{-1}\right)(1.3 \times 10^{-7} \text{ Ci})\left(10^{12} \frac{\text{pCi}}{\text{Ci}}\right)$$
$$= 1.95 \times 10^{-6} \quad (2.0 \times 10^{-6})$$

The answer is (B).

7. No, the exposure level is not acceptable. The whole-body dose of 1.2 mrem is less than the acceptable occupational exposure of 5 rem, but the risk of 1.8 in one million is greater than the acceptable risk of one in one million.

The answer is (C).

8. Find the sound pressure level, measured in decibels.

Noise Pollution
$$\text{SPL (dB)} = 10 \log_{10}(P^2/P_0^2)$$

Use a form of this equation modified for sound power to find the sound power level relative to the reference sound power of 10^{-12} W. Modifying for sound power transforms the P^2 aspects of the equation representing sound pressure to W representing sound power. This is to account for the difference between how sound pressure changes as it travels outward from its origin point while sound power does not.

$$L_W = 10\log_{10}\left(\frac{W_{\text{Watt}}}{W_{0,\text{Watt}}}\right)$$
$$= 10\log_{10}\left(\frac{50 \text{ W}}{10^{-12} \text{ W}}\right)$$
$$= 137 \text{ dB} \quad (140 \text{ dB})$$

The answer is (D).

9. With isotropic radiation, sound radiates from the source in all directions. Find the area of air receiving noise normal to the direction of propagation.

$$A = 4\pi r^2 = 4\pi(100 \text{ m})^2$$
$$= 125\,600 \text{ m}^2$$

Find the sound intensity in the residential area.

$$I = \text{sound intensity, W/m}^2$$
$$W = \text{sound power} = 50 \text{ W}$$

Environmental Health & Safety

$$I = \frac{W}{A} = \frac{50 \text{ W}}{125\,600 \text{ m}^2}$$
$$= 3.98 \times 10^{-4} \text{ W/m}^2 \quad (4.0 \times 10^{-4} \text{ W/m}^2)$$

The answer is (A).

10. Find the sound pressure level, measured in decibels.

Noise Pollution

$$\text{SPL (dB)} = 10 \log_{10}(P^2/P_0^2)$$

Use a form of this equation modified for sound intensity to find the sound intensity level relative to the reference sound intensity of 10^{-12} W/m^2.

$$\text{SPL} = 10 \log_{10}\left(\frac{I}{I_0}\right)$$
$$= 10 \log_{10}\left(\frac{4.0 \times 10^{-4} \dfrac{\text{W}}{\text{m}^2}}{10^{-12} \dfrac{\text{W}}{\text{m}^2}}\right)$$
$$= 86 \text{ dB}$$

The answer is (A).

11. From an estimated average trend curve for net hearing loss, for a frequency of 1000 Hz, 5 dB net hearing loss along the 88 dB curve occurs at about 8 yr of exposure. [Hearing]

The answer is (C).

12. Use the equation for effective half-life.

τ_e = effective half-life, yr
τ_r = radioactive half-life = 28.8 yr
τ_b = biological half-life = 49.3 yr

Effective Half-Life

$$\frac{1}{\tau_e} = \frac{1}{\tau_r} + \frac{1}{\tau_b}$$
$$= \frac{1}{28.8 \text{ yr}} + \frac{1}{49.3 \text{ yr}}$$
$$= 0.055 \text{ yr}^{-1}$$
$$\tau_e = \frac{1}{0.055 \dfrac{1}{\text{yr}}} = 18.2 \text{ yr}$$

The answer is (A).

58 Indoor Air Quality

Content in blue refers to the *NCEES Handbook*.

PRACTICE PROBLEMS

1. What are the symptoms of low-level indoor formaldehyde exposure?

   (A)   lung cancer

   (B)   cancer of the mucosa of the eyes, mouth, and nose

   (C)   headaches, sinus congestion, depression

   (D)   hyperactivity, increased appetite, hair loss

2. What is the primary source of asbestos and formaldehyde in indoor air?

   (A)   movement of individuals into and out of buildings

   (B)   materials used in building construction

   (C)   ventilation systems with outside air intakes

   (D)   furniture and office supplies

3. When does asbestos present a health risk?

   (A)   in any form when accessible for human contact

   (B)   in friable form

   (C)   in nonfriable form

   (D)   in unprocessed mineral form only

4. What are the effects of nonoccupational asbestos exposure?

   (A)   lung cancer and mesothelioma

   (B)   cancer of the mucosa of the eyes, mouth, and nose

   (C)   asbestosis

   (D)   headache, sinus congestion, depression

Environmental Health & Safety

5. The ventilation system for a 40,000 ft² floor area manufacturing space with 20 ft ceilings theoretically replaces the air volume in the space once every 10 minutes. The removal reaction rate constant for the space is 0.64 hr⁻¹. Most nearly, what is the actual air volume replacement interval for the space?

(A) 6.6 min

(B) 7.4 min

(C) 9.0 min

(D) 15 min

6. What is the primary source of radon gas in most buildings?

(A) masonry building materials

(B) soil surrounding foundations and crawl spaces

(C) concrete walls and slabs

(D) lead containing paints and sealers

7. At what concentration in the indoor air does the U.S. Environmental Protection Agency recommend immediate action to mitigate radon exposure?

(A) 4 pCi/L

(B) 20 pCi/L

(C) 100 pCi/L

(D) 200 pCi/L

8. What is the dominant mechanism for radon transport into buildings?

(A) It is carried with individuals as they move from one location to another.

(B) It is the density differential between radon gas and air.

(C) It is the presence of forced air ventilation systems.

(D) It is the pressure differential between inside and outside air.

9. What radioactive particles are associated with radon?

(A) alpha particles

(B) beta particles

(C) gamma particles

(D) delta particles

SOLUTIONS

1. Symptoms of low-level formaldehyde exposure include headaches, sinus congestion, and depression.

The answer is (C).

2. Building materials are the primary source of asbestos and formaldehyde in indoor air.

The answer is (B).

3. Asbestos presents a health risk when it is in friable form, allowing it to potentially become airborne.

The answer is (B).

4. Nonoccupational exposure to asbestos can cause lung cancer and mesothelioma. Asbestosis is typically associated with occupational exposure.

The answer is (A).

5. Use the equation for the time constant.

$$Q = \text{ventilation rate, ft}^3/\text{hr}$$
$$V = \text{room volume, ft}^3$$
$$\tau = \text{system time constant, hr}$$
$$k = \text{removal reaction rate constant} = 0.64 \text{ hr}^{-1}$$

Indoor Air Quality

$$\tau = \left(\frac{Q}{V} + k\right)^{-1}$$

t = theoretical air volume replacement interval = 10 min

$$Q = \frac{V}{t}$$

$$\frac{Q}{V} = \frac{1}{t}$$

$$\tau = \left(\frac{1}{t} + k\right)^{-1} = \cfrac{1}{\cfrac{1}{10 \text{ min}} + \cfrac{0.64 \, \dfrac{1}{\text{hr}}}{60 \, \dfrac{\text{min}}{\text{hr}}}}$$

$$= 9.04 \text{ min} \quad (9.0 \text{ min})$$

The answer is (C).

6. The primary source of radon gas in most buildings is soil surrounding foundations and crawl spaces.

The answer is (B).

7. The USEPA recommends immediate action to mitigate radon exposure when concentrations exceed 4 pCi/L. [Standards and Regulations for Radon in Air]

The answer is (A).

8. The dominant mechanism for radon transport into buildings is the pressure differential between inside and outside air.

The answer is (D).

9. The radioactive particles associated with radon are alpha particles.

The answer is (A).

Topic VI Associated Engineering Principles

59 Statistics

Content in blue refers to the *NCEES Handbook*.

PRACTICE PROBLEMS

1. How many permutations are possible for all single-digit positive integers?

(A) 40,320

(B) 45,360

(C) 181,440

(D) 362,880

2. There are 10 competitors in an event. How many ways can the competitors finish in first, second, and third place?

(A) 3.3

(B) 30

(C) 720

(D) 5040

3. How many distinguishable ways can the letters in "engineer" be written?

(A) 336

(B) 3360

(C) 5040

(D) 40,320

4. Most nearly, how many different ways can two items be selected from a population of eight?

(A) 4

(B) 16

(C) 28

(D) 56

5. Out of a population of 1200 students, 875 are residents, and 80 have a GPA of 3.5 or greater. Among the 80 students with a GPA of 3.5 or greater, 52 are residents. If a student is selected at random, the probability that they will be a resident, or have a GPA of 3.5 or greater, or be both a resident and have a GPA of 3.5 or greater is most nearly

(A) 0.73

(B) 0.75

(C) 0.80

(D) 0.84

6. What is the expected value for the data set in the table?

x_k	$f(x_k)$
141	0.1
277	0.5
518	0.2
759	0.1

(A) 188

(B) 332

(C) 349

(D) 397

7. The table shown gives the probable costs of a necessary repair to a vital system in a wastewater treatment plant.

repair cost ($)	probability
73	0.2
187	0.4
312	0.3
501	0.1

What is most nearly the probability that the cost of repairs will be less than $312?

(A) 0.2

(B) 0.4

(C) 0.6

(D) 0.7

8. The probability density function of a component failure in 10^4 operating cycles is given by the function shown.

$$f(x) = \frac{2}{(x+1)^3}$$

The probability that the component will fail after 2×10^4 operating cycles is most nearly

(A) 0.074

(B) 0.25

(C) 0.89

(D) 0.93

9. The length of a given group of samples is normally distributed with a mean of 182 cm and a standard deviation of 25 cm. The probability that a sample from the group will be longer than 209 cm is most nearly

(A) 14%

(B) 15%

(C) 23%

(D) 28%

10. A system for the storage and handling of hazardous materials is characterized by the reliability ratings shown. Components occur in process sequence from A to B to C.

component	type	reliability
A	series	0.93
B	series	0.97
C	series	0.96

The overall reliability of the system is most nearly

(A) 0.87

(B) 0.90

(C) 0.93

(D) 0.95

11. A system for the storage and handling of hazardous materials is characterized by the reliability ratings shown. Components occur in process sequence from A to B to C.

component	type	reliability
A	series	0.98
B1	parallel	0.83
B2	parallel	0.81
C	series	0.98

The overall reliability of the system is most nearly

(A) 0.83

(B) 0.86

(C) 0.89

(D) 0.93

12. A system for the storage and handling of hazardous materials is characterized by the reliability ratings shown. Components occur in process sequence from A to B.

component	type	reliability
A	series	0.95
B1	parallel	0.78
B2	parallel	0.80
B3	parallel	0.79

If the reliabilities for the system are based on 1000 h of operation, what is the failure rate for the system?

(A) 6.1×10^{-5} failures/h

(B) 9.2×10^{-5} failures/h

(C) 1.1×10^{-4} failures/h

(D) 1.8×10^{-4} failures/h

13. A hydrologic basin is characterized by a historical rainfall record that is thought to be normally distributed with a mean of 48 in and a standard deviation of 11.2 in. What are the three ideal coordinates that would represent the fit of the data to a normal distribution?

(A) (0.5, 48), (0.159, 59.2), (0.841, 36.8)

(B) (0.5, 48), (0.841, 59.2), (0.159, 36.8)

(C) (1.0, 48), (1.682, 59.2), (0.318, 36.8)

(D) (1.0, 48), (1.682, 70.4), (0.318, 25.6)

Assoc. Eng. Principles

STATISTICS

14. The expected value for the data in the table shown is 332.

x_k	$f(x_k)$
141	0.1
277	0.5
518	0.2
759	0.1

The standard deviation for the data is most nearly

(A)   155

(B)   174

(C)   196

(D)   216

SOLUTIONS

1. The single-digit positive integers are 1, 2, 3, 4, 5, 6, 7, 8, and 9.

$$P(n,n) = n!$$
$$n = 9$$
$$P(n,n) = 9!$$
$$= 362{,}880$$

The answer is (D).

2. The number of permutations of the top three positions is

$$n = 10$$
$$r = 3$$

Permutations and Combinations

$$P(n,r) = \frac{n!}{(n-r)!}$$
$$= \frac{10!}{(10-3)!}$$
$$= 720$$

The answer is (C),

3. The number of possible permutations is

$$n = 8$$
$$n_e = 3$$
$$n_n = 2$$
$$n_g = 1$$
$$n_i = 1$$
$$n_r = 1$$

Permutations and Combinations

$$P(n; n_1, n_2, \ldots n_k) = \frac{n!}{n_1! \, n_2! \ldots n_k!}$$
$$= \frac{8!}{(3!)(2!)(1!)(1!)(1!)}$$
$$= 3360$$

The answer is (B).

4. The number of different combinations is

$$n = 8$$
$$r = 2$$

Assoc. Eng.
Principles

PPI ● **ppi2pass.com**

Permutations and Combinations

$$C(n,r) = \frac{n!}{[r!\,(n-r)!]}$$

$$= \frac{8!}{2!(8-2)!}$$

$$= 28$$

The answer is (C).

5. The total probability is

$$n_r = 875$$
$$n_{3.5} = 80$$
$$n_{r,3.5} = 52$$
$$n_S = 1200$$

Laws of Probability: Property 2. Law of Total Probability

$$P(A+B) = P(A) + P(B) - P(A,B)$$

$$P(\text{resident}) = \frac{875}{1200} = 0.729$$

$$P(3.5\ \text{GPA}) = \frac{80}{1200} = 0.0667$$

$$P\binom{\text{resident and}}{3.5\ \text{GPA}} = \frac{52}{1200} = 0.0433$$

$$P\begin{pmatrix}\text{resident, 3.5 GPA,}\\ \text{or resident and}\\ 3.5\ \text{GPA}\end{pmatrix} = 0.729 + 0.0667 - 0.0433$$

$$= 0.752 \quad (0.75)$$

The answer is (B).

6. The expected value is

Expected Values

$$\mu = E[X] = \sum_{k=1}^{n} x_k f(x_k)$$

$$= (0.1)(141) + (0.5)(277) + (0.2)(518) + (0.1)(759)$$
$$= 332$$

The answer is (B).

7. The probability the cost of repairs will be less than $312 is a cumulative distribution function.

Cumulative Distribution Functions

$$F(x_m) = P(X \le x_m) \text{ for } -\infty < x_m < \infty$$

$-\infty < x_m < \$73$	$F(x_m) = P(\text{cost} \le x_m) = 0$
$\$73 \le x_m < \$187$	$F(x_m) = P(\text{cost} \le x_m) = 0.2$
$\$73 \le x_m < \$312$	$F(x_m) = P(\text{cost} \le x_m) = 0.2 + 0.4 = 0.6$
$\$73 \le x_m < \$501$	$F(x_m) = P(\text{cost} \le x_m) = 0.2 + 0.4 + 0.3 = 0.9$
$\$73 \le x_m < \infty$	$F(x_m) = P(\text{cost} \le x_m) = 0.2 + 0.4 + 0.3 + 0.1 = 1.0$

The probability that the cost of repairs will be less than $312 is 0.6.

The answer is (C).

8. The probability that the component will fail after 2×10^4 operating cycles is

Probability Density Function

$$P(a \le X \le b) = \int_a^b f(x)\,dx$$

$$P(0 < X < 2) = \int_0^2 \frac{2}{(x+1)^3} = 0.89$$

The answer is (C).

9. Find the standard normal value, Z.

$$\mu = \text{population mean} = 182\ \text{cm}$$
$$\sigma = \text{standard deviation} = 25\ \text{cm}$$

Normal Distribution (Gaussian Distribution)

$$Z = \frac{x - \mu}{\sigma} = \frac{209 - 182}{25} = 1.08$$

The equation for the probability that Z is greater than or equal to 1.08 (i.e., the probability that a sample is longer than 209 cm) is

$$P(Z \ge 1.08) = 1 - F(1.08)$$

From a unit normal distribution table, the value of $F(1.08)$ falls between 0.8413 for an x-value of 1.0, and 0.8643 for an x-value of 1.1. [Unit Normal Distribution]

Find the exact value using interpolation.

$$0.8413 + \left(\frac{1.0 - 1.08}{1.0 - 1.1}\right)(0.8643 - 0.8413) = 0.8597$$

Calculate the probability.

$$P(Z \geq 1.08) = 1 - F(1.08)$$
$$= 1 - 0.8597$$
$$= 0.1403 \quad (14\%)$$

The answer is (A).

10. R_A, R_B, and R_C are the reliability of components A, B, and C, respectively. The overall system reliability is

$$R_{\text{system}} = R_A R_B R_C$$
$$= (0.93)(0.97)(0.96)$$
$$= 0.87$$

The answer is (A).

11. R_A, R_B, and R_C are the reliability of components A, B, and C, respectively. The reliability of stage B is

$$R_B = 1 - (1 - R_{B1})(1 - R_{B2})$$
$$= 1 - (1 - 0.83)(1 - 0.81)$$
$$= 0.97$$

Find the overall system reliability.

$$R_{\text{system}} = R_A R_B R_C$$
$$= (0.98)(0.97)(0.98)$$
$$= 0.93$$

The answer is (D).

12. Use the equation for reliability, and rearrange to solve for the failure rate.

$$R_t = \text{reliability at time } t$$
$$\lambda = \text{failure rate}$$

$$R_t = e^{-\lambda t}$$
$$\lambda = -\frac{\ln R_t}{t}$$

For component A,

$$\lambda = -\frac{\ln 0.95}{1000 \text{ h}} = 5.13 \times 10^{-5} \text{ failures/h}$$

The reliability of component B is

$$R_B = 1 - (1 - R_{B1})(1 - R_{B2})(1 - R_{B3})$$
$$= 1 - (1 - 0.78)(1 - 0.80)(1 - 0.79)$$
$$= 0.99$$

The failure rate for component B is

$$\lambda = -\frac{\ln 0.99}{1000 \text{ h}} = 1.00 \times 10^{-5} \text{ failures/h}$$

The system failure rate is

$$5.13 \times 10^{-5} \frac{\text{failures}}{\text{h}} + 1.00 \times 10^{-5} \frac{\text{failures}}{\text{h}}$$
$$= 6.13 \times 10^{-5} \text{ failures/h} \quad (6.1 \times 10^{-5} \text{ failures/h})$$

The answer is (A).

13. The probability scale plots at the standardized mean and one standard deviation above and below the mean.

From a unit normal distribution table, for $x = 1.0$, the value of the unit standard deviation above and below the mean is 0.6827. Therefore, one unit standard deviation either above or below the mean is 0.341.

The x-values for coordinates are

$$0.5$$
$$0.5 + 0.341 = 0.841$$
$$0.5 - 0.341 = 0.159$$

The rainfall scale plots at the mean for the data and one standard deviation for the data above and below the mean.

The y-values for coordinates are

$$48$$
$$48 + 11.2 = 59.2$$
$$48 - 11.2 = 36.8$$

The plotting coordinates are

$$(0.5, 48)$$
$$(0.841, 59.2)$$
$$(0.159, 36.8)$$

The answer is (B).

14. Calculate the variance of X, $V[X]$.

$$\mu = \text{expected value of } X, E[X]$$
$$\sigma^2 = \text{variance}$$

Expected Values

$$\sigma^2 = V[X] = \sum_{k=1}^{n}(x_k - \mu)^2 f(x_k)$$

$$= (141 - 332)^2(0.1) + (277 - 332)^2(0.5)$$

$$+ (518 - 332)^2(0.2) + (759 - 332)^2(0.1)$$

$$= 30{,}313$$

Calculate the standard deviation.

σ = standard deviation

Expected Values

$$\sigma = \sqrt{V[X]} = \sqrt{30{,}313} = 174$$

The answer is (B).

60 Sustainability

Content in blue refers to the *NCEES Handbook*.

PRACTICE PROBLEMS

1. What type of water right is most common in areas of the U.S. where water availability is limited?

- (A) contestable
- (B) littoral
- (C) prior appropriation
- (D) riparian

SOLUTIONS

1. Prior appropriation water rights are most common in the western U.S., where water availability is limited. Prior appropriation rights give precedence to whomever was the first to put the water to beneficial use. These rights exist as quantifiable shares and, as such, may be bought and sold.

Riparian or littoral water rights are most common in areas of the U.S. where water availability is not limited. They are typical of the eastern U.S. Riparian rights allow anyone whose property abuts a waterway to use the water. These rights are conveyed with the land.

The answer is (C).

61 Economics: Cash Flow, Interest, Decision-Making

PRACTICE PROBLEMS

1. Bench studies have determined that aluminum sulfate (alum) promotes acceptable floc formation at a dose of 23 mg/L when applied to a surface water source. The water demand is $19\,000$ m³/d. Alum is available from one supplier at 17% purity for $234/1000 kg or from a second supplier at 26% purity at $319/1000 kg. The project has a 10-year life. The interest rate is 4% from the first supplier and 6% from the second supplier. By comparing costs at the end of the project life, the price difference between the two suppliers is most nearly

(A) less than $30,000

(B) between $30,000 and $50,000

(C) between $50,000 and $70,000

(D) more than $70,000

2. A precious metals mine produces $20\,000$ m³/d of wastewater containing free cyanide (CN^-) at 1400 mg/L. The wastewater meets discharge criteria if the free cyanide is oxidized to cyanate (CNO^-). The daily mass of chlorine required to treat the wastewater is $77\,000$ kg. Chlorine gas at a purity of 99.8% is available at $584 for 1000 kg. The price includes transportation to the site. The oxidation treatment process will operate continuously 24 h per day and 365 d per year. The project uses a fixed amount of chlorine each year. This year, the cost of the chlorine used is $16,446,212. The cost of chlorine is expected to increase each year by an amount equal to 4% of this year's cost. For a net financing cost of 6%, the total present worth of the chlorine cost over 12 years is most nearly

(A) $27,000,000

(B) $138,000,000

(C) $164,000,000

(D) $201,000,000

3. A precious metals mine produces $20\,000$ m³/d of wastewater containing free cyanide (CN^-) at 1400 mg/L. The wastewater can meet discharge criteria if the free cyanide is oxidized to cyanate (CNO^-). The daily mass of ozone required to treat the water is $52\,000$ kg. The ozone generator used for treatment produces 3.5% O_3, with a power requirement of 14 kW·h/kg of ozone produced. Electrical power costs $0.042/kW·h. The oxidation treatment process will operate continuously 24 h per day and 365 d per year. For a net annual financing cost of 6%, the uniform annual cost for ozone over the first three years of the project is most nearly

(A) $7,700,000/yr

(B) $11,000,000/yr

(C) $17,000,000/yr

(D) $22,000,000/yr

4. To offset energy costs, natural gas use at a wastewater treatment plant will be augmented with methane gas recovered from the plant's anaerobic digesters. A mixture of 70% methane and 30% natural gas will be used. The plant currently pays $0.020/ft³ for natural gas. The expected operating and amortized capital cost of digester gas to scrub the methane before blending is $0.0080/ft³, and the blending will continue full time 365 days per year. Total gas production by the digesters is expected to peak at 38,000 ft³/day, 65% of which will be methane. Most nearly, what will be the approximate annual present worth savings over a 10–year period at 4% annual inflation rate over the current natural gas costs if all the available methane gas, when blended with the natural gas, precisely meets the needs of the plant?

(A) $560,000

(B) $830,000

(C) $890,000

(D) $1,300,000

5. Bench studies have determined that aluminum sulfate (alum) promotes acceptable floc formation at a dose of 23 mg/L when applied to a surface water source. Alum is available at 17% purity for $234/1000 kg and electrical power costs $0.05/kW·h. The power required at 85% efficiency is 2.6×10^4 N·m/s. If the motor efficiency is 85%, the total monthly cost for electricity is most nearly

(A) $80/mo

(B) $470/mo

(C) $790/mo

(D) $940/mo

SOLUTIONS

1. Find the monthly demand for alum from supplier 1.

$$C = \left(19\,000\ \frac{m^3}{d}\right)\left(23\ \frac{mg}{L}\right)\left(1000\ \frac{L}{m^3}\right)$$
$$\times \left(30\ \frac{d}{mo}\right)\left(10^{-6}\ \frac{kg}{mg}\right)$$
$$= 13\,110\ kg/mo$$

Find the monthly cost for alum.

$$\left(13\,110\ \frac{kg}{mo}\right)\left(\frac{\$234}{1000\ kg}\right)\left(\frac{100\%}{17\%}\right)$$
$$= \$18,046/mo$$

Find the monthly cost of alum from supplier 2.

$$\left(13\,110\ \frac{kg}{mo}\right)\left(\frac{\$319}{1000\ kg}\right)\left(\frac{100\%}{26\%}\right) = \$16,085/mo$$

Calculate the total future worth for each supplier using the uniform series compound factor. [Economics]

$$F = (F/A,\ i\%,\ n)A$$

For supplier 1, $i = 4\%$, $n = 10$ yr. [Factor Table]

$$F_1 = (12.0061)\left(\frac{\$18,046}{mo}\right)\left(12\ \frac{mo}{yr}\right) = \$2,599,945$$

For supplier 2, $i = 6\%$, $n = 10$ yr. [Factor Table]

$$F_2 = (13.1808)\left(\frac{\$16,085}{mo}\right)\left(12\ \frac{mo}{yr}\right) = \$2,544,158$$

The difference in price over the life of the project is

$$\$2,599,945 - \$2,544,158 = \$55,787$$

The difference in price is $57,787 in favor of supplier 2.

The answer is (C).

2. The annual cost of chlorine is

$$\left(77\,000\ \frac{kg}{d}\right)\left(\frac{\$584}{1000\ kg}\right)\left(365\ \frac{d}{yr}\right)\left(\frac{100\%}{99.8\%}\right)$$
$$= \$16,446,212/yr$$

In order to use standard cash flow factors, divide the annual cost into two portions, one portion being an annual cost of $16,446,212 and the other portion being the increase in cost, which is zero in year 1 and increases

Assoc. Eng.
Principles

by 4% of $16,446,212 each year. Calculate the present worth of these two portions separately, then add them together to get the total present worth. The amount of the annual cost increase at 4% of the current cost is

$$\left(\frac{4\%}{100\%}\right)(\$16,446,212) = \$657,848$$

Calculate the total present worth of the cost increase using the uniform gradient present worth compound factor. [Economics]

$$P = (P/G,\ i\%,\ n)\,G$$

Find the present worth for $i = 6\%$ and $n = 12$ yr. [Factor Table]

$$P = (40.3369)(\$657,848) = \$26,535,549$$

Calculate the total present worth of the annual cost using the uniform series present worth compound factor. [Economics]

$$P = (P/A,\ i\%,\ n)\,A$$

Find the present worth for $i = 6\%$ and $n = 12$ yr. [Factor Table]

$$P = (8.3838)(\$16,446,212) = \$137,881,752$$

The total present worth of the chlorine cost is

$$\$137,881,752 + \$26,535,549$$
$$= \$164,417,301 \quad (\$164,000,000)$$

The answer is (C).

3. The annual cost of ozone is

$$\left(52\,000\ \frac{\text{kg}}{\text{d}}\right)\left(14\ \frac{\text{kW·h}}{\text{kg}}\right)\left(\frac{\$0.042}{\text{kW·h}}\right)\left(365\ \frac{\text{d}}{\text{yr}}\right)$$
$$= \$11,160,240/\text{yr}$$

Calculate the total present worth of the costs for the first three years using the single payment present worth factor. Values for the factor can be found in a table of equivalent cash flow factors. [Factor Table]

Economics

$$P = (P/F, i\%, n)$$
$$P_1 = (P/F, 6\%, 1)F$$
$$= (0.9434)(\$11,160,240)$$
$$= \$10,528,570$$
$$P_2 = (P/F, 6\%, 2)F$$
$$= (0.8900)(\$11,160,240)$$
$$= \$9,932,614$$
$$P_3 = (P/F, 6\%, 3)F$$
$$= (0.8396)(\$11,160,240)$$
$$= \$9,370,138$$
$$P_{\text{total}} = P_1 + P_2 + P_3$$
$$= \$10,528,570 + \$9,932,614$$
$$\qquad + \$9,370,138$$
$$= \$29,831,322$$

Calculate the uniform annual cost from the total present worth using the capital recovery factor. Values for the factor can be found in a table of equivalent cash flow factors. [Factor Table]

Economics

$$A = (A/P, i\%, n)$$
$$A = (A/P, 6\%, 3)P_{\text{total}}$$
$$= (0.3741)(\$29,831,322)$$
$$= \$11,159,898 \quad (\$11,000,000/\text{yr})$$

The answer is (B).

4. The natural or blended gas needed is

$$\frac{\left(38,000\ \dfrac{\text{ft}^3}{\text{day}}\right)(0.65)}{0.70} = 35,286\ \text{ft}^3/\text{day}$$

The current natural gas cost is

$$\left(35,286\ \frac{\text{ft}^3}{\text{day}}\right)\left(\frac{\$0.020}{\text{ft}^3}\right) = \$706/\text{day}$$

The blended natural gas needed is

$$\left(35,286\ \frac{\text{ft}^3}{\text{day}}\right)(0.30) = 10,586\ \text{ft}^3/\text{day}$$

The blended natural gas cost is

$$\left(10,586\ \frac{\text{ft}^3}{\text{day}}\right)\left(\frac{\$0.020}{\text{ft}^3}\right) = \$212/\text{day}$$

Assoc. Eng. Principles

The scrubber cost is

$$\left(38{,}000 \ \frac{\text{ft}^3}{\text{day}}\right)\left(\frac{\$0.0080}{\text{ft}^3}\right) = \$304/\text{day}$$

The total cost for blended gas is

$$\frac{\$212}{\text{day}} + \frac{\$304}{\text{day}} = \$516/\text{day}$$

The gas blending continues full time for 365 day/yr, so the annual savings realized is

$$\left(\frac{\$706}{\text{day}} - \frac{\$516}{\text{day}}\right)\left(365 \ \frac{\text{day}}{\text{yr}}\right)$$
$$= \$69{,}350/\text{yr} \quad (\$69{,}000/\text{yr})$$

Calculate the present worth for the annual cost using the capital recovery factor.

$$i = \text{annual interest rate} = 4\%$$
$$n = \text{term} = 10 \ \text{yr}$$
$$A = \text{first year annual cost (savings)} = \$69{,}000$$

Economics

$$A = (P/A, i\%, n)$$

From a table of cash flow equivalent factors, the value of P/A when $n = 10$ is 8.1109. [Factor Table]

$$(P/A, 4\%, 10) = 8.1109$$
$$P = A(P/A, 4\%, 10)$$
$$= (\$69{,}000)(8.1109)$$
$$= \$559{,}652 \quad (\$560{,}000)$$

The answer is (A).

5. Find the total monthly cost for electricity. As 1 N·m/s $= 1$ W,

$$C = \frac{(2.6 \times 10^4 \ \text{W})\left(\left(\dfrac{\$0.05}{\text{kW·h}}\right)\left(24 \ \dfrac{\text{h}}{\text{d}}\right)\left(30 \ \dfrac{\text{d}}{\text{mo}}\right)\right)}{1000 \ \dfrac{\text{W}}{\text{kW}}}$$
$$= \$936/\text{mo} \quad (\$940/\text{mo})$$

The answer is (D).

62 Economics: Capitalization, Depreciation, Accounting

PRACTICE PROBLEMS

1. Electrical use by a pump operating at 100% efficiency is 2.6×10^4 N·m/s. The cost of electricity is $0.10/kW·h. For a motor efficiency of 87% and a net interest rate of 8%, the amount of money that should be reserved today to cover the electrical costs over the next five years is most nearly

(A) $91,000

(B) $105,000

(C) $110,000

(D) $135,000

2. A city purchased refuse collection trucks for $1.5M. The purchase is financed over an eight-year term at an annual interest rate of 6%. The trucks will have $130,000 salvage value at the end of the eight-year term. The annual maintenance costs are shown in the following table.

year	cost
1	$45,000
2	$48,000
3	$51,000
4	$54,000
5	$57,000
6	$60,000
7	$63,000
8	$66,000

How much money should the city borrow to finance the project over the eight-year term?

(A) $1,750,000

(B) $1,800,000

(C) $1,850,000

(D) $1,950,000

3. A developer is offered property at a price of $560,000. She can finance the property with a 10% down payment and equal monthly payments of $12,700 over a five-year period. The effective interest rate of the transaction is most nearly

(A) 1.08%

(B) 1.33%

(C) 1.47%

(D) 2.65%

4. Earth moving equipment is purchased for $280,000 with no money for the down payment. The buyer pays for the equipment in 60 equal monthly payments at an annual interest rate of 6% compounded continuously. The monthly payment is most nearly

(A) $4,810

(B) $4,950

(C) $5,440

(D) $8,060

5. A maintenance facility is constructed at a cost of $7,800,000 with financing at 4% interest over a five-year term. The facility has the following anticipated annual operating costs also financed at a 4% annual interest rate.

year	cost
1	$430,000
2	$440,000
3	$480,000
4	$510,000
5	$560,000

The equivalent uniform annual cost of the facility is most nearly

(A) $701,000

(B) $894,000

(C) $1,185,000

(D) $2,233,000

6. Compare two alternatives. The first alternative has a capital cost of $1,200,000 and an annual operation and maintenance cost of $140,000. The second alternative has a capital cost of $1,600,000 and an annual operation and maintenance cost of $95,000. After the first year, the operation and maintenance cost will increase annually by 5%. For a 15-year project term and an annual interest rate of 4%, which alternative results in the least present worth cost?

(A) alternative 1 has the least present worth by $40,000

(B) alternative 1 has the least present worth by $63,000

(C) alternative 2 has the least present worth by $232,000

(D) alternative 2 has the least present worth by $275,000

7. An investment of $230,000 will pay an annual annuity of $45,600 over a six-year period. If the MARR is 8%, is the investment acceptable?

(A) acceptable, $i > 8\%$

(B) not acceptable, $6\% < i < 8\%$

(C) not acceptable, $4\% < i < 6\%$

(D) not acceptable, $i < 4\%$

SOLUTIONS

1. The total annual cost for electricity is

$$\frac{\left(2.6 \times 10^4 \; \dfrac{\text{N·m}}{\text{s}}\right)\left(\dfrac{\$0.10}{\text{kW·hr}}\right)\left(8760 \; \dfrac{\text{hr}}{\text{yr}}\right)}{\left(\dfrac{1000 \; \dfrac{\text{N·m}}{\text{s}}}{\text{kW}}\right)\left(\dfrac{87\%}{100\%}\right)} = \$26{,}179/\text{yr}$$

Calculate the total present worth of the annual cost using the uniform series present worth factor. [Economics]

$$P = (P/A, \; i\%, \; n)A$$

Find the present worth for $i = 8\%$ and $n = 5$ yr. [Factor Table]

$$P = (3.9927)(\$26{,}179) = \$104{,}525 \quad (\$105{,}000)$$

The answer is (B).

2. The gradient is uniform at $3000/yr with a recurring annual cost of $42,000. [Economics] [Factor Table]

$$
\begin{aligned}
P &= (P/A, \; 6\%, \; 8 \; \text{yr})A + (P/G, \; 6\%, \; 8 \; \text{yr})G \\
&\quad + \text{first year cost} + \text{capital cost} \\
&\quad - (P/F, \; 6\%, \; 8 \; \text{yr})F \\
&= (6.2098)(\$42{,}000) + (19.8416)(\$3000) \\
&\quad + \$42{,}000 + \$1{,}500{,}000 \\
&\quad - (0.6274)(\$130{,}000) \\
&= \$1{,}786{,}980 \quad (\$1{,}800{,}000)
\end{aligned}
$$

The answer is (B).

3. Calculate the financed amount.

$$\$560{,}000 - (0.10)(\$560{,}000) = \$504{,}000$$

Use the uniform series present worth factor to find the effective interest rate. [Economics]

$$(P/A, \; i\%, \; 60 \; \text{mo})$$

$$\frac{P}{A} = \frac{\$504{,}000}{\$12{,}700} = 39.6850$$

Use factor tables to find i for $P/A = 39.6850$ and $n = 60$.

Using a factor table, $i = 1.50\%$ at 60 months gives $P/A = 39.3803$.

Using a factor table, $i = 1.00\%$ at 60 months gives $P/A = 44.9550$.

The desired value lies between 1.00% and 01.50 %. By interpolation,

$$i = 1.00\% + \frac{(1.50\% - 1.00\%)(39.6850 - 44.9550)}{39.3803 - 44.9550}$$

$$= 1.4727\% \quad (1.47\%)$$

The answer is (C).

4. For continuous compounding, calculate the annual effective interest rate.

$$i_e = e^i - 1 = e^{0.06} - 1 = 0.06184$$

Calculate the monthly interest rate.

$$i_{\text{monthly}} = \frac{0.06184}{12} = 0.005153$$

Use the capital recovery formula to calculate the annual cost. [Economics]

$$A = \frac{i(1 + i)^n}{(1 + i)^n - 1} P$$

$$= \left(\frac{(0.005153)(1 + 0.005153)^{60}}{(1 + 0.005153)^{60} - 1} \right)(\$280,000)$$

$$= \$5,437 \quad (\$5,440)$$

The answer is (C).

5. The gradient is not uniform, so the uniform annual cost needs to be calculated for each year. Calculate the total present value then calculate the uniform annual cost based on the total present value. The first-year operating cost is not realized until the end of the year. [Economics]

$$P = (P/F, 4\%, x \text{ yr}) F$$

Using a factor table, for $i = 4\%$,

year 1: $P = (0.9615)(\$430,000) = \$413,445$

year 2: $P = (0.9246)(\$440,000) = \$406,824$

year 3: $P = (0.8890)(\$480,000) = \$426,720$

year 4: $P = (0.8548)(\$510,000) = \$435,948$

year 5: $P = (0.8219)(\$560,000) = \$460,264$

Calculate total present value.

$$P = \$7,800,000 + \$413,445 + \$406,824$$
$$+ \$426,720 + \$435,948 + \$460,264$$
$$= \$9,943,200$$

Calculate the uniform annual cost. [Economics]

$$F = (F/A, 4\%, 5 \text{ yr}) A$$

Using a factor table, for $i = 4\%$,

$$A = (0.2246)(\$9,943,200) = \$2,233,240 \quad (\$2,233,000)$$

The answer is (D).

6. For alternative 1, the present worth can be calculated using the following equation. [Economics]

$$P = \text{capital cost} + (P/A, 4\%, 15 \text{ yr}) A$$
$$+ (P/G, 4\%, 15 \text{ yr}) G$$

The first-year operation and maintenance cost is reduced by 5% to represent the base annual cost.

Using a factor table, for $i = 4\%$,

$$P = \$1,200,000 + (11.1184)$$
$$\times \big(\$140,000 - (0.05)(\$140,000) \big)$$
$$+ (69.7355) \big((0.05)(\$140,000) \big)$$
$$= \$3,166,900$$

For alternative 2, the present worth can be calculated using the following equation. [Economics]

$$P = \text{capital cost} + (P/A, 4\%, 15 \text{ yr}) A$$
$$+ (P/G, 4\%, 15 \text{ yr}) G$$

The first-year operation and maintenance cost is reduced by 5% to represent the base annual cost.

Using a factor table, for $i = 4\%$,

$$P = \$1,600,000 + (11.1184) \big(\$95,000 - (0.05)(\$95,000) \big)$$
$$+ (69.7355) \big((0.05)(\$95,000) \big)$$
$$= \$2,934,680$$

The cost difference between the two alternatives is

$$\$3,166,900 - \$2,934,680 = \$232,220 \quad (\$232,000)$$

The answer is (C).

7. Calculate the interest rate for a term of six years using the uniform series present worth. [Economics]

$$P/A = (P/A, i\%, 6) = \frac{\$230,000}{\$45,600} = 5.04386$$

Assoc. Eng.
Principles

Using a factor table, for $i = 4\%$,

$$P/A = 5.2421$$

Using a factor table, for $i = 6\%$,

$$P/A = 4.9173$$

The interest rate is between 4% and 6%, which is less than the MARR. The investment is not acceptable.

The answer is (C).

Project Management

Content in blue refers to the *NCEES Handbook*.

PRACTICE PROBLEMS

1. A project has the base costs and component risk factors shown in the table.

project component	base cost	risk factor
site work	$1,932,000	5.33%
concrete	$2,414,000	6.84%
mechanical	$892,000	9.71%
electrical	$296,000	6.62%

The overall project contingency is most nearly

(A) 5.09%

(B) 6.76%

(C) 7.13%

(D) 7.55%

2. Two projects are being evaluated for implementation at an industrial facility. Project 1 requires upgrading chemical storage and conveyance equipment at a cost of $3 million. Project 2 requires developing and maintaining a hazardous waste reduction program at a cost of $2.6 million. The budget will cover the cost of only one of the projects. Project characteristics are defined as follows.

Project 1:

risk component	cost of current risk	cost of risk at project completion
A	2×10^6	2×10^4
B	6×10^5	6×10^4
C	9×10^5	4.5×10^3

Project 2:

risk component	cost of current risk	cost of risk at project completion
A	1.8×10^6	1.8×10^4
B	1×10^6	1×10^4
C	3×10^5	3×10^4

What is the risk reduction potentially realized by implementation of each project?

(A) 8.50×10^4, 5.80×10^4

(B) 1.14×10^6, 1.01×10^6

(C) 3.42×10^6, 3.04×10^6

(D) 3.50×10^6, 3.10×10^6

3. Two projects are being evaluated for implementation at an industrial facility. Project 1 requires upgrading chemical storage and conveyance equipment at a cost of $3 million. Project 2 requires developing and maintaining a hazardous waste reduction program at a cost of $2.6 million. The budget will cover the cost of only one of the projects. The risk reduction potential for project 1 is 3.42×10^6, and the risk reduction for project 2 is 3.04×10^6. The risk reduction to cost ratios for project 1 and project 2, respectively, are most nearly

(A) 0.028, 0.022

(B) 0.38, 0.39

(C) 1.14, 1.17

(D) 1.17, 1.19

4. Two projects are being evaluated for implementation at an industrial facility. Project 1 requires upgrading chemical storage and conveyance equipment at a cost of $3 million. Project 2 requires developing and maintaining a hazardous waste reduction program at a cost of $2.6 million. The budget will cover the cost of only one of the projects. Project 1 has a risk reduction to cost ratio of 1.14 and project 2 has a risk reduction to cost ratio of 1.17. Which project should be funded first?

(A)   Project 1, because it has the smaller risk reduction to cost ratio.

(B)   Project 1, because it has the greater risk reduction to cost ratio.

(C)   Project 2, because it has the smaller risk reduction to cost ratio.

(D)   Project 2, because it has the greater risk reduction to cost ratio.

5. Performance measures of a project under construction are as follows.

$$BCWP = \$172{,}480$$
$$ACWP = \$185{,}530$$
$$BCWS = \$173{,}860$$
$$BAC = \$483{,}100$$

What is the project budget and schedule status?

(A)   ahead of schedule, under budget

(B)   ahead of schedule, over budget

(C)   behind schedule, under budget

(D)   behind schedule, over budget

SOLUTIONS

1. Calculate the project contingency.

$$\text{contingency} = \sum_{i=1}^{n} \frac{(\text{risk}_i)(\text{cost}_i)}{\text{cost}_i}$$

$$= \frac{\begin{matrix}(5.33\%)(\$1{,}932{,}000) \\ +(6.84\%)(\$2{,}414{,}000) \\ +(9.71\%)(\$892{,}000) \\ +(6.62\%)(\$296{,}000)\end{matrix}}{\begin{matrix}\$1{,}932{,}000 + \$2{,}414{,}000 \\ +\$892{,}000 + \$296{,}000\end{matrix}}$$

$$= 6.76\%$$

The answer is (B).

2. The equation for cost benefit from total risk reduction is

$$R_{\text{red}} = \sum R_{\text{current}} - \sum R_{\text{final}}$$

$R_{\text{red}} =$ cost benefit from total risk reduction

$R_{\text{current}} =$ cost of each risk component under current conditions

$R_{\text{final}} =$ cost of each risk component after project completion

For project 1,

$$R_{\text{red}} = (\$2 \times 10^6 + \$6 \times 10^5 + \$9 \times 10^5)$$
$$- (\$2 \times 10^4 + \$6 \times 10^4 + \$4.5 \times 10^3)$$
$$= \$3.42 \times 10^6$$

For project 2,

$$R_{\text{red}} = (\$1.8 \times 10^6 + \$1 \times 10^6 + \$3 \times 10^5)$$
$$- (\$1.8 \times 10^4 + \$1 \times 10^4 + \$3 \times 10^4)$$
$$= \$3.04 \times 10^6$$

The answer is (C).

3. For project 1,

$$\frac{R_{\text{red}}}{\text{project cost}} = \frac{\$3.42 \times 10^6}{\$3.0 \times 10^6} = 1.14$$

For project 2,

$$\frac{R_{red}}{project\ cost} = \frac{\$3.04 \times 10^6}{\$2.6 \times 10^6} = 1.17$$

The answer is (C).

4. The project with the greater risk reduction to cost ratio should be funded first, because it provides a greater benefit per unit of cost. Project 2 has the greater risk reduction to cost ratio.

The answer is (D).

5. Check the cost variance.

CPM Precedence Relationships: Variances

$$\begin{aligned} CV &= BCWP - ACWP \\ &= \$172,480 - \$185,530 \\ &= -\$13,050 \quad [< \$0] \end{aligned}$$

The project is over budget by \$13,050.

Check the schedule variance.

CPM Precedence Relationships: Variances

$$\begin{aligned} SV &= BCWP - BCWS \\ &= \$172,480 - \$173,860 \\ &= -\$1380 \quad [< \$0] \end{aligned}$$

The project is behind schedule, having spent \$1380 more than should be spent by this date.

The answer is (D).

Assoc. Eng.
Principles

64 Mass and Energy Balance

PRACTICE PROBLEMS

1. A small natural lake occupies an area of 73 ha to an average depth of 30 m. The stream feeding the lake has an average flow of 0.21 m^3/s with a total phosphorus concentration of 0.13 mg/L. The total phosphorus deposition rate in the lake is 10 m/year. The total phosphorus concentration in the lake is most nearly

(A) 0.062 mg/L

(B) 0.12 mg/L

(C) 3.6 mg/L

(D) 7.0 mg/L

2. A city generates 72 tons of municipal solid waste per day that is to be incinerated. The waste has a heating value of 11 000 kJ/kg, is 12% ash, and has a total combined moisture and hydrogen water of 48%. Heat loss to ash is 400 kJ/kg, and radiation losses are expected to be 0.0035 kJ/kg. The net heat produced from incinerating the waste is most nearly

(A) 2.9×10^8 kJ/d

(B) 6.4×10^8 kJ/d

(C) 7.1×10^8 kJ/d

(D) 1.4×10^9 kJ/d

SOLUTIONS

1. Use a mass balance to determine phosphorus loading. [Conversion Factors]

$$Q = \text{inlet and outlet flow rate}$$
$$\text{(assume they are equal)}$$
$$= 0.21 \text{ m}^3/\text{s}$$
$$C_{o\text{TP}} = \text{inflow total phosphorus concentration}$$
$$= 0.13 \text{ mg/L}$$
$$v_s = \text{total phosphorus deposition rate}$$
$$= 10 \text{ m/yr}$$
$$A_s = \text{lake surface area} = 73 \text{ ha}$$
$$C_P = \text{total phosphorus concentration in}$$
$$\text{the lake}$$

$$QC_p + C_p v_s A_s = QC_{o\text{TP}}$$
$$C_p = \frac{QC_{o\text{TP}}}{Q + v_s A_s}$$

$$= \frac{\left(0.21 \ \dfrac{\text{m}^3}{\text{s}}\right)\left(0.13 \ \dfrac{\text{mg}}{\text{L}}\right)}{\left(\begin{array}{c}0.21 \ \dfrac{\text{m}^3}{\text{s}} + \left(10 \ \dfrac{\text{m}}{\text{yr}}\right)(73 \text{ ha})\left(\dfrac{1 \text{ yr}}{365 \text{ d}}\right) \\ \times \left(\dfrac{1 \text{ d}}{86\,400 \text{ s}}\right)\left(10\,000 \ \dfrac{\text{m}^2}{\text{ha}}\right)\end{array}\right)}$$

$$= 0.062 \text{ mg/L}$$

The answer is (A).

2. Find the net heat.

$\dot{m}$ = waste mass flow rate, kg/d

q = net heat, kJ/d

q_a = heat loss to ash, kJ/d

q_r = heat loss to radiation, kJ/d

q_T = total heat, kJ/d

q_v = latent heat of vaporization of water
   = 2420 kJ/kg

q_w = heat loss to water, kJ/d

$$\dot{m} = \left(72 \, \frac{\text{tons}}{\text{day}}\right)\left(2000 \, \frac{\text{lbm}}{\text{ton}}\right)\left(\frac{1 \, \text{kg}}{2.204 \, \text{lbm}}\right)$$

$$= 65\,336 \, \text{kg/d}$$

$$q_T = \left(65\,336 \, \frac{\text{kg}}{\text{d}}\right)\left(11\,000 \, \frac{\text{kJ}}{\text{kg}}\right)$$

$$= 7.2 \times 10^8 \, \text{kJ/d}$$

$$q_a = \left(65\,336 \, \frac{\text{kg}}{\text{d}}\right)(0.12)\left(400 \, \frac{\text{kJ}}{\text{kg}}\right)$$

$$= 3.1 \times 10^6 \, \text{kJ/d}$$

$$q_w = \left(65\,336 \, \frac{\text{kg}}{\text{d}}\right)(0.48)\left(2420 \, \frac{\text{kJ}}{\text{kg}}\right)$$

$$= 7.6 \times 10^7 \, \text{kJ/d}$$

$$q_r = \left(65\,336 \, \frac{\text{kg}}{\text{d}}\right)\left(0.0035 \, \frac{\text{kJ}}{\text{kg}}\right)$$

$$= 229 \, \text{kJ/d} \quad \text{[negligible]}$$

$$q = q_T - q_a - q_w - q_r$$

$$= 7.2 \times 10^8 \, \frac{\text{kJ}}{\text{d}} - 3.1 \times 10^6 \, \frac{\text{kJ}}{\text{d}} - 7.6 \times 10^7 \, \frac{\text{kJ}}{\text{d}}$$

$$= 6.4 \times 10^8 \, \text{kJ/d}$$

The answer is (B).

Data Management

PRACTICE PROBLEMS

1. The U.S. Environmental Protection Agency has adopted guidelines that provide a methodology to municipalities for the implementation of asset management strategies. Which of the following guidelines apply to sanitary wastewater collection systems?

- (A) Asset Management and Compliance Strategy (AMCS)

- (B) Capacity, Management, Operations, and Maintenance (CMOM)

- (C) Capital, Infrastructure, and Maintenance Framework (CIMF)

- (D) Operations Management, Tracking, and Reporting (OMTR)

2. What ISO series provides asset management guidance?

- (A) ISO 17000

- (B) ISO 28000

- (C) ISO 40500

- (D) ISO 55000

SOLUTIONS

1. USEPA guidelines administered through the Office of Enforcement and Compliance provide a methodology for municipalities to implement asset management strategies. These guidelines are commonly referred to as Capacity, Management, Operations, and Maintenance (CMOM) and apply to sanitary wastewater collection systems.

The answer is (B).

2. The International Organization for Standardization developed asset management guidance is the ISO 55000 series standard. The ISO 55000 series contains three standards: ISO 55000 Asset Management—Overview, Principles and Terminology; ISO 55001 Asset Management—Requirements; and ISO 55002 Asset Management—Guidelines for the Application of ISO 55001.

The answer is (D).